KW-208-753

OXFORD REVIEWS OF REPRODUCTIVE BIOLOGY

Volume 16
1994

OXFORD REVIEWS OF REPRODUCTIVE BIOLOGY
EDITORIAL BOARD

Professor R B Heap (chairman), Institute of Animal Physiology and Genetics Research, Cambridge Research Station, Babraham Hall, Cambridge CB2 4AT, UK.

Professor D T Baird, Department of Obstetrics and Gynaecology, University of Edinburgh, 37 Chalmers Street, Edinburgh EH3 9EW, UK.

Professor F Bazer, 442D Kleberg Center, Department of Animal Science, Texas A & M University, College Station, TX 77843–2471, USA.

Professor H Beier, Institut für Anatomie und Reproduktionsbiologie Medizinische Fakultät der RWTW Aachen, Wendlingweg 2, D–W–5100 Aachen, Germany.

Professor A R Bellvé, Department of Anatomy and Cell Biology and Center of Cellular and Molecular Urology, Columbia Presbyterian Medical Center, New York NY 10032, USA.

Dr H M Charlton (editor), Department of Human Anatomy, University of Oxford, South Parks Road, Oxford OX1 3QX, UK.

Dr Hilary Dobson, University of Liverpool, Department of Veterinary Preclinical Sciences, Leahurst, Neston, Wirral L64 7TE, UK.

Professor A P F Flint, Animal Physiology Section, Dept of Physiology and Environmental Science, University of Nottingham, Sutton Bonington, Loughborough, Leics, LE12 5RD, UK.

Professor M Johnson, University of Cambridge, Department of Anatomy, Downing Street, Cambridge, CB2 3DY, UK.

Professor D de Kretser (director), Institute of Reproduction and Development, Monash University Level 3, Block E, Monash Medical Centre, 246, Clayton Road, Clayton, Victoria 3168, Australia.

Professor A McNeilly, MRC Reproductive Biology Unit, Centre for Reproductive Biology, 37 Chalmers Street, Edinburgh, EH3 9EH, UK.

Dr S R Milligan, Physiology Section, Division of Biomedical Sciences, King's College, Strand, London WC2R 2LS, UK.

Professor R Ozon, Laboratoire de Physiologie de la Reproduction, Université Pierre et Marie Curie, Bâtiment B, 7ème étage–Case 13, 9 quai St Bernard, 75005 Paris, France.

Professor R V Short, Department of Physiology, Monash University, Clayton, Victoria 3168, Australia.

Professor S K Smith, University of Cambridge Clinical School, Department of Obstetrics and Gynaecology, The Rosie Maternity Hospital, Robinson Way, Cambridge, CB2 2SW, UK.

B/QH 471 CHA

OXFORD REVIEWS OF REPRODUCTIVE BIOLOGY

EDITED BY

H.M. CHARLTON

Volume 16
1994

Oxford   New York   Tokyo

OXFORD UNIVERSITY PRESS

1994

SOCIETY OF MEDICINE
Royal Society of
Medicine Library
3 0 NOV 1998
WITHDRAWN STOCK
LIBRARY

Oxford University Press, Walton Street, Oxford OX2 6DP
Oxford   New York   Toronto
Delhi   Bombay   Calcutta   Madras   Karachi
Kuala Lumpur   Singapore   Hong Kong   Tokyo
Nairobi   Dar es Salaam   Cape Town
Melbourne   Auckland   Madrid
and associated companies in
Berlin   Ibadan

Oxford is a trade mark of Oxford University Press

Published in the United States
by Oxford University Press Inc., New York

© Oxford University Press, 1994

All rights reserved. No part of this publication may be
reproduced, stored in a retrieval system, or transmitted, in any
form or by any means, without the prior permission in writing of Oxford
University Press. Within the UK, exceptions are allowed in respect of any
fair dealing for the purpose of research or private study, or criticism or
review, as permitted under the Copyright, Designs and Patents Act, 1988, or
in the case of reprographic reproduction in accordance with the terms of
licences issued by the Copyright Licensing Agency. Enquiries concerning
reproduction outside those terms and in other countries should be sent to
the Rights Department, Oxford University Press, at the address above.

This book is sold subject to the condition that it shall not,
by way of trade or otherwise, be lent, re-sold, hired out, or otherwise
circulated without the publisher's prior consent in any form of binding
or cover other than that in which it is published and without a similar
condition including this condition being imposed
on the subsequent purchaser.

A catalogue record for this book is available from the British Library

Library of Congress Cataloging in Publication Data
Library of Congress Card No 80–648347

ISBN 0 19 2624261

Typeset by The Electronic Book Factory Ltd, Fife, Scotland
Printed in Great Britain by
Bookcraft (Bath) Ltd, Midsomer Norton

Contents

Contributors

Daniel D. Carson: Department of Biochemistry and Molecular Biology, M.D. Anderson Cancer Center, 1515 Holcombe Boulevard, Houston, TX 77030, USA.

Ann G. Clarke: The Thymus Laboratory, Cellular Physiology, AFRC Babraham Institute, Cambridge, UK, and Department of Obstetrics and Gynaecology, University of Nottingham, UK.

John P. Hearn: Wisconsin Regional Primate Research Center, Madison, Wisconsin 153715, USA.

Lyn A. Hinds: CSIRO Division of Wildlife and Ecology, PO Box 84, Lyneham, ACT 2602, Australia.

Lawrence A. Johnson: US Department of Agriculture, Agricultural Research Service, Germplasm and Gamete Physiology Laboratory, Beltsville, MD 20705, USA.

Nathalie Josso: Unité de Recherches sur l'Endocrinologie du Développement (INSERM), Ecole Normale Supérieure, 1 rue Maurice-Arnoux, 92120 Montrouge, France.

Marion D. Kendall: The Thymus Laboratory, Cellular Physiology, AFRC Babraham Institute, Cambridge, UK, and Thymus Research Laboratory, The Rayne Institute, UMDS, London, UK.

John K.H. Lu: Departments of Obstetrics/Gynecology and Anatomy/Cell Biology, University of California at Los Angeles, Los Angeles, California, USA.

Martin R. Luck: Animal Physiology Section, Department of Physiology and Environmental Science, University of Nottingham, Sutton Bonington, Loughborough, Leics LE12 5RD, UK.

Joseph Meites: Departments of Physiology, Michigan State University, East Lansing, Michigan, USA.

Georgina E. Webley: Institute of Zoology, Regent's Park, London NW1 4RY, UK.

Carole C. Wegner: Department of Biochemistry and Molecular Biology, M.D. Anderson Cancer Center, 1515 Holcombe Boulevard, Houston, TX 77030, USA.

Abbreviations

17α-OH-P	17α-hydroxyprogesterone
5-HT	5-hydroxytryptamine (serotonin)
ACTH	adrenocorticotrophic hormone
AFP	α-fetoprotein
AMH	anti-Müllerian hormone
bAMH	bovine AMH
Bt$_2$cAMP	dibutyryl cAMP
CA(s)	catecholamine(s)
CAM	cell adhesion molecules
cAMP	cyclic adenosine monophosphate
cDNA	complementary DNA
CG	chorionic gonadotrophin
CHO	Chinese hamster ovary
CIA	collagen-induced arthritis
CL	corpus luteum
CMJ	corticomedullary junction
ConA	concanavalin A
CRH	corticotrophin-releasing hormone
DA	dopamine
dbcAMP	dibutyryl cyclic adenosine monophosphate
DHEA	dehydroepiandrosterone
DHT	dihydrotestosterone
DN	double negative
DP	double positive
ECM	extracellular matrix
EHS	EHS mouse subcutaneous tumour
FCM	flow cytometry
FSH	follicle-stimulating hormone
GH	growth hormone
GnRH	gonadotrophin-releasing hormone
hAMH	human AMH
hCG	human chorionic gonadotrophin
hPL	human placental lactogen
IDC(s)	interdigitating cell(s)
IFN	interferon
IL-x.	interleukin-x.
IVF	*in vitro* fertilization
LDL	low density lipoprotein
LH	luteinizing hormone

LHRH	LH-releasing hormone
LNF-I	lacto-*N*-fucopentase-I
LRP	lipoprotein receptor-related gene
MAO	monoamine oxidase
MER(s)	medullary epithelial ring(s)
MFCM	modified FCM
MHC	major histocompatibility complex
mRNA	messenger RNA
MSH	melanocyte-stimulating hormone
NA	noradrenaline
NK	natural killer (cells)
NS	natural suppressor (cells)
PA	plasminogen activator
PAF	platelet-activating factor
PAI	PA inhibitor
PAPP	pregnancy-associated plasma proteins
$PGF_{2\alpha}$	prostaglandin $F_{2\alpha}$
PI	phosphoinositide
PMDS	persistent Müllerian duct syndrome
PMSG	pregnant mare's serum gonadotrophin
PNA	peanut antigen
POMC	pro-opiomelanocortin
PRL	prolactin
PVS	perivascular space
RA	rheumatoid arthritis
RGD	Arginine, Glycine, Aspartate
RIA(s)	radioimmunoassay(s)
SDS-PAGE	sodium dodecylsulphate-polyacrylamide gel electrophoresis
SP	single positive
SPARC	secreted protein, acidic and rich in Cysteine
STH	somatotrophic hormone
TCR	T-cell receptor
TGF	transforming growth factor
THF	thymic humoral factor
TIMP2	tissue inhibitor of metalloproteinase type 2
tPA	tissue-type PA
TRH	thyrotrophin-releasing hormone
TSH	thyroid-stimulating hormone
TSP	thrombospondin
uPA	urokinase-type PA

1 Embryo–maternal interactions during the establishment of pregnancy in primates

GEORGINA E. WEBLEY and JOHN P. HEARN

I   INTRODUCTION

The endocrine regulation of implantation and the recognition of embryonic signals by the primate corpus luteum is quite different from that in other mammalian species. The corpus luteum will normally decline at the end of the ovarian cycle. It requires a luteotrophic rescue, stimulated by chorionic gonadotrophin (CG) derived from the peri-implantation embryo. Consequently, there is a relatively short time for the differentiating embryo to achieve competence in its secretion of CG, to attach and invade the maternal endometrium and for CG to be transmitted, presumably through the maternal circulation, in order to stimulate the corpus luteum and allow pregnancy to continue.

In contrast, in many non-primate species, the corpus luteum is controlled by prostaglandin $F_{2\alpha}$ ($PGF_{2\alpha}$) secreted by the uterus towards the end of the ovarian cycle; a luteolysin which is neutralized when an early embryo is present in the uterus and inhibits the release of $PGF_{2\alpha}$.

The primate corpus luteum is not immune from the luteolytic effects of $PGF_{2\alpha}$ but appears more resistant. While luteolysis is not understood, the most feasible hypothesis is that $PGF_{2\alpha}$ also controls primate luteolysis but is derived from within the corpus luteum itself.

In this chapter, we consider the embryo–maternal interactions during implantation and the establishment of pregnancy in the marmoset monkey, rhesus monkey, and human. While all three species conform to a basic primate pattern, there are species-specific variations in the timing of events around implantation and of the first release of CG. There are also variations in the responses of the corpus luteum between these species. Some of these differences may be trivial and some may be due to the various analytical methods that have been employed. We aim at a synthesis and focus on some of the questions that are now amenable to study, based on the knowledge gained to date.

II   REGULATION OF CORPUS LUTEUM REGRESSION

The mechanism of luteolysis in the primate corpus luteum is still debated. There are those who consider it unnecessary to evoke an active luteolytic mechanism with a luteolytic agent (Lenton and Woodward 1988; Fisch et al. 1989). Instead they suggest that luteolysis is an autonomous pre-programmed event which is initiated at the time of ovulation and only prevented in early pregnancy by the secretion of CG by the implanting embryo. We prefer the hypothesis that there is an active luteolytic process in the primate corpus luteum induced by a luteolytic agent. While oestradiol is considered a possible candidate (Schoonmaker et al. 1982; Hahlin et al. 1986) we believe that the data shown below from ours and others studies are consistent with a luteolytic role for $PGF_{2\alpha}$.

1   The luteolytic action of $PGF_{2\alpha}$

In non-primate species luteolysis is induced by the action of $PGF_{2\alpha}$ released from the uterus (McCracken et al. 1972, Horton and Poyser 1976). In the primate, *uterine* $PGF_{2\alpha}$ does not appear to initiate luteolysis since hysterectomy has no effect on cycle length in the rhesus monkey (Neill *et al.* 1969) and in women (Beling *et al.* 1970). Instead intraovarian production of $PGF_{2\alpha}$ may provide the source for induction of luteolysis in the primate. The production of prostaglandins by luteal tissue has been demonstrated in the rhesus monkey and in women (Challis *et al.* 1976; Swanston *et al.* 1977; Balmaceda *et al.* 1979; Valenzuela *et al.* 1983). We have detected $PGF_{2\alpha}$ production by luteal tissue from the marmoset monkey (Abayasekara and Webley, unpublished observation); $PGF_{2\alpha}$ concentrations of between 1.0 and 1.2 ng/ml were produced by marmoset luteal cells (1×10^4) cultured for 24 h in serum supplemented serum. An intraluteal action for $PGF_{2\alpha}$ is further supported by the presence of $PGF_{2\alpha}$ receptors in human corpora lutea (Powell *et al.* 1977; Tanaka *et al.* 1983).

The ability of $PGF_{2\alpha}$ to cause luteolysis has been demonstrated in a variety of *in vitro* and *in vivo* systems. $PGF_{2\alpha}$ has a luteolytic action on human ovarian cells *in vitro* (McNatty *et al.* 1975; Dennefors *et al.* 1982), intraluteal administration of $PGF_{2\alpha}$ to the rhesus monkey *in vivo* results in luteolysis (Auletta *et al.* 1984) and perfusion of $PGF_{2\alpha}$ through the corpus luteum of the marmoset monkey *in vivo* is associated with a decrease in progesterone production (Hearn and Webley 1987). Systemic administration of $PGF_{2\alpha}$ is less clear in its effect. A single intramuscular injection (0.5 µg) of a $PGF_{2\alpha}$ analogue (cloprostenol) to the marmoset monkey has a marked and rapid luteolytic action after day 8 of the luteal phase or in early pregnancy (Summers *et al.* 1985; Webley *et al.* 1991*a*). In contrast, similar dose/weight injections of cloprostenol to the baboon are not luteolytic at any stage of the cycle or pregnancy (Eley *et al.* 1987). Furthermore, $PGF_{2\alpha}$ is only transient in effect when administered systemically to women (Karim and Hillier 1979; Wentz and Jones 1979) and requires multiple injections to be effective in the rhesus monkey (Wilks 1983). Therefore, while $PGF_{2\alpha}$ appears to have a consistent luteolytic effect *in vitro* its action *in vivo* in primates is variable.

Although the response to systemic administration of $PGF_{2\alpha}$ may vary between primate species the action of $PGF_{2\alpha}$ at the luteal cell is the same in all the primate species studied and similar to that described in non-primate species. $PGF_{2\alpha}$ acts at the luteal cell by inhibiting the luteotrophic actions of luteinizing hormone (LH) or human chorionic gonadotrophin (hCG). This has been demonstrated in mid-luteal phase tissue from the human (Dennefors *et al.* 1982; Patwardhan and Lanthier 1984), rhesus monkey (Stouffer *et al.* 1979), marmoset monkey (Webley *et al.* 1989; Michael and Webley 1993), and in cultured human granulosa cells (McNatty *et al.* 1975; Michael and Webley 1991*a*; Webley *et al.* 1991*c*). Furthermore, perfusion of the marmoset corpus luteum *in vivo* with $PGF_{2\alpha}$ prevents the stimulatory action of hCG perfused subsequently (Hearn and Webley 1987).

Our hypothesis for the antigonadotrophic action of $PGF_{2\alpha}$ at the primate

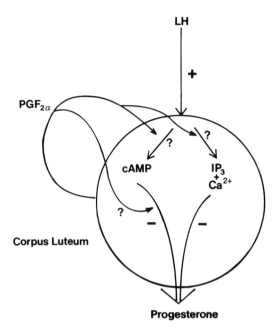

Fig. 1.1.   Diagram of our hypothesis for the antigonadotrophic action of $PGF_{2\alpha}$ at the primate luteal cell and the second messengers involved in decreasing the production of progesterone. IP_3, inositol triphosphate; Ca^{2+} free calcium.

luteal cell is shown in Fig. 1.1 $PGF_{2\alpha}$ can inhibit LH/hCG stimulated cyclic adenosine monophosphate (cAMP) (Hamberger *et al.* 1979; Dennefors *et al.* 1982). We have demonstrated in the marmoset monkey, from incubation of luteal tissue pieces *in vitro*, that the $PGF_{2\alpha}$ analogue cloprostenol inhibits both the stimulation of cAMP by hCG and the stimulation of progesterone by dibutyryl cAMP (Michael and Webley 1993). $PGF_{2\alpha}$ can therefore act to inhibit the ability of LH/hCG to stimulate progesterone at sites before and after the generation of cAMP. The action of $PGF_{2\alpha}$ has also been shown to involve another second messenger system, the generation of inositol phosphates via increased phosphoinositide (PI) turnover, in non-primate luteal cells (Davis *et al.* 1989; Lahav *et al.* 1989). A role for increased PI turnover in the luteolytic action of $PGF_{2\alpha}$ has now been demonstrated in primate luteal cells (Houmard *et al.* 1992; Michael and Webley 1993) and will be discussed below (see p. 6).

The consistent antigonadotrophic action of $PGF_{2\alpha}$ in luteal tissue of various primate and non-primate species suggests that variations in the effects of systemic administration may be due to differences between species in access and delivery of $PGF_{2\alpha}$ to luteal tissue. In support of this hypothesis there are known to be large differences between species in the rate or velocity of $PGF_{2\alpha}$ metabolism. For example, in sheep 99% of injected $PGF_{2\alpha}$ is metabolized in a single passage through the lungs compared with 65% in cows (Hansel and Dowd 1986). Since $PGF_{2\alpha}$ appears to be predominantly of intraovarian origin in primates, the induction of luteolysis may involve a paracrine action within the corpus luteum. Systemic concentrations of $PGF_{2\alpha}$ may therefore not be of relevance to the endogenous control of luteolysis.

If $PGF_{2\alpha}$ of luteal origin controls the onset of luteal regression in primates, an increase in luteal concentrations of $PGF_{2\alpha}$ might be expected to occur at the onset of luteolysis as described previously for domestic species (Silvia et al. 1991). Reports of changes in primate luteal concentrations of $PGF_{2\alpha}$ are, however, inconsistent. A mid-luteal peak in $PGF_{2\alpha}$ concentrations has been recorded in human corpus luteum (Patwardhan and Lanthier 1985), which would coincide with the onset of luteolysis and the first decrease in progesterone production. However, others found no correlation between $PGF_{2\alpha}$ levels and the regression of the human corpus luteum (Challis et al. 1976; Swanston et al. 1977). In the rhesus monkey higher luteal concentrations of $PGF_{2\alpha}$ were recorded in the early and late luteal phase with a decrease in the mid-luteal phase (Houmard and Ottobre 1989).

2   Changing sensitivity to $PGF_{2\alpha}$

Recently, we have obtained evidence to suggest that it may not be necessary to evoke a clear change in luteal concentrations of $PGF_{2\alpha}$ across the luteal phase. Instead, luteal cells of increasing age may change their response to a given concentration of $PGF_{2\alpha}$. In the marmoset monkey, the action of $PGF_{2\alpha}$ to inhibit gonadotrophic-stimulated cAMP is not evident at all times in the luteal phase. Cloprostenol cannot induce luteolysis in vivo if administered prior to day 8 after ovulation (Summers et al. 1985; Michael and Webley 1993). This corresponds to the inability of cloprostenol to influence hCG-stimulated progesterone and cAMP by luteal tissue in vitro obtained in the early luteal phase (Michael and Webley 1993). Conversely, on day 14 after ovulation when cloprostenol is luteolytic in vivo, it also inhibits hCG action in vitro.

Study of the involvement of increased PI turnover in the antigonadotrophic action of $PGF_{2\alpha}$ at the primate corpus luteum has revealed a different pattern of activation to that of cAMP. We have demonstrated, from incubation of marmoset luteal tissue with tritiated inositol (Michael and Webley, 1993), that cloprostenol can stimulate a significant increase in PI turnover. This action of cloprostenol is, however, only evident on day 3 after ovulation, a stage in the luteal phase when it is unable to influence hCG action in vitro or induce luteolysis in vivo. In contrast, on day 14 after ovulation when cloprostenol is antigonadotrophic *in vitro* and luteolytic *in vivo*, it has no effect on PI turnover. The action of cloprostenol on hCG-stimulated progesterone, PI turnover, and *in vivo* progesterone response by the marmoset corpus luteum in the early and mid-luteal phase is shown in Fig. 1.2.

The results in the marmoset monkey indicate that the luteolytic action of $PGF_{2\alpha}$ involves its ability to inhibit the luteotrophic stimulus of LH/hCG at sites before and after cAMP accumulation. This action of $PGF_{2\alpha}$ is only evident in the luteal phase after the corpus luteum has reached a certain age (Table 1.1). An increase in PI turnover does not appear to be involved in the luteolytic action of $PGF_{2\alpha}$ since it occurs only in the early luteal phase when $PGF_{2\alpha}$ is not luteolytic *in vivo*. However, increased PI turnover in the early luteal phase may instead be necessary to render the corpus luteum sensitive to the antigonadotrophic actions of $PGF_{2\alpha}$ in the mid-luteal phase when luteolysis is initiated. The absence of a PI response to $PGF_{2\alpha}$ when the corpus luteum reaches the mid-luteal phase may be a trigger to the onset of an antigonadotrophic action at the time of luteolysis. Such changes in second messenger responses would allow the corpus luteum to exhibit altered responses to $PGF_{2\alpha}$ without requiring increases in intraluteal concentrations of $PGF_{2\alpha}$.

The effect of $PGF_{2\alpha}$ on PI turnover has also recently been described in the rhesus corpus luteum of increasing age and in simulated early pregnancy (Table 1.1; Houmard *et al.* 1992). While the results may not be dissimilar to those in the marmoset monkey the interpretation by Houmard *et al.* (1992) is very different. They found that in dispersed luteal cells from the rhesus, $PGF_{2\alpha}$ increases the accumulation of inositol phosphates in the early but not the mid-luteal phase as in intact tissue from the marmoset monkey. However, Houmard *et al.* (1992) also measured increases in PI turnover in the late luteal phase and in simulated early pregnancy. They suggest that there is a $PGF_{2\alpha}$-induced increase in PI turnover, which coincides with an increase in $PGF_{2\alpha}$ concentrations at the end of the luteal phase (Houmard and Ottobre 1989) to cause luteolysis. Furthermore, they propose that the increased PI turnover in simulated early pregnancy results from exposure to gonadotrophins and that luteolysis does not occur in early pregnancy because there is no increase in luteal $PGF_{2\alpha}$ concentrations. While we have not yet measured PI turnover at the end of the luteal phase or in early pregnancy in the marmoset the most important time for the process of luteolysis must occur during the mid-luteal phase. A decrease in progesterone is first recorded in the marmoset from day 14 after ovulation (Webley *et al.* 1991*b*) (luteal phase length 19 ± 0.6 days, Harlow *et al.* 1983) and in the rhesus monkey from day 11 of a 14-day luteal phase (Atkinson *et al.* 1975), indicating that luteolysis will have already begun on these days. The second messenger system activated by $PGF_{2\alpha}$ at these times is therefore likely to be implicated in the initiation of luteolysis. This together with the

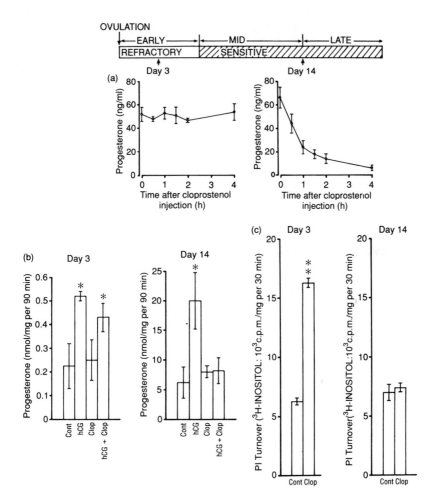

Fig. 1.2 (a) Plasma progesterone concentrations (mean ± SEM) in blood samples taken from marmoset monkeys following intramuscular injection of cloprostenol (0.5 µg) on day 3 (*n*=5) and day 14 (*n*=4) after ovulation. (b) Progesterone production (mean ± SEM; (*n*=3 tubes per treatment) by marmoset luteal tissue on days 3 and 14 after ovulation and treated for 90 min with medium alone (control), cloprostenol (clop), hCG and cloprostenol + hCG. Representative results of 5 (day 3) and 3 (day 14) experimental replicates. *$P<0.01$, **$P<0.001$ *vs* control. (c) PI turnover (^{3}H inositol phosphate generation; mean ± SEM, *n*=3 tubes per treatment) by marmoset luteal tissue on days 3 and 14 after ovulation and treated for 30 min *in vitro* in the presence of 10 mM LiCl with medium alone or with cloprostenol. Representative results of three experimental replicates for each day. **$P<0.001$ *vs* control.

Table 1.1 *Demonstrates the presence (√) or absence (×) of in vivo and in vitro responses to cloprostenol by marmoset (Michael and Webley 1993) and rhesus (Stouffer et al. 1979; Houmard et al. 1992) monkey luteal tissue at different stages in the luteal phase and in simulated early pregnancy (rhesus monkey only). Material not assayed. (−)*

		Stage of the Luteal phase			
Action of cloprostenol	Species	Early	Mid	Late	Stimulated early pregnancy
Luteolytic *in vivo*	Marmoset monkey	×	√	√	—
Antigonadotrophic *in vitro*	Marmoset monkey	×	√	√	—
	Rhesus monkey	×	√	×	√
Stimulates PI turnover *in vitro*	Marmoset monkey	√	×	×	—
	Rhesus monkey	√	×	√	√

discrepancies in measured luteal $PGF_{2\alpha}$ concentrations during the cycle supports our hypothesis that PI turnover may be involved in the timing of luteolysis but not the actual process. However, we agree with Houmard *et al.* (1992) that exposure to gonadotrophins may increase $PGF_{2\alpha}$-stimulated PI turnover since the increase in PI accumulation we measured in the early luteal phase of the marmoset was just 3 days after the ovulatory LH surge (Michael and Webley 1993). Furthermore, $PGF_{2\alpha}$ and cloprostenol stimulate PI turnover in human granulosa cells *in vitro* following exposure to high concentrations of human menopausal gonadotrophin and hCG *in vivo* (Davis *et al.* 1989; Abayasekara *et al.* 1993).

Variations in the response to $PGF_{2\alpha}$ can also be seen in the culture systems used to study its action. When luteal cells are dispersed and incubated in suspension for short periods of 2 or 3 h, instead of an inhibitory action, $PGF_{2\alpha}$ can stimulate progesterone production. This paradoxical luteotrophic response is evident in dispersed luteal cells from human (Richardson and Mason 1980), marmoset (Webley *et al.* 1989), and rhesus monkey (Stouffer *et al.* 1979). In contrast, when luteal cells are incubated in tissue pieces (Hamberger *et al.* 1979; Dennefors *et al.* 1982; Webley *et al.* 1989; Michael and Webley 1993) or as a monolayer of cells cultured over several days (Fisch *et al.* 1989; Webley and Michael 1992) the antigonadotrophic action of $PGF_{2\alpha}$ is evident. This suggests that a degree of tissue integrity and cell to cell contact is required in order for primate luteal cells to respond to the luteolytic action of $PGF_{2\alpha}$.

3   Is there an active luteolysin?

Further evidence to support the presence of an endogenous luteolytic agent requiring cell to cell contact for its action can be seen when basal progesterone production is compared across the luteal phase for luteal cells in different *in vitro* systems. Changes in basal progesterone production by incubated tissue pieces show a fall from early to late luteal phase for human (Fisch *et al.* 1989) and marmoset (Michael and Webley 1993) luteal tissue. This fall in basal progesterone production across the luteal phase is also evident when luteal cells are cultured as a monolayer for 3 days in serum supplemented medium (Webley and Michael 1993). In contrast when luteal cells, taken at different times in the luteal phase, are incubated in suspension, basal progesterone production does not change across the luteal phase of the rhesus (Brannian and Stouffer 1991) and marmoset monkeys (Webley *et al.* 1989; Webley and Michael 1992). Therefore, one of the requirements for a putative endogenous luteolysin to decrease progesterone production is cell to cell contact. Brannian and Stouffer (1991) also suggest that dispersal of cells and placement in fresh medium may reduce luteolysins to insufficient concentrations to elicit a response.

The requirement for a degree of cell contact, either to other cells or to culture wells, in order for the antigonadotrophic action of $PGF_{2\alpha}$ to be evident may reflect the proposed involvement of intracellular free calcium. Increasing intracellular calcium with calcium ionophores inhibits basal and hCG-stimulated progesterone and stimulates $PGF_{2\alpha}$ production by rhesus luteal cells (Houmard and Ottobre 1989; Houmard et al. 1991). $PGF_{2\alpha}$ has itself been shown to increase intracellular free calcium in rhesus luteal cells, the proportion of cells showing an increase rising from early to mid/late luteal phase (Houmard et al. 1992). Studies on ovine luteal cells have shown the increase in intracellular free calcium induced by $PGF_{2\alpha}$ is primarily due to influx of calcium from extracellular sources (Wiltbank et al. 1989). The latter authors suggest that this induced influx by $PGF_{2\alpha}$ may be by mechanisms other than increased PI turnover. This would agree with the results in the rhesus (Houmard et al. 1992) and marmoset (Michael and Webley 1993) in that luteolysis is initiated in the mid-luteal phase when there is no significant increase in PI turnover but there is an increase in the number of cells responding to $PGF_{2\alpha}$ with increases in intracellular calcium. Cell contact may be necessary for the spread of a calcium signal as described in mast cells (Osipchuk and Cahalan 1992). Alternatively, the change in shape of a cell associated with different in vitro cell systems may influence the operation of ligand-gated and/or voltage-dependent calcium channels. Alternatively, the increase in calcium in one cell may influence the membrane of neighbouring cells. A role for intracellular calcium in the luteolytic actions of $PGF_{2\alpha}$ is implicated by the involvement of the calcium-dependent enzymes, cAMP phosphodiesterase (Michael and Webley 1991b), and protein kinase C (Abayasekara et al. 1993) in the antigonadotrophic action of $PGF_{2\alpha}$ on human luteinized granulosa cells.

A further consideration is the presence of different cell types described in the corpus luteum of a number of species including human (Ohara et al. 1987) and rhesus monkey (Hild-Petito et al. 1989). We have not been able to distinguish two populations of cells in terms of cell size in the marmoset corpus luteum (Webley et al. 1990), although cells of a similar size range may differ in their origins. The large luteal cells in the ovine corpus luteum contain the high affinity receptors for $PGF_{2\alpha}$ (Braden et al. 1988) and increases in intracellular calcium in response to $PGF_{2\alpha}$ occur in the large and not the small luteal cells (Wiltbank et al. 1989). Although in the rhesus monkey there is a difference in the profile of the calcium response in large and small luteal cells there is no difference in the peak concentration of calcium achieved in response to $PGF_{2\alpha}$ in the two cell types (Houmard et al. 1992). There may be a species difference in the responses of different cell types to $PGF_{2\alpha}$ but since the contribution of different cell types in the corpus luteum is known to change across the luteal phase (Farin et al. 1988), there might also be variation in response to $PGF_{2\alpha}$ in the different cells.

While the evidence described above supports a role for $PGF_{2\alpha}$ as the endogenous luteolysin in the primate corpus luteum, conclusive evidence from *in vivo* inhibition of $PGF_{2\alpha}$ action or production has been difficult to obtain, partly because of the auto or paracrine nature of $PGF_{2\alpha}$. We attempted passively to immunize marmoset monkeys (Webley, Kelly and Flint, unpublished observation) against $PGF_{2\alpha}$ and obtained high titres of antibody in the peripheral circulation. While luteal phases were extended in some treatment cycles the marmoset can spontaneously have luteal phases of extended duration (Harding *et al.* 1982) and no clear difference in treatment and control cycles was obtained. An attempt to inhibit prostaglandin synthesis with sodium meclofenamate (Sargent *et al.* 1988) in the rhesus monkey resulted in a shortening of the luteal phase suggesting a predominantly luteotrophic role for prostaglandins. However, subsequent *in vitro* studies showed that sodium meclofenamate also inhibits gonadotrophic support of luteal cell function (Zelinski- Wooten *et al.* 1990) so that shortening of the luteal phase may have resulted from inhibition of LH support rather than a decrease in luteal prostaglandin production. Further confirmation of the luteolytic role of $PGF_{2\alpha}$ in the primate corpus luteum awaits *in vivo* manipulation of the mechanism through which $PGF_{2\alpha}$ acts rather than inhibition of production.

III   THE DEVELOPMENT OF EMBRYONIC SIGNALS

CG is recognized as the principal embryonic signal that takes over the luteotrophic support of the corpus luteum and extends its life into early pregnancy.

Pregnancy is thought to depend on progesterone from the corpus luteum for the first 6–7 weeks in the human (Sapo and Pulkkinen 1978) and marmoset (Hearn 1983), while in the rhesus monkey pregnancy appears to become independent of the corpus luteum by weeks 4–5. In our assumption of the role of CG within these limits many questions are still left unanswered. Among these is the role of CG during later stages of pregnancy since the circulating levels of this hormone in maternal plasma are still relatively low in the human at 6 weeks, reaching a peak at 10–12 weeks and declining to 19 weeks but remaining at substantial levels for the rest of pregnancy. Functions such as the immunological protection of the fetoplacental unit, initiation of differentiation of the fetal gonads and the regulation of placental steroid production have all been suggested but are hard to prove.

1   The start of pregnancy

We focus here only on the first 2–3 weeks of pregnancy. During this time the embryo enters the uterus as a morula late on day 3 or early day 4 after ovulation. While some variation must be allowed, it appears that in Old World primates including the human embryo attachment is initiated on days 7–9 after ovulation whereas in the marmoset monkey, the only New World primate studied in detail to date, attachment is on days 11–12.

In the past 15 years, there have been numerous candidates identified as potential embryo–maternal signals. These include the early pregnancy factor (Morton *et al.* 1977), platelet-activating factor (O'Neil 1985), Schwangershaftsprotein (Sinosich *et al.* 1985), histamine-releasing factor (Cocchiara *et al.* 1987), and CG. Apart from the last of these, there has been a distinct lack of definitive results, although there are intriguing data that may yet confound the sceptic. We therefore concentrate here on the activation, secretion and biological activity of CG. Our hypotheses are that CG is:

(1) an embryo-derived glycoprotein responsible for the luteotrophic support of the corpus luteum during early pregnancy;
(2) with a possible intraembryonic role in the differentiation of the periimplantation trophectoderm; and
(3) a probable role at the embryo–endometrial interface to facilitate trophoblastic invasion.

The first of these hypotheses is widely accepted as proven although the mechanisms are not fully elucidated, while the second and third have yet to be proven and are still only partially tested through the indirect lines of evidence that we summarize below.

2   Activation of the embryonic genome

The maternal embryonic transition, with activation of the embryonic genome, occurs early in the mouse but is initiated by the 16-cell stage in the majority of mammalian species studied to date. The β-subunit of CG was detected in three of seven 6–8-cell human embryos by *in situ* hybridization (Bonduelle *et al.* 1988). These were triploid embryos the normality of which may be questioned (as may that of many of the 'spare' embryos available for study of human embryo differentiation). The suggestion that transcription and expression of CG is already attained at this early stage, before any clear cell lineage specialization has been identified, requires further definition. In addition we need to know if the α-subunit of CG is also transcribed this early, since the biological activity of CG depends on its function as a dimer, with the α-subunit holding structural and the β-subunit the biological effects.

One of the confusing variables at present in establishing the expression and secretion of biologically active CG is that the data available from

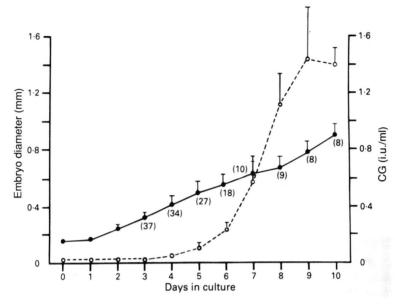

Fig. 1.3   The growth in embryo diameter ●——● and the onset of secretion of CG ○——○ in embryos recovered on day 8 after ovulation (blastocyst) and cultured for 10 days *in vitro*. Values are mean ± SEM for the number of embryos in parentheses.

the human have relied to a large extent on immunoassay kits, while the marmoset and rhesus work has relied substantially on biological assays of CG. However, as we show below, there is no evidence for the secretion of CG by preimplantation embryos until the blastocyst stages when it may be detectable at very low levels only before attachment commences. As yet, the stage specific activation of genes in the primate embryo is virtually unstudied, so that the ontogeny of protein expression and secretion is almost unknown.

3   The first secretion of CG

i   *The marmoset monkey*

In the marmoset monkey, embryos recovered by surgical flushing from the uterus and grown *in vitro* either on fibroblast monolayers, or on plastic, hatched, attached and expanded to spread trophoblast across the culture dish while the embryo itself routinely differentiated to yolk sac stages (Hearn *et al.* 1988*a*, 1991). Under these conditions, there was little or no evidence for detectable CG in the incubation medium until the time

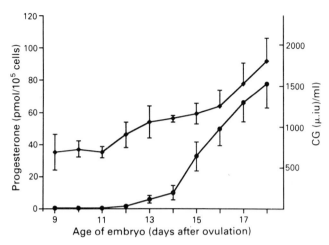

Fig. 1.4   Progesterone concentrations ◆ produced by marmoset luteal cells in monolayer culture and incubated for 24 h (after incubation for 48 h) in the presence of embryo-conditioned medium pooled from four embryos and cultured for 10 days from day 8 after ovulation to day 18. Concentrations of CG (●) in the embryo-conditioned medium were measured by mouse Leydig cell bioassay.

of attachment when an exponential rise occured. In a few samples, very low levels of CG were detected above background from preattachment embryos, but whether this is an abnormality of degenerating embryos has yet to be determined. Figure 1.3 shows an average growth curve for embryos cultured *in vitro* and the profile of CG secretion measured in these preparations. The clear pattern is for embryos to attach to the dish about 4 days after they are flushed from the uterus as 8-day blastocysts, giving a close approximation to our estimation of embryo attachment *in vivo* on days 11–12 after ovulation (Moore *et al.* 1985). Figure 1.4 shows the results of culturing dispersed marmoset luteal cells with incubation medium collected every 48 h from marmoset embryos. There was no confirmed elevation of luteal progesterone production until after the characteristic postattachment rise in CG in the media. Both the above experiments suggest, at least in the marmoset using a mouse Leydig cell bioassay adapted for this species (Hearn *et al.* 1988*b*) that substantial production of CG is not a characteristic of the preattachment embryo in the marmoset. However, as reported below, there may be conflicting evidence *in vivo* that the corpus luteum is recognizing an embryonic message (CG?) by day 14 after ovulation, i.e. 2 days after embryo attachment, but before any significant levels of CG can be detected in the peripheral plasma. There are also our data, summarized below, to show that elevated α-inhibin is detectable up to 4 days before embryo attachment in the conception cycle of the marmoset monkey. The recent report that preattachment human embryos secrete inhibin (Phocas *et al.* 1992) is perhaps a portent of future papers that will greatly extend the repertoire of embryonic products before implantation.

ii   The rhesus monkey

Recent studies in the rhesus monkey (Seshagiri and Hearn 1993; Hearn *et al.* 1993) show that bioactive CG can be measured in an *in vitro* embryo culture preparation to have a similar, postattachment, exponential rise as we described earlier in the marmoset. In the case of the rhesus, however, we report low but distinct and measurable CG secretion by the embryo into incubation media before attachment. Table 1.2 shows the levels monitored, suggesting that CG is already expressed and secreted by the blastocyst before hatching. This finding may suggest that CG has a local role at the implantation site. The recent reports that CG receptors may be present in the endometrial epithelium lend credence to such an effect.

iii   The human

Studies in the human, based on embryos grown in culture from *in vitro* fertilization through to blastocyst stages, suggest also that low levels of CG may be secreted immediately before attachment, while exponentially increasing levels are likewise a characteristic of postattachment development (Hay and Lopata 1988; Dokras *et al.* 1991). In current studies (Hay

Table 1.2   In vitro *secretion of CG by cultured rhesus monkey morula* (n=4) *and blastocysts* (n=3) *during the peri-attachment period*

Stage of embryo development	CG secretion[a] (ng/ml) (mean ± SEM; $n=7$)
Expanded and not hatched	0.014 ± 0.0
Hatched and not attached	1.7 ± 0.5
Attached (5–11 days)	122.7 ± 45.5
Attached (10–17 days)	5108.7 ± 1706
Attached (16–40 days)	317.0 ± 201.4

[a]Aliquots of spent medium (serum-supplemented-CMRL-1066) from cultured embryos were analysed by a sensitive mouse Leydig cell bioactive CG assay. The values are expressed as equivalents of hCG (CR 125), used for the standard curve. Data presented are after subtracting the non-specific value due to the blank culture medium.

and Lopata 1990), embryos are shown to respond to the addition of human cord serum or growth factors or insulin and selenium with stimulated rises in CG production. Although it may be likely that some of these embryos grown for several days *in vitro* are abnormal, the most likely interpretation from all of the above evidence is that CG is secreted by primate embryos at low levels before attachment and at rapidly increasing levels after attachment. The event of attachment itself may have consequences in the activation of CG regulating genes, the production and expression of attachment linked proteins, and the availability of bioactive hormone.

The function of preattachment CG may relate to the attachment process itself, acting via the corpus luteum and/or directly on the endometrium. It may be expressed in low levels for an as yet unknown function within the embryo: this is a dynamic stage in differentiation with rapid segregation into inner cell mass, endoderm and trophectoderm, presumably requiring careful intraembryonic signalling. However, we should keep in mind that elevated preattachment CG levels might also be an indicator of embryonic dedifferentiation and therefore of embryonic demise with an associated, deregulated release of CG.

4   Immunization against CG

It is now well established that active or passive immunization against the β-subunit of CG prevents implantation or terminates early stages of pregnancy (reviewed by Hearn 1979). The results were similar in marmosets (Hearn 1976), baboons (Stevens 1976), and rhesus monkeys (Talwar *et al.* 1976). Active immunization during early pregnancy resulted in raised antibody levels and consequent abortions 3–4 weeks later, while passive immunization produced similar results for 10 days. In addition, when marmoset embryos were incubated individually with anti-hCGβ antiserum drawn from the actively immunized marmosets noted above, the embryos failed to attach and outgrow (Hearn *et al.* 1988*a*). This effect *in vitro* requires confirmation with larger numbers of embryos, but it suggests that antisera to CG can have a direct inhibitory effect on the attachment process *in vitro*, as it has *in vivo*. The corollary is that CG may be required locally at the attachment site in addition to its more conventionally accepted role in stimulating the corpus luteum.

5   The regulation of CG secretion

The intraembryonic regulation of CG secretion and release has yet to be investigated in detail and our working hypothesis at present is an analogue of the hypothalamopituitary release of gonadotrophins, i.e. that CG release from the peri-implantation embryo is controlled by intraembryonic gonadotrophin-releasing hormone (GnRH). The rationale for this hypothesis derives from:

(1) the similar molecular structure between CG and LH, with CG being essentially an LH molecule with a 33 amino acid addition at the C-terminal end; and
(2) the demonstrated presence of CG and GnRH in placental trophoblast later in pregnancy.

Our preliminary findings show that GnRH may be located on the preimplantation embryo by immunocytochemistry and that low levels of GnRH may be measured in the incubation media of individual embryos cultured *in vitro*. Table 1.3 shows the production of CG and GnRH from individual embryos. While, at present, this data merely show an association between these two hormones we are progressing to test their causal relationship through manipulation with analogues and with antisera.

Recent advances in our and others' laboratories give us some confidence that progress can now be made. We have shown that morula and blastocyst stage embryos can be recovered routinely by invasive and non-invasive embryo flushing techniques in the marmoset and rhesus monkeys (Seshagiri and Hearn 1993). We have been able to analyse protein production in single rhesus monkey embryos, using two-dimensional gel systems, and we are now able to co-culture blastocysts with endometrial epithelial cell lines. We have yet to achieve stage specific *in situ* hybridization to locate precisely the sites of production of these hormones in the blastocyst; and we are close to achieving the polymerase chain reaction amplification of messenger RNAs

Table 1.3   In vitro *secretion of GnRH and CG by cultured rhesus monkey morulae* (n=10) *and blastocysts* (n=10) *during the peri-attachment period*

Stage of embryo development	GnRH secretion[a] (pg/ml) (mean $\pm$ SEM; n=20)	CG secretion[a] (ng/ml) (mean $\pm$ SEM; n=10)
Expanded and not hatched	0.32 ± 0.05	$<0.01 \pm 0.0$
Hatched and not attached	0.52 ± 0.04	1.6 ± 0.4
Post-attached (1–5 days)	0.62 ± 0.07	101.10 ± 37.0
Post-attached (6–12 days)	0.70 ± 0.08	2604.8 ± 1036.3
Post-attached (13–60 days)	1.30 ± 0.23	603.04 ± 245.1

[a]Aliquots of spent-media samples of cultured embryos were analysed by a sensitive RIA for GnRH or mouse Leydig cell bioassay for CG. Data are presented after substracting the non-specific value due to the blank culture medium. (from Seshagiri *et al.* in Press).

(mRNAs) for these proteins in morulae and blastocysts. When successful, our objective in these studies is further to question the intraembryonic signals that control this critical period of embryonic differentiation and their responsibility for the embryo–maternal signals on which embryo survival itself depends.

IV   THE ESTABLISHMENT OF EARLY PREGNANCY

It is widely accepted that CG secreted by the embryo is responsible for 'rescue' of the corpus luteum and maintenance of progesterone production in early pregnancy. The luteotrophic action of CG has been well described in several primate species including the rhesus monkey and human (Knobil 1973; Baird 1985; Stouffer *et al.* 1989). In the marmoset monkey, direct perfusion of hCG through the corpus luteum with a microcannula *in vivo* stimulated an immediate increase in progesterone production (Hearn and Webley 1987; Webley and Hearn 1987).

The three primate species listed above differ in the pattern of plasma progesterone associated with an increase in plasma CG concentrations in early pregnancy. In women, the increase in hCG concentrations detected in the circulation from day 9 after ovulation paralleled an increase in progesterone concentrations above mid-luteal levels (Lenton and Woodward 1988). In contrast, in the marmoset monkey plasma progesterone concentrations in early pregnancy are maintained at similar levels to those in the mid-luteal phase (Webley *et al.* 1991*b*) whereas in the rhesus monkey, the increase in CG concentrations in early pregnancy or following hCG treatment leads to an initial increase followed by a decline in plasma progesterone concentrations, although CG levels continue to rise (Wilks and Noble 1983; Stouffer *et al.* 1987).

1   Rescue of the corpus luteum

i   Progesterone

How is CG able to prevent luteolysis and maintain these varying patterns of progesterone production in early pregnancy? We have studied the interaction of hCG and cloprostenol on progesterone production by human granulosa cells luteinized *in vitro* (Michael and Webley 1991*a*; Webley *et al.* 1991*c*). Using a protocol of 3 days of pre-culture followed by 24 h of exposure to treatment, cloprostenol had an antigonadotrophic action on progesterone production (Michael and Webley, 1991*a*; Webley *et al.* 1991*b*). This was, however, prevented if the cells were exposed to hCG (or human LH) for the 3 days prior to treatment. Exposure to CG at a critical time during the luteal phase may be a mechanism by which CG can prevent the subsequent luteolytic action of endogenous $PGF_{2\alpha}$. The inability of cloprostenol to cause luteolysis in the early luteal phase of the marmoset monkey may similarly result from pre-exposure to high endogenous LH concentrations at the LH surge (Michael and Webley 1993).

The variable progesterone responses to increased CG concentrations may be due to species variation in the requirement by the CL for luteotrophic support. There is a marked difference in the ability of hCG to stimulate progesterone production by dispersed human and marmoset luteal cells incubated in suspension (Webley et al. 1989). Progesterone production by human cells is stimulated by up to eightfold whereas production by marmoset cells only increases by 50%. While this may be due to species variation in gonadotrophin structure or the sensitivity of receptors, basal peripheral concentrations of progesterone are considerably higher in marmosets than in women (Hodges et al. 1983).

The transient increase in progesterone concentrations during early pregnancy in the rhesus monkey is considered to be due to a change in steroidogenic response of luteal cells to gonadotrophin (Stouffer et al. 1989). The initial rise in progesterone coincides with the onset of CG secretion at the time of implantation on days 9–11 postovulation. The ability of CG to stimulate cAMP then decreases with continued exposure, although other luteotrophins such as PGE_2 are still able to stimulate cAMP production (Vandevoort et al. 1988) suggesting a homologous desensitization of the LH/hCG–cAMP pathway. We have seen a change in the progesterone response of luteal cells from the marmoset monkey in early pregnancy (Webley et al. 1989). On day 14 after ovulation incubated luteal cells from pregnant animals showed a higher basal production of progesterone and a decreased response to stimulation as compared with cells from non-pregnant animals. This was within 2 days of embryo attachment at a time when CG cannot yet be detected in the peripheral circulation (Hearn et al. 1988a) and before there is any morphological distinct vascular connection (Smith et al. 1987). By day 20, plasma CG concentrations were increasing rapidly yet luteal cells from day 20 pregnant animals have a lower basal production of progesterone than cells from day 14 pregnant animals and respond to stimulation by luteotrophins (Webley et al. 1989). Although different in nature the apparent changes in the luteal cells of early pregnancy in the rhesus and marmoset monkeys show that luteal 'rescue' is more complex than a simple relationship between increased CG and progesterone production.

Our studies in the marmoset monkey show that, just 2 days after implantation, luteal cells appear to have been exposed to an endogenous luteotrophin (Webley et al. 1989). This suggests that the corpus luteum may be able to respond to CG at very low concentrations before it can be detected in the peripheral circulation. Alternatively, the embryo may secrete a different message to prevent luteolysis before CG reaches sufficient concentration in the circulation to assume the luteotrophic role.

Our data so far suggest the first option since we have demonstrated in the marmoset monkey that embryo- conditioned medium from embryos cultured between days 8 and 18 after ovulation can stimulate progesterone production by luteal cells in culture. Progesterone concentrations increased with increase in CG concentrations in the embryo-conditioned medium from day 12 after ovulation, the time of embryo attachment *in vivo* (Moore *et al.* 1985) and *in vitro*.

We have yet to detect CG secretion by the marmoset embryo prior to attachment but our recent data from rhesus embryos in culture show low levels of preattachment secretion of CG (Seshageri and Hearn 1993). There have also been previous reports in the rhesus monkey of CG production by preimplantation blastocysts with detection of CG in ovarian venous blood prior to detection in the peripheral circulation (Hodgen and Itskovitz 1988). Our *in vitro* data in the marmoset monkey have not shown a preattachment influence of embryo secretions on progesterone production. However, measurement of other luteal products in particular the peptide hormone, inhibin, indicate that there are changes in the corpus luteum of pregnancy before implantation.

ii   Inhibin

We have measured immunoreactive inhibin concentrations in the peripheral circulation of the marmoset monkey using a radioimmunoassay based on antisera against a synthetic fragment of the α-subunit of human inhibin (Webley *et al.* 1991*b*; Knight *et al.* 1992). As described in women (McLachlan *et al.* 1987*c*), stump-tailed macaque (Fraser *et al.* 1989), and by others in the marmoset monkey (Smith *et al.* 1990) immunoreactive inhibin concentrations increase in the luteal phase of the cycle. We detected a further increase in the luteal phase after conception compared with non-conception cycles (Webley *et al.* 1991*b*). Significantly ($P<0.05$) higher concentrations of inhibin were recorded from days 8/9 after ovulation (Fig. 1.5), 4 days before the expected time of implantation, and were not related to any detectable increase in peripheral concentrations of CG. An increase in immunoreactive inhibin concentrations following conception has also been recorded in women (McLachlan *et al.* 1987*b*). However, in women, the first increase in plasma immunoreactive inhibin concentrations coincided with the time of implantation and the first increase in hCG concentrations such that parallel rises in inhibin, hCG, and progesterone concentrations were recorded after day 9.

The corpus luteum appears to be the major source of inhibin in the luteal phase of conception and nonconception cycles in the primate. Administration of a luteinizing hormone–releasing hormone (LHRH) antagonist during the luteal phase of women (McLachlan *et al.* 1989), stump-tailed macaques (Smith and Fraser, 1991), and marmoset monkeys (Webley *et al.* 1991*b*) is associated with a decrease in inhibin as well as progesterone

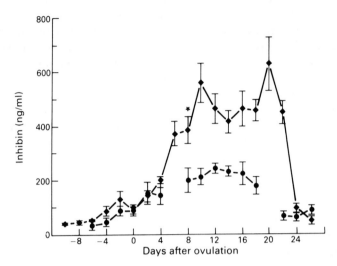

Fig. 1.5.   Concentrations of immunoreactive inhibin (mean ± SEM) in plasma samples from non-pregnant ($n=5$, ●) and pregnant ($n=6$, ◆). *$P<0.05$, first time in the cycle when immunoreactive inhibin concentrations in pregnant animals were significantly higher than in non-pregnant animals. Embryo implantation commences on days 11–12.

concentrations. Further disruption of luteal function with $PGF_{2\alpha}$ in the marmoset (Webley *et al.* 1991*b*) or lutectomy in cynomologous monkeys (Basseti *et al.* 1990) also results in a decrease in inhibin concentrations. The marmoset corpus luteum has a high content of inhibin (Knight *et al.* 1992) and human luteinized granulosa cells secrete inhibin in culture (Tsonis *et al.* 1987). Furthermore, the human and marmoset corpus luteum contain mRNA for the α-subunit of inhibin (Davies *et al.* 1987; Hillier *et al.* 1989). In fact, luteal production of immunoreactive inhibin in the marmoset appears to be predominantly inhibin α-subunit since the luteal increase in production could not be detected using a two-site immunoradiometric assay specific for dimeric inhibin (α- and β- subunit; Knight *et al.* 1992). Since it is only dimeric inhibin which is known to have bioactivity in suppressing follicle-stimulating hormone release from the pituitary (Knight 1991), the role of increased α-inhibin is unclear. Our preliminary data suggest that α-inhibin may have a role in potentiating the progesterone response to CG.

The embryo itself has been suggested as a source of inhibin in early pregnancy. In a study in which pregnancy was maintained in women with non-functional ovaries after oocyte donation, inhibin concentrations rose within 2–4 weeks of transfer (McLachlan *et al.* 1987*a*). We were, however, unable to detect inhibin production by marmoset embryos in culture although inhibin production by human pre-embryos before formation of the cytotrophoblast has been described (Phocas *et al.* 1992). Our results in the marmoset (Webley *et al.* 1992), obtained by flushing embryos from the uterus between days 5 and 9 after ovulation and measuring subsequent concentrations of peripheral immunoreactive inhibin concentrations, suggest that the marmoset embryo provides a stimulus on days 7/8 after ovulation, which triggers an increase in luteal immunoreactive inhibin production. The nature of this stimulus is unclear but its continued presence was not necessary to maintain increased immunoreactive inhibin production.

iii   Regulation of inhibin and progesterone

It appears that there may be differential control of progesterone and inhibin production by the corpus luteum of early pregnancy. The significant increase in immunoreactive inhibin concentrations from day 8 after ovulation in conception cycles in the marmoset monkey is not associated with a similar increase in progesterone concentrations (Webley *et al.* 1992). The stimulus provided by the embryo is therefore not effective in stimulating a further increase in progesterone production at this time in the luteal phase. The ability of hCG to stimulate inhibin production has been demonstrated on dispersed human luteal cells (Han-zheng *et al.* 1992) and human luteinized granulosa cells (Tsonis *et al.* 1987). *In vivo*, hCG can prevent the decline in inhibin after administration of LHRH antagonist (McLachlan *et al.* 1989; Smith and Fraser 1991; Webley *et al.* 1991*b*), and 'rescue' of the corpus luteum with physiological concentrations of hCG to women stimulated a significant increase in inhibin production (Illingworth *et al.* 1990). In a preliminary study, dispersed marmoset luteal cells taken on day 8 after ovulation were incubated for 24 h in serum-free medium. Increasing concentrations of hCG stimulated immunoreactive inhibin production whereas PGE_2 inhibited production in a dose-dependent manner. In contrast, under these conditions, hCG had no significant effect whereas PGE_2 stimulated a dose–dependent increase in progesterone production. Differential control of progesterone and inhibin production has also been described in macaque luteinized granulosa cells (Brannian *et al.* 1992). Although both hCG and PGE_2 stimulated progesterone and inhibin the pattern of response and the optimum culture conditions required varied for the two hormones.

iv   Relaxin

Another peptide hormone produced by the primate corpus luteum which appears to show a different pattern of production to progesterone is relaxin. In the rhesus monkey, circulating levels of relaxin immunoreactivity increased in early pregnancy at a time when progesterone concentrations were declining. The increase in relaxin could be induced by administration of hCG but the response depended on the age of the corpus luteum (Ottobre *et al*. 1984). In women, relaxin has been shown to be exclusively derived from the corpus luteum (Johnson *et al*. 1991*a*) and to be stimulated by exogenous administration of hCG (Quagliarello *et al*. 1980; Johnson *et al*. 1991*b*). Relaxin is detectable in the peripheral circulation during the non-pregnant luteal phase with the peak in concentrations occurring 10–12 days after ovulation when luteal function is declining (Stewart *et al*. 1990). The relaxin secretion pattern is therefore delayed by 6–9 days compared with progesterone production.

The differences in the pattern of production of progesterone, relaxin, and inhibin provide additional ways of assessing luteal function and, in particular, provide insight into the mechanism of luteal 'rescue' in early pregnancy. It would appear that all three hormones are stimulated by CG but they show very different patterns of response to the increasing concentrations of CG detected in early pregnancy and can indeed increase production prior to detection of CG, implicating production of alternative embryo signals.

V   CONCLUSIONS

The establishment of early pregnancy in primates is quite different to that of non-primate species. It appears to depend on different mechanisms, including the secretion of CG by the peri-implantation embryo and the recognition of this signal by the corpus luteum, which is then extended into pregnancy (Fig. 1.6).

In the primates studied so far, these mechanisms follow a similar pattern albeit with minor species variation in the timing of secretion of CG and in the relative susceptibility of the corpus luteum to the luteolytic effects of $PGF_{2\alpha}$.

The control of luteolysis in primates depends on an intraovarian, probably an intraluteal paracrine system, in which $PGF_{2\alpha}$ is synthesized potentially at any stage during the luteal phase. The luteolytic action evoked may be due to a change in the second messenger response to $PGF_{2\alpha}$ rather than to a change in its concentration.

The regulation of CG release is still hypothetical but may be controlled by intraembryonic GnRH. The ability to identify the ontogeny of stage-specific gene expression in individual primate embryos, together with the resultant proteins including embryo-maternal signals may have relevance to the future regulation of fertility.

In the corpus luteum of pregnancy, there are characteristic increases in the secretion of progesterone, inhibin, and relaxin. The temporal relations between these three hormones and CG differ among themselves and between primate species. Therefore we are still unsure of the definitive

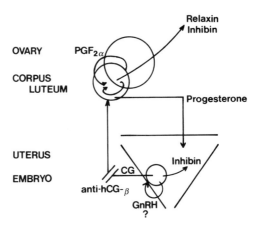

Fig. 1.6   The major interactions during the rescue of the corpus luteum at the start of pregnancy in primates. Embryo-derived CG, regulated by GnRH(?) overcomes the PGF $_{2\alpha}$ induced decline of the corpus luteum. The corpus luteum continues to produce progesterone. There is an association between the exponential rise in CG and the production of inhibin and relaxin from the corpus luteum. This association varies temporally in different primate species, with the ovarian inhibin rise preceding by 4 days embryo attachment in the marmoset. CG release from the embryo is at low levels before attachment, rising exponentially once attachment is achieved.

mechanisms involved in the process of luteolysis and its reversal at a critical stage in the early establishment of pregnancy.

ACKNOWLEDGEMENTS

We acknowledge the collaborative contributions of Drs P.B. Seshagiri, P.G. Knight, and A.E. Michael in aspects of this research. We thank Philip Caron, Steve Eisele, Scott Kudia, Denny Mohr, Joan Scheffler, and Pippa Marsden for their excellent technical assistance. We acknowledge programme grant support form UK Medical Research and Agriculture and Food Research Councils, programme support to J.P.H. from NIH Grant RR-00167 and project support from the WHO Special Programme of Research, Training and Research Development in Human Reproduction.

REFERENCES

Abayasekara, D.R.E., Jones, P.M., Persaud, S.J., Michael, A.E., and Flint, A.P.F. (1993). Prostaglandin F2α activates protein kinase C in human ovarian cells. *Molecular and Cellular Endocrinology*, **97**, 81–91.

Atkinson, L.E., Hotchkiss, J., Fritz, G.R, Surve, A.H., Newill, J.D., and Knobil, E. (1975). Circulating levels of steroids and chorionic gonadotrophin during pregnancy in the rhesus monkey, with special attention to the rescue of the corpus luteum in early pregnancy. *Biology of Reproduction*, **12**, 335–45.

Auletta, F.J., Kamps, D.L., Pories, S., Bisset, J., and Gibson, M. (1984). An intra-corpus luteum site for the luteolytic action of prostaglandin F2α in the rhesus monkey. *Prostaglandins*, **27**, 285–98.

Baird, D.T. (1985). Control of luteolysis. In *The luteal phase* (ed. S.L. Jeffcoate), pp. 25–43. John Wiley and Sons, Chichester.

Balmaceda, J., Asch, R.H., Fernandez, E.O., Velenzuela, G., Eddy, C.A., and Pauerstein, C.J. (1979). Prostaglandin production by rhesus monkey corpora lutea *in vitro*. *Fertility and Sterility*, **31**, 214–16.

Basseti, S.G., Winters, S.J., Keeping, H.S., and Zeleznik, A.J. (1990). Serum immunoreactive inhibin levels before and after lutectomy in the cynomolgus monkey (*Macaca fascicularis*). *Journal of Clinical Endocrinology and Metabolism*, **70**, 590–4.

Beling C.G., Marcus, S.L., and Markham, S.M. (1970). Functional activity of the corpus luteum following hysterectomy. *Journal of Clinical Endocrinology*, **30**, 30–9.

Bonduelle, M.L., Dodd, R., Liebaers, I., Steirteghem, A. Van., Williamson, R., and Akhurst, R. (1988). Chorionic gonadotrophin-β mRNA, a trophoblast marker, is expressed in human 8-cell embryos derived from tripronucleate zygotes. *Human Reproduction*, **3**, 909–14.

Braden, T.D., Gamboni, F., and Niswender, G.D. (1988). Effect of prostaglandin F$_{2\alpha}$-induced luteolysis on populations of cells in the ovine corpus luteum. *Biology of Reproduction*, **39**, 245–53.

Brannian, J.D. and Stouffer, R.L. (1991). Progesterone production by monkey luteal cell subpopulations at different stages of the menstrual cycle: changes in agonist responsiveness. *Biology of Reproduction*, **44**, 141–9.

Brannian, J.D., Stouffer, R.L., Molskness, T.A., Chandrasekher, Y.A., Sarkissian, A., and Dahl, K.D. (1992). Inhibin production by macaque granulosa cells from pre- and periovulatory follicles: regulation by gonadotrophins and prostaglandin E2. *Biology of Reproduction*, **46**, 451–7.

Challis, J.R.G., Calder, A.A., Dilley, A., Forster, S.S., Hillier, K., Hunter, D.J.S., *et al.* (1976). Production of prostaglandins E and F2α by corpus luteum, corpora albicantes and stroma from the human ovary. *Journal of Endocrinology*, **68**, 401–8.

Cocchiara, R., Di Trapani, G., Azzolina, A., Albeggiani, G., Ciriminna, R., Cefalu, E., *et al.* (1987). Isolation of a histamine releasing factor from human embryo culture medium after *in vitro* fertilization. *Human Reproduction*, **2**, 341–4.

Csapo, A.I. and Pulkkinen, M.O. (1978). Indispensability of the human corpus luteum in the maintenance of early pregnancy: lutectomy evidence. *Obstetrics and Gynecology Survey*, **110**, 69–89.

Davies, S.R., Krozowski, Z., McLachlan, R.I., and Burger, H. G. (1987). Inhibin gene expression in the human corpus luteum. *Journal of Endocrinology*, **115**, R21–3.

Davis, J.S., Alila, H.W., West, L.A., Corradino, R.A., Weakland, L.L., and

Hansel, W. (1989). Second messenger systems and progesterone secretion in the small cells of the bovine corpus luteum: effects of gonadotrophin and prostaglandin F2α. *Journal of Steroid Biochemistry*, **32**, 643–9.

Dennefors, B.L., Sjorgen, A., and Hamberger, L. (1982). Progesterone and adenosine 3,5-monophosphate formation by isolated human corpora lutea of different ages: influence of human chorionic gonadotrophin and prostaglandins. *Journal of Clinical Endocrinology and Metabolism*, **55**, 102–7.

Dokras, A., Sargent, I.L., Ross, C., Gardner, R.L., and Barlow, D.H. (1991). The human blastocyst: morphology and human chorionic gonadotrophin secretion *in vitro*. *Human Reproduction*, **6**, 1143–51.

Eley, R.M., Summers, P.M., and Hearn, J.P. (1987). Failure of the prostaglandin F2α analogue, cloprostenol, to induce functional luteolysis in the olive baboon (*Papio cynocephlaus*). *Journal of Medical Primatology*, **16**, 1–12.

Farin, C.E., Moeller, C.L., Mayan, H., Gamboni, F., Sawyer, H.R., and Niswender, G.D. (1988). Effect of luteinizing hormone and human chorionic gonadotropin on cell populations in the ovine corpus luteum. *Biology of Reproduction*, **38**, 413–21.

Fisch, B., Margara, R.A., Winston, R.M.L., and Hillier, S.G. (1989). Cellular basis of luteal steroidogenesis in the human ovary. *Journal of Endocrinology*, **122**, 303–11.

Fraser, H.M., Robertson, D.M., and De Kretser, D.M. (1989). Immunoreactive inhibin concentrations in serum throughout the menstrual cycle in the macaque: suppression of inhibin during the luteal phase after treatment with an LHRH antagonist. *Journal of Endocrinology*, **121**, R9–12.

Hahlin, M., Bennegard, B., and Dennefors, B. (1986). Human luteolysis-interaction between HCG and oestradiol-17β in an *in-vitro* model. *Human Reproduction*, **1**, 75–9.

Hamberger, L., Nilsson, L., Dennefors, B., Khan, I., and Sjorgen, A. (1979). Cyclic AMP formation of isolated human corpora lutea in response to hCG–interference by PGF2α. *Prostaglandins*, **17**, 615–21.

Hansel, W. and Dowd, J.P. (1986). New concepts of the control of corpus luteum function. *Journal of Reproduction and Fertility*, **78**, 755–68.

Han-zheng, W., Shu-hua, L., Xiang-jun, H., Zhi-da, S., Wei-xiong, S., Wei, Z., et al. (1992). Control of inhibin production by dispersed human luteal cells *in vitro*. *Reproduction, Fertility and Development*, **4**, 67–75.

Harding, R.D., Hulme, M.J., Lunn, S.F., Henderson, C., and Aitken, R.J. (1982). Plasma progesterone levels throughout the ovarian cycle of the common marmoset monkey (*Callithrix jacchus*). *Journal of Medical Primatology*, **11**, 43–51.

Harlow, C.R., Gems, S., Hodges, J.K., and Hearn, J.P. (1983). The relationship between plasma progesterone and the timing of ovulation and early embryonic development in the marmoset monkey (*Callithrix jacchus*). *Journal of Zoology*, **210**, 273–82.

Hay, D.L. and Lopata, A. (1988). Chorionic gonadotrophin secretion by human embryos *in vitro*. *Journal of Clinical Endocrinology and Metabolism*, **67**, 1322–4.

Hay, D.L. and Lopata, A. (1990). Factors modifying early hCG production by the human blastocyst *in vitro*. *Assisted Reproductive Technology and Andrology*, **1**, 9–23.

Hearn, J.P. (1976). Immunization against pregnancy. *Proceedings of the Royal Society of London, Series B*, **195**, 149–60.

Hearn, J.P. (1979). Immunological interference with the maternal recognition of pregnancy in primates. In *Maternal recognition of pregnancy* (Ciba Foundation Symposium 64), pp. 353–75. Excerpta Medica, Amsterdam.

Hearn, J.P. (1983). The common marmoset monkey (*Callithrix jacchus*). In *Reproduction in New World primates* (ed. J.P. Hearn), pp. 181–216. MTP Press, Lancaster.

Hearn, J.P. and Webley, G.E. (1987). Regulation of the corpus luteum of early pregnancy in the marmoset monkey: local interactions of luteotrophic and luteolytic hormones *in-vivo* and their effects on the secretion of progesterone. *Journal of Endocrinology*, **114**, 231–9.

Hearn, J.P., Gidley-Baird, A.A., Hodges, J.K., Summers, P.M., and Webley, G.E. (1988a). Embryonic signals during the peri-implantation period in primates. *Journal of Reproduction and Fertility*, Suppl. **36**, 49–58.

Hearn, J.P., Hodges, J.K., and Gems, S. (1988b). Early secretion of chorionic gonadotrophin by marmoset embryos *in-vitro* and *in-vivo*. *Journal of Endocrinology*, **119**, 249–55.

Hearn, J.P., Webley, G.E., and Gidley-Baird, A.A. (1991). Chorionic gonadotrophin and embryo–maternal recognition during the peri-implantation period in primates. *Journal of Reproduction and Fertility*, **92**, 497–509.

Hearn, J.P., Seshagiri, P.B. and Webley, G.E. (1993). *The Physiology of Implantation in Primates*. Serono Symposium, Oregon, (in press).

Hilde-Petito, S.A., Shighi, S.M., and Stouffer, R.L. (1989). Isolation and characterization of cell subpopulations from the rhesus monkey corpus lutem of the menstrual cycle. *Biology of Reproduction*, **40**, 1975–85.

Hillier, S.G., Wickings, E.J., Saunders, P.T.K., Dixson, A.F., Shimasaki, S., Swanston, I.A., *et al.* (1989). Control of inhibin secretion by primate granulosa cells. *Journal of Endocrinology*, **123**, 65–73.

Hodgen, G.D. and Itskovitz, J. (1988). Recognition and maintenance of pregnancy. In *'The Physiology of Reproduction'* (ed. E. Knobil and J. Neill), pp. 1995–2021. Raven Press, New York.

Hodges, J.K., Henderson, C., and Hearn, J.P. (1983). Relationship between ovarian and placental steroid production during early pregnancy in the marmoset monkey *(Callithrix jacchus)*. *Journal of Reproduction and Fertility*, **69**, 613–21.

Horton, E.W. and Poyser, N.l. (1976). Uterine luteolytic hormone: a physiological role for prostaglandin F2α. *Physiological Reviews*, **56**, 595–651.

Houmard, B.S. and Ottobre, J.S. (1989). Progesterone and prostaglandin production by primate luteal cells collected at various stages of the luteal phase: modulation by calcium ionophore. *Biology of Reproduction*, **41**, 401–8.

Houmard, B.S., Guan, Z., Kim-Lee, M., Stokes, B.T., and Ottobre, J.S. (1991). The effects of elevation and depletion of intracellular free calcium on progesterone and prostaglandin production by the primate corpus luteum. *Biology of Reproduction*, **45**, 560–5.

Houmard, B.S., Guan, Z., Stokes, B.T., and Ottobre, J.S. (1992). Activation of the phosphatidylinositol pathway in the primate corpus lutem by prostaglandin F2α. *Endocrinology*, **131**, 743–8.

Illingworth, P.J., Reddi, K., Smith, K., and Baird, D.T. (1990). Pharmacological 'rescue' of the corpus lutem results in increased inhibin production. *Clinical Endocrinology*, **33**, 323–32.

Johnson, M.R., Abdalla, H., Allman, A.C.J., Wren, M.E., Kikland, A., and Lightman, S.L. (1991*a*). Relaxin levels in ovum donation pregnancies. *Fertility and Sterility*, **56**, 59–61.

Johnson, M.R., Okokon, E., Collins, W.P., Sharma, V., and Lightman, S.L. (1991*b*). The effect of human chorionic gonadotrophin and pregnancy on the circulating level of relaxin. *Journal of Clinical Endocrinology and Metablism*, **72**, 1042–7.

Karim, S.M. and Hillier, K. (1979). Prostaglandins in the control of animal and human reproduction. *British Medical Bulletin*, **35**, 173–80.

Knight, P.G. (1991). Identification and purification of inhibin and inhibin-related proteins. *Journal of Reproduction and Fertility, Suppl.* **43**, 111–23.

Knight, P.G., Muttukrishna, S., Groome, N., and Webley, G.E. (1992). Evidence that most of the radioimmunoassayable inhibin secreted by the corpus luteum of the common marmoset monkey is of a non-dimeric form. *Biology of Reproduction*, **47**, 554–60.

Knobil, F. (1973). On the regulation of the primate corpus luteum. *Biology of Reproduction*, **8**, 246–58.

Lahav, M., Davis, J.S., and Rennert, H. (1989). Mechanism of the luteolytic action of prostaglandin F2α in the rat. *Journal of Reproduction and Fertility, Suppl.* **37**, 233–40.

Lenton, E.A. and Woodward, A.J. (1988). The endocrinology of conception cycles and implantation in women. *Journal of Reproduction and Fertility, Suppl.* **36**, 1–15.

McCracken, J.A., Carlson, J.C., Glew, M.E., Goding, J.R., Baird, D.T., Green, K., and Samuelsson, B. (1972). Prostaglandin F2α is identified as a luteolytic hormone in sheep. *Nature*, **238**, 129–34.

McLachlan, R.I., Healy, D.L., Lutjen, P.L., Findlay, J.K., De Kretser, D.M., and Burger, H.G. (1987*a*). The maternal ovary is not the source of circulating inhibin levels during human pregnancy. *Clinical Endocrinology*, **27**, 663–8.

McLachlan, R.I., Healy, D.L., Robertson, D.M., Burger, H.G., and De Kretser, D.M. (1987*b*). Circulating immunoreactive inhibin in the luteal phase and early gestation of women undergoing ovulation induction. *Fertility and Sterility*, **48**, 1001–5.

McLachlan, R.I., Robertson, D.M., Healy, D.L., Burger, H.G., and De Kretser, D.M. (1987*c*). Circulating immunoreactive inhibin levels during the normal human menstrual cycle. *Journal of Clinical Endocrinology and Metabolism*, **65**, 954–61.

McLachlan, R.I., Cohen, N.I., Vale, W.W., Rivier, J.E., Burger, H.G., Bremner, W.J., and Soules, M.R. (1989). The importance of luteinizing hormone in the control of inhibin and progesterone secretion by the human corpus luteum. *Journal of Clinical Endocrinology and Metabolism*, **68**, 1078–85.

McNatty, K.P., Henderson, K.M., and Sawers, R.S. (1975). Effects of prostaglandin F2α and E2 on the production of progesterone by cultured human granulosa cells in tissue culture. *Journal of Endocrinology*, **67**, 231–40.

Michael, A.E. and Webley, G.E. (1991*a*). Prior exposure to gonadotrophins prevents the subsequent antigonadotrophic effects of cloprostenol by a cyclic AMP-dependent mechanism in cultured human granulosa cells. *Journal of Endocrinology*, **131**, 319–25.

Michael, A.E. and Webley, G.E. (1991*b*). Prostaglandin F2α stimulates cAMP phosphodiesterase via protein kinase C in cultured human granulosa cells. *Molecular and Cellular Endocrinology*, **82**, 207–14.

Michael, A.E. and Webley, G.E. (1993). Roles of cyclic AMP and inositol phosphates in the luteolytic action of cloprostenol, a prostaglandin F2α analogue, in marmoset monkeys (*Callithrix jacchus*). *Journal Reproduction and Fertility*, **97**, 425–31.

Moore, H.D.M., Gems, S., and Hearn, J.P. (1985) Early implantation stages in the marmoset monkey (*Callithrix jacchus*). *American Journal of Anatomy*, **172**, 265–78.

Morton, H., Rolfe, B., Clunie, G.J.A., Anderson, M.J., and Morrisson, J. (1977). An early pregnancy factor detected in human serum by the rosette inhibition test. *Lancet*, **i**, 394–7.

Neill, J.D., Johansson, E.D.B., and Knobil, E. (1969). Failure of hysterectomy to influence the normal pattern of cyclic progesterone secretion in the rhesus monkey. *Endocrinology*, **84**, 464–5.

O'Neill, C. (1985). Examination of causes of early pregnancy associated thrombocytopenia in mice. *Journal of Reproduction and Fertility*, **73**, 567–77.

Ohara, A., Mori, T., Taii, S., Ban, C., and Narimoto, K. (1987). Functional differentiation in steroidogenesis of two types of luteal cells isolated from mature human corpora lutea of menstrual cycle. *Journal of Clinical Endocrinology and Metabolism*, **65**, 1192–200.

Osipchuk, Y. and Cahalan, M. (1992). Cell-to-cell spread of calcium signals mediated by ATP receptors in mast cells. *Nature*, **359**, 241–4.

Ottobre, J.S., Nixon, W.E., and Stouffer, R.L. (1984). Induction of relaxin secretion in rhesus monkeys by hCG: dependence on the age of the corpus luteum of the menstrual cycle. *Biology of Reproduction*, **31**, 1000–6.

Patwardhan, V.V. and Lanthier, A. (1984). Effect of prostaglandin F2α on the hCG-stimulated progesterone production by human corpora lutea. *Prostaglandins*, **27**, 465–73.

Patwardhan, V.V. and Lanthier, A. (1985). Luteal phase variation in endogenous concentrations of prostaglandin E2 and PGF2α and in the capacity for their *in vitro* formation in the human corpus luteum. *Prostaglandins*, **30**, 91–8.

Phocas, I., Sarandakou, A., Rizos, D., Dimitriadou, F., Mantzavinos, T., and Zourlas, P.A. (1992). Secretion of α-immunoreactive inhibin by human pre-embryos cultured *in vitro*. *Human Reproduction*, **7**, 545–9.

Powell, W.S., Hammarstrom, S., Samuelsson, B., and Sjoberg, B. (1977). Prostaglandin F2α receptor in human corpora lutea. *Lancet*, **i**, 1120.

Quagliarello, J., Goldsmith, L., Steinetz, B., Lustig, D.S., and Weiss, G. (1980). Induction of relaxin secretion in nonpregnant women by human chorionic gonadotrophin. *Journal of Clinical Endocrinology and Metabolism*, **51**, 74–7.

Richardson, M.C. and Mason, G.M. (1980). Progesterone production by dispersed cells from human corpus luteum: stimulation by gonadotrophins and prostaglandin F2α; a lack of response to adrenaline and isoprenaline. *Journal of Endocrinology*, **87**, 247–54.

30 Georgina E. Webley and John P. Hearn

Sargent, E.L., Baughman, W.L., Novy, M.J., and Stouffer, R.L. (1988). Intraluteal infusion of a prostaglandin synthesis inhibitor, sodium meclofenamate, causes premature luteolysis in the rhesus monkey. *Endocrinology*, **123**, 2261–9.

Schoonmaker, J.N., Bergman, K.S., Steiner, R.A., and Karsch, F.J. (1982). Estradiol-induced luteal regression in the rhesus monkey; evidence for an extraovarian site of action. *Endocrinology*, **110**, 1708–15.

Seshagiri, P.B. and Hearn, J.P. (1993). *In vitro* development of *in vivo* produced rhesus monkey morulae and blastocysts to hatched, attached and post-attached blastocyst stages: morphology and early secretion of chorionic gonadotrophin. *Human Reproduction*, **8**, 278–87.

Seshagiri, P.B., Terasawa, E., and Hearn, J.P., The secretion of gonadotrophim releasing hormone by peri-implantation embryos of the Rhesus monkey: comparison with the secretion of chorionic gonadotrophin. *Human Reproduction* (in press).

Silvia, W.J., Lewis, G.S., McCracken, J.A., Thatcher, W.W., and Wilson, L., Jr. (1991). Hormonal regulation of uterine secretion of prostaglandin F2α during luteolysis in ruminants. *Biology of Reproduction*, **45**, 655–63.

Sinosich, M.J., Ferrier, A., and Saunders, D.M. (1985). Monitoring of postimplantation embryo viability following successful *in vitro* fertilization and embryo transfer by measurements of placental proteins. *Fertility and Sterility*, **44**, 70–4.

Smith, C.A., Moore, H.D.M., and Hearn, J.P. (1987). The ultrastructure of early implantation in the marmoset monkey (*Callithrix jacchus*). *Anatomical Embryologica*, **175**, 399–410.

Smith, K.B. and Fraser, H.M. (1991). Control of progesterone and inhibin secretion during the luteal phase in the macaque. *Journal of Endocrinology*, **128**, 107–13.

Smith, K.B., Lunn, S.F., and Fraser, H.M. (1990). Inhibin secretion in the common marmoset monkey. *Journal of Endocrinology*, **126**, 489–95.

Stevens, V.C. (1976). Perspectives of development of a fertility control vaccine from hormonal antigens of the trophoblast. In Development of vaccines for fertility regulation (World Health Organization meeting), pp. 93–110. Scriptor, Copenhagen.

Stewart, D.R., Celniker, A.C., Taylor, C.A. Jr., Cragun, J.R., Overstreet, J.W., and Lasley, B.L. (1990). Relaxin in the peri-implantation period. *Journal of Clinical Endocrinology and Metabolism*, **6**, 1771–3.

Stouffer, R.L., Nixon, W.E., and Hodgen, G.D. (1979). Disparate effects of prostaglandins on basal and gonadotrophin-stimulated progesterone production by luteal cells isolated from rhesus monkeys during the menstrual cycle and pregnancy. *Biology of Reproduction*, **20**, 897–903.

Stouffer, R.L, Ottobre, J.S., and Vandervoort, C.A. (1987). Regulation of the primate corpus luteum during early pregnancy. In *The primate ovary* (ed. R.L. Stouffer), pp. 207–20. Plenum Press, New York.

Stouffer, R.L., Ottobre, J.S., Molskness, T.A., and Zelinski-Wooten, M.B. (1989). The function and regulation of the primate corpus luteum during the fertile menstrual cycle. In *Development of preimplantation embryos and their environment* (ed. K. Yoshinaga and T. Mori), pp. 129–42. Alan R. Liss, New York.

Summers, P.M., Wennink, C.J., and Hodges, J.K. (1985). Cloprostenol-induced luteolysis in the marmoset monkey (*Callithrix jaccus*). *Journal of Reproduction and Fertility*, **73**, 133–8.

Swanston, I., McNatty, K.P., and Baird, D.T. (1977). Concentrations of prostaglandin F2α and steroids in the human corpus luteum. *Journal of Endocrinology*, **73**, 115–22.

Talwar, G.P., Dubey, S.K., Salahuddin, M., and Shastri, N. (1976). Kinetics of antibody response in animals injected with processed beta hCG conjugated to tetanus toxoid. *Contraception*, **13**, 153–62.

Tanaka, S., Shimoya, Y., Hamamatsu, M., and Hashimoto, M. (1983). Binding sites for prostaglandin F2α in human corpora lutea. *Asia-Oceania Journal of Obstetrics and Gynaecology*, **9**, 445–51.

Tsonis, C.G., Hillier, S.G., and Baird, D.T. (1987). Production of inhibin bioactivity by human granulosa cells: stimulation by LH and testosterone *in vitro*. *Journal of Endocrinology*, **112**, R11–14.

Valenzuela, G., Balmaceda, J.P., Harper, M.J.K., and Asch, R.H. (1983). Platelets bind to rhesus monkey corpora lutea and stimulate its prostaglandin synthesis. *Fertility and Sterility*, **39**, 370–3.

Vandevoort, C.A., Stouffer, R.L., Molskness T.A., and Ottobre, J.S. (1988). Chronic exposure of the developing corpus luteum in monkeys to chorionic gonadotrophin: persistent progestrone production despite desensitization of adenylate cyclase. *Endocrinology*, **122**, 1876–82.

Webley, G.E. and Hearn, J.P. (1987). Local production of progesterone by the corpus luteum of the marmoset monkey in response to perfusion with chorionic gonadotrophin and melatonin *in vivo*. *Journal of Endocrinology*, **112**, 449–57.

Webley, G.E. and Michael, A.E. (1992). The paracrine action of an endogenous luteolysin in the marmoset corpus luteum requires cell to cell contact. *Journal of Reproduction and Reproduction*, Abstr. Ser. **10**, 7.

Webley, G.E., Richardson, M.C., Summers, P.M., Given, A., and Hearn, J.P. (1989). Changing responsiveness of luteal cells of the marmoset monkey (*Callithrix jacchus*) to luteotrophic and luteolytic agents during normal conception cycles. *Journal of Reproduction and Fertility*, **87**, 301–10.

Webley, G.E., Richardson, M.C., Smith, C.A., Masson, G.M., and Hearn, J.P. (1990). Size distribution of luteal cells from pregnant and non-pregnant marmoset monkeys and a comparison of the morphology of marmoset luteal cells with those from the human corpus luteum. *Journal of Reproduction and Fertility*, **90**, 427–37.

Webley, G.E., Hodges, J.K., Given, A., and Hearn, J.P. (1991a). Comparison of the luteolytic action of gonadotrophin-releasing hormone antagonist and cloprostenol, and the ability of human chorionic gonadotrophin and melatonin to override their luteolytic effects in the marmoset monkey. *Journal Endocrinology*, **128**, 121–9.

Webley, G.E., Knight, P.G., Given, A., and Hodges, J.K. (1991b). Increased concentrations of immunoreactive inhibin during conception cycles in the marmoset monkey: suppression with an LHRH antagonist and cloprostenol. *Journal of Endocrinology*, **128**, 465–473.

Webley, G.E., Richardson, M.C., Given, A., and Hearn, J.P. (1991c). Pre-incubation of human granulosa cells with gonadotrophin prevents the cloprostenol-induced inhibition of progesterone production. *Human Repro-duction*, **6**, 779–82.

Webley, G.E., Knight, P.G., and Hearn, J.P. (1992). Stimulation of immunoreactive inhibin production by preimplantation embryos during early pregnancy in the marmoset monkey (*Callithrix jacchus*). *Journal of Reproduction and Fertility*, **96**, 385–93.

Wentz, A.C. and Jones, G.S. (1973). Transient luteolytic effect of prostaglandin F2α in the human. *Obstetrics and Gynaecology*, **42**, 172–81.

Wilks, J.W. (1983). Pregnancy interception with a combination of prostaglandins: studies in monkeys. *Science*, **221**, 1407–9.

Wilks, J.W. and Noble, A.S. (1983). Steroidogenic responsiveness of the monkey corpus luteum to exogenous chorionic gonadotrophin. *Endocrinology*, **112**, 1256–66.

Wiltbank, M.C., Guthrie, P.B., Mattson, M.P., Kater, S.B., and Niswender, G.D. (1989). Hormonal regulation of free intracellular calcium concentrations in small and large ovine luteal cells. *Biology of Reproduction*, **41**, 771–8.

Zelinski-Wooten, M.B., Sargent, E.L., Molskness, T.A., and Stouffer, R.L. (1990). Disparate effects of the prostaglandin synthesis inhibitors, meclofenamate, and flurbiprofen on monkey luteal tissue *in vitro*. *Endocrinology*, **126**, 1380–7.

2 The Gonadal Extracellular Matrix

MARTIN R. LUCK

I   INTRODUCTION

1   Aim

All eukaryotic cells, including those of the gonads, occur in association with an extracellular matrix (ECM). The components of the ECM include proteins and proteoglycans and vary widely in size, form, and concentration. Their role was traditionally considered to be the provision of structural integrity to tissues through specific anchorage and communicative functions. More recent studies in a variety of tissues show that they also have a ubiquitous role in regulating cellular differentiation, especially during embryological development, in the immune system and in tissues undergoing pathological change.

The tissues of the testis and ovary exhibit complex sequences of cellular differentiation during development. They are also remarkable for the cycles of cellular activity and tissue remodelling which characterize their adult function. These events are as dramatic as any which occur in non-pathological tissues and provide good reasons for supposing that the ECM is crucial to gonadal function. The purpose of this review is to assemble information on the gonadal ECM and to locate areas where our knowledge is incomplete. It considers the role of the matrix in gonadal embryology, in maintaining the integrity of adult tissue structures, and in facilitating and co-ordinating cycles of cellular activity. It also considers the cellular origin of specific components, their regulation by enzymes and hormones and the specific mechanisms by which they affect endocrine and germ cell function. In attempting to understand the role of the ECM, it appeared prudent to consider the testis and ovary in the same chapter so that embryological, histological, and functional parallels may be drawn.

2   Terms of reference

This review deals with the protein (collagen, fibronectin, laminin) and glycoprotein (proteoglycan) components of extracellular fractions of the mammalian testis and ovary. It also covers the enzymes and cofactors concerned with their production, turnover, and remodelling during embryology and cycles of adult gonadal function. The review does not deal with accessory gonadal tissue or the extragonadal circumstances of released gametes.

Various authors have used the terms 'basement membrane' and 'basal lamina' interchangeably. To avoid ambiguity, and following Dym and Fawcett (1970), Hadley *et al.* (1985), and Alberts *et al.* (1989), *basal lamina* here refers to an extracellular membrane, usually consisting of collagen type IV, proteoglycans, and laminin. The term *basement membrane* refers to more complex or ill-defined structures, for example, where the basal lamina has an additional collagenous matrix ('lamina reticularis') and an associated cell layer. Alberts *et al*. (1989) and Stryer (1988) provide comprehensive and accessible biochemical descriptions of the various ECM components. The nature and variety of *collagens* are described by van der Rest and Garrone (1991). *Matrigel* is a biomatrix, secreted by a tumour cell line, comprising collagen type IV, laminin, entactin, and heparan sulphate proteoglycan; it therefore represents a typical basal lamina of epithelial origin. *Peptide sequences* are referred to by single letter codes (Stryer 1988).

II   THE ECM IN GONADAL EMBRYOLOGY

1   General

A fundamental observation about the adult gonad is that the germ cells, together with subsets of somatic endocrine cells, occur within discrete tissue compartments. Thus spermatogonia and Sertoli cells reside within the seminiferous tubules of the testis, while oocytes and granulosa cells are confined within the follicles of the ovary. Other types of steroidogenic cells, respectively the Leydig and theca or theca–interstitial cells, together with non-endocrine cells such as endothelial cells, fibroblasts, and the peritubular myoid cells, remain outside these compartments. In both sexes, the germinal compartments are delineated by basement membranes which serve to isolate them from the vascular regions of the organ. This isolation by ECM presumably has endocrinological, immunological, or other physiological significance for the maturation of the germ cells and subsequent fertility. Of particular interest is the timing of appearance of the basement membrane relative to the gathering of germ cells and somatic cells in the fetal gonad and the role which such membranes may play in the sexual differentiation of the organ. Byskov and Høyer (1988) provide a well-illustrated overview of gonadal embryology.

2   Pre-differentiation

The primitive gonad develops from primordial germ cells in association with somatic cells derived, initially at least, from mesonephric tissue. Prior to sexual differentiation, germ cells migrate towards the epithelium of the gonadal ridge, guided by a pathway of fibronectin (Ffrench-Constant *et al*. 1991). They become closely associated with clusters of somatic cells which may extend from the mesonephros to the gonadal ridge (Zamboni and Upadhyay 1982). The origin of the somatic cells has been much debated and may vary between species. They seem principally to derive from the mesonephros, but may include cells derived from the surface (coelomic) epithelium of the primitive gonad in amounts related to the genetic sex of the embryo (Zamboni and Upadhyay 1982; Byskov and Høyer 1988; Yoshinaga *et al*. 1988; Patek *et al*. 1991). A basal lamina, apparently derived initially from mesenchymal cells and containing fibronectin and laminin on a backbone of collagen, subsequently encloses these clusters. Proliferation of the somatic cells leads to the formation of the 'sex cords' from which seminiferous tubules and follicles will eventually develop (Paranko *et al*. 1983; Paranko 1987; Gelly *et al*. 1989; Wartenberg *et al*. 1991).

The sequence of early developmental events is described in the pig (see Pelliniemi 1976) in which the primitive gonad can be distinguished between days 21 and 22 and sexual differentiation occurs at day 24 of gestation. Germ cells accumulate within the epithelium and cords of cells, distinguished by a discontinuous basal lamina, and start to develop back towards the underlying mesenchyme. The basal lamina permits colonization of the epithelium by increasing numbers of germ cells, but remains associated with the developing cords. Some somatic cells lose contact with the basal lamina as a result of proliferation. By day 24, three cell 'layers' are apparent: the surface epithelium, a central gonadal blastema made up of primitive cords, and the underlying mesenchyme. The surface epithelium consists of columnar and cuboidal cells, the former resting on a basal lamina with a few collagen fibrils in the interstitium. This basal lamina continues from the surface into the blastema along with the cords but does not penetrate the deeper layers. The cuboidal cells do not have contact with the basal lamina but follow the cords towards the centre of the gonad. The basal lamina eventually separates the cords from the surrounding stromal tissue. The stroma will include mesodermal cells, connective tissue, and interstitial collagen (type III) fibrils, together with blood capillaries of mesonephric origin which have their own basal laminae (Merchant 1975; Pelliniemi 1976; Paranko 1987).

A similar picture, of cord formation alongside a discontinuous basal lamina, appears in other species (Jeppeson 1975; Merchant 1975; Dang and Fouquet 1979, Pelliniemi *et al*. 1979; Paranko 1987; Gelly *et al*. 1989; Wartenberg *et al*. 1991). Although completion of the basal lamina is apparently delayed until germ cell migration is complete (Paranko *et al*. 1983), cord development proceeds normally in the absence of germ cells, suggesting that germ cells play little part in controlling cord formation during the early stages of gonadal development (Merchant 1975).

3   Sexual differentiation

The early separation, by basal lamina, of cord cells from those of the stroma defines the two types of gonadal somatic cells. Thus the cells internal to the cords, which will give rise to Sertoli and granulosa cells, are epithelial (mesonephric and surface epithelial) in origin while those external to the cord, which will form Leydig, peritubular myoid, and theca/interstitial cells, are mesenchymal in origin (Jeppesen 1975; Merchant 1975; Merchant-Larios 1979*a*; Palotie *et al.* 1984). The subsequent behaviour of the cord, and its extension or division by basal lamina structures, defines the tissue arrangement which will develop at the time of sexual differentiation.

i   The male

Differentiation of the testis is marked by the separation of the cord cells from the surface epithelium by the tunica albuginea (Zamboni and Upadhyay 1982). In most species, the cords of the differentiating male gonad develop more rapidly and completely than those of the female. A superficial membrane, described in one report as containing laminin, fibronectin, heparan sulphate proteoglycan, and collagen types IV and V (Pelliniemi *et al.* 1984) and in another as comprising collagen type III (Paranko 1987), is secreted by presumptive myoid cells, with collagen type I being deposited in the intercellular space (Paranko 1987). Proliferation of somatic (presumptive Sertoli) cells continues so that the cords develop radially, eventually separating from the surface epithelium. As a result of proteolysis of the epithelial membrane, the completed cord membrane eventually comes to lie just beneath the periphery of the organ (Merchant 1975; Dang and Fouquet 1979; Paranko *et al.* 1983; Pelliniemi *et al.* 1984 Paranko 1987).

The role of the basal lamina in directing cord formation is not entirely clear. In the male pig fetus, the cords have fully enclosed the germ cells by day 26. The basal lamina is still incomplete but there is sufficient structure to give the interstitial cells a cord-like appearance also. Basal lamina continues to be secreted to enclose the cords, suggesting that it follows rather than initiates cord formation (Pelliniemi *et al.* 1979). This view is supported by evidence from the rat fetus in which cord formation, defined by presumptive Sertoli cells, can be observed some days before basal lamina deposition and before membrane fibronectin becomes immunohistologically detectable (Jost *et al.* 1981; Paranko *et al.* 1983). Indeed, it appears that the very earliest stages of cord formation are characterized by the disappearance of fibronectin from groups of cells running from the cortex to the medulla of the undifferentiated gonad (Pelliniemi *et al.* 1984). On the other hand, evidence from *in vitro* studies suggests a crucial role for the basal lamina in cord formation and the maintenance of cord integrity. Cultured explants of fetal testis will spontaneously form cord-like structures in which Sertoli cells are surrounded by basal lamina and peritubular cells (Agelopoulou and Pelliniemi *et al.* 1984; Tung and Fritz 1986*a*). Chemical inhibition of collagen synthesis disrupts these structures and also prevents the expression of laminin and fibronectin and the deposition of basal lamina (Jost *et al.* 1985). Fibronectin, secreted by peritubular cells, may provide a pathway to direct the otherwise random migration of Sertoli cells and their associated germ cells during cord formation and lengthening (Tung and Fritz 1986*a*).

ii   The female

At the time of sexual differentiation the cords of the female gonad are much less differentiated than those of the male. They remain contiguous with the surface epithelium, forming a 'rete' system (Byskov 1978), and lack a pronounced centrifugal differentiation. Germ cells proliferate with less evidence of surrounding connective tissue (Paranko 1987). In the sheep, epithelial colonization by germ cells occurs between days 24 and 58 of gestation, with sexual differentiation taking place at day 31. However, in the female gonad the first appearance of the basal lamina does not occur until about day 34. Clusters of cells, with surrounding basal laminae, appear within the cords at day 45, and between days 49 and 58 the tissue undergoes remodelling to form concentric layers of collagenous connective tissue (Zamboni *et al*. 1979). A similar pattern is described for the mouse ovary (Upadhyay *et al*. 1979). Thus the somatic cells may be entirely responsible for the early organization of the ovary, with basal lamina formation and compartmentalization occurring at a relatively late stage. Subsequent development occurs radially so that follicular structures nearer the centre of the ovary show greater organization than those towards the periphery.

One consequence of the difference in the extent and timing of cord development between male and female gonads is that male germ cells initially show a greater enrichment with somatic cells than do their female counterparts (Dang and Fouquet 1979). This difference is seen particularly clearly in the horse where the steroidogenic differentiation of somatic cells occurs prior to sexual differentiation, and the basal lamina appears to exclude them from (male; presumptive Leyding cells) or include them within (female) the cord structure (Merchant-Larios 1976, 1979*b*). By the time of birth in humans (Makabe *et al*. 1991) and of zona pellucida formation in mouse oocytes (days 1–2 postnatal; Odor and Blandau 1969) the majority of follicles have achieved only a single layer of somatic cells within the confines of the basal lamina. This indicates a slow prenatal proliferation rate but is consistent with evidence that the cells of the mature mouse follicle arise by oligoclonal proliferation of a small number, perhaps five, of progenitor cells (Telfer *et al*. 1988; Patek *et al*. 1991).

4   Postnatal maturation of the testicular ECM

The wall of the adult seminiferous tubule comprises several cellular and non-cellular layers (Fig. 2.1). The neonatal structure is relatively immature and undergoes thickening during maturation. The inner non-cellular layer develops distinct collagen fibrils so that a basement membrane eventually arises consisting of collagen sandwiched between two basal laminae. In the rat, this process coincides with the muscular differentiation of the peritubular cells into myoid cells and results in the completion of the tubular membrane by the time spermatogenesis has been established

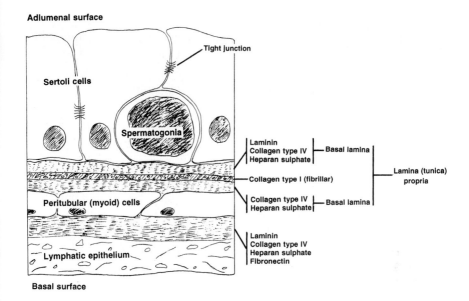

Fig. 2.1   Diagram of the wall of the seminiferous tubule to illustrate the juxtaposition of cellular and extracellular components. (Composite from Dym and Fawcett 1970, Hadley *et al.* 1985, 1988; Not to scale.)

(Leeson and Leeson 1963). The production of collagen, as a fraction of total protein, is greater in immature (25%) than in pubertal (5%) somatic cells, and the ratio of type III (secreted) to type I (secreted and cell-associated) falls as maturation proceeds. The peritubular cell secretion of plasminogen activators, regulators of the plasmin cascade, also varies as development proceeds (Rosselli and Skinner 1992). Immature cells produce larger quantities of proteoglycans than mature cells, with a higher ratio of hyaluronic acid to chondroitin sulphate (Rodriguez and Minguell 1989; Rodriguez *et al.* 1991). The level of fibronectin gene expression in the seminiferous tubule also increases, to reach a maximum at puberty from which it slowly declines (Skinner *et al.* 1989). Taken together, these observations demonstrate that the testis is a site of considerable ECM synthesis during the prepubertal period.

III   THE TESTIS

1   Leydig cells

The groups of Leydig cells which populate the interstitial areas of the testis are surrounded by a complex basement membrane (Kuopio and

Pelliniemi 1989). In primates (Hatakeyama 1965; Belt and Cavazos 1971; Kuopio *et al*. 1989) and dogs (Connell and Christensen 1975), there is a collagenous membrane surrounding individual and groups of interstitial cells which is laid down after steroidogenesis has begun and excludes blood capillaries and fibroblasts from contact with the steroidogenic cells. This membrane is unconnected with that of the seminiferous tubules and in the rat may appear as a patchy lamina on the free surfaces of Leydig cells or between Leydig cells and macrophages (Kuopio and Pelliniemi 1989).

The secretory origin of this membrane is uncertain. In humans, the fibrillar collagens, types I and III, have been identified (Takaba 1990). In the rat, the presence of heparan sulphate proteoglycan (Hayashi *et al*. 1987), laminin, and collagen type IV (Kuopio and Pelliniemi 1989; Kuopio *et al*. 1989) suggests that Leydig cells are of epithelial rather than mesenchymal origin. Alternatively, the migration and clustering of mesenchymal Leydig cells in the fetal testis may be mediated by ECM originating from an epithelial source (Kuopio *et al*. 1989). *In vitro* studies demonstrate that the morphology and differentiation of Leydig cells in culture are dramatically affected by the ECM. They adhere weakly to collagens I, V, and VIII but attach firmly to type IV, fibronectin, laminin, and matrigel. On thick matrigel they form aggregates and assume a rounded shape. On thin matrigel they elongate and form cords. The rounded habit is associated with low proliferation and the maintenance of steroidogenesis (3β-hydroxysteroid dehydrogenase activity). Elongation is associated with dedifferentiation as indicated by rapid proliferation, low steroidogenesis, and the expression of a protein (SPARC) associated with tissue remodelling (Vernon *et al*. 1991). The functional significance of Leydig cell isolation remains to be established.

2   The seminiferous tubule: microanatomy

The boundary layer (lamina propria or tunica propria) of the seminiferous tubule is an interesting, complex and highly organized structure comprising cellular and acellular components (Fig. 2.1). The Sertoli cells of the germinal compartment (seminiferous epithelium) rest on a complex basement membrane. This comprises a basal lamina supported by collagen fibrils, with an additional glycoprotein layer on the outside. The latter forms an inner coating for a layer of peritubular myoid cells which may be one or several cells thick (Dym and Fawcett 1970); Connell and Christensen 1975). External to the myoid cells is a further layer of glycoprotein which also contains collagen fibrils and is in contact with the endothelial cells of the peritubular lymphatics. The peritubular cell layer occasionally displays intercellular clefts, filled with dense filamentous material (Dym and Fawcett 1970).

3   Seminiferous tubule: ECM components and their cellular origins

Studies of human (Takaba *et al*. 1991), rat (Enders *et al*. 1986; El Ouali *et al*. 1991), and other mammalian and non-mammalian testicular material (Christl 1990) indicate that the tubular basement membrane contains collagen, fibronectin, and laminin together with the heparan sulphate and chondroitin sulphate proteoglycans. In the human, the basement membrane contains collagen types I, III, IV, and V (Takaba 1990). In the rat, the basal lamina on the lumenal side contains laminin, collagen type IV, and heparan sulphate, while the lamina on the tubular side of the myoid cells contains only collagen type IV and heparan sulphate (Hadley and Dym 1987*a*; Gelly *et al*. 1989; El Ouali *et al*. 1991). Fibronectin occurs between germ cells, between the myoid cells and the external lymphatics (Hadley and Dym 1987*a*), in the interstitial areas (Davis *et al*. 1990) and between the 'supporting cells' of the bull tubule (Strübing 1989), but has not been detected in the area between the myoid and Sertoli cells (Hadley and Dym 1987*a*).

The compartmentalization of these components suggests that they may be secreted by particular cell types and/or that secretion occurs in a directional manner within the tissue laminae. The cellular source of a number of components has been debated. The weight of evidence, from *in vitro* studies with rat material, suggests that fibronectin originates exclusively from the myoid cells (Tung *et al*. 1984*a*; Skinner *et al*. 1985; Skinner *et al*. 1989; Davis *et al*. 1990), although one study (Borland *et al*. 1986) reports the presence of fibronectin in a testicular cell culture comprising 95% pure Sertoli cells. It is possible that the ages of the rats used, or the inclusion of hormones such as epidermal growth factor and insulin in the culture medium, may have led to differences between this and other studies. Laminin was thought to be secreted exclusively by Sertoli cells (Tung *et al*. 1984*b*; Skinner *et al*. 1985; Borland *et al*. 1986), although more recent reports suggest that it is also expressed by peritubular cells (Davis *et al*. 1990; El Ouali *et al*. 1991). Peritubular cells secrete collagen type I while both cell types appear to secrete collagen type IV (Skinner *et al*. 1985 Davis *et al*. 1990). Heparan sulphate appears to be the product of Sertoli cells (Borland *et al*. 1986) but other proteoglycan components of the basal lamina, such as chondroitin sulphate, may be produced by both cell types (Elkington and Fritz 1980; Skinner and Fritz 1985).

4   Simple monoculture models of cell–ECM interaction

The possibility of cell-specific origins for various ECM components in the testis has suggested a hypothesis of cooperation between cell types in the synthesis and organization of the tubular membrane (Tung and Fritz 1980).

However, it is clear from simple cell culture studies that the morphology and secretory behaviour of Sertoli and peritubular cells are themselves influenced by the ECM.

Sertoli cells cultured on plastic or on collagen type I adopt a squamous habit which contrasts strongly with their columnar appearance *in vivo* (Suárez-Quian *et al.* 1984; Tung and Fritz 1984; Hadley *et al.* 1985; Hadley and Dym 1987*b*). When cultured on collagen type IV or laminin they show greater viability, retain a columnar shape, and have a polarity of organelle distribution which resembles that seen in the tubule (Suárez-Quian *et al.* 1984). Cultured Sertoli cells resemble their *in vivo* counterparts most closely when a biomatrix, such as Matrigel or a reconstituted seminiferous tubule matrix (a fibrillar material containing 50% protein, mostly collagen and glycoprotein, together with fibronectin and laminin), is used as substratum (Tung and Fritz 1984). Such cells are cuboidal and form basal tight junctions; they also have a lower proliferation rate than cells grown on plastic and characteristic patterns of protein glycosylation (Tung and Fritz 1984; Tung *et al.* 1984*b*; Hadley *et al.* 1985; Hadley and Dym 1987*b*; Anthony and Skinner 1989; Page *et al.* 1990). Changes in the expression of cytokeratin and vimentin by these cells suggest that the intermediate filaments of the cytoskeleton mediate the differentiation response to complex extracellular environments (Guillou *et al.* 1990). Sertoli cells grown on collagen type I showed a membrane-associated pattern of proteoglycan secretion in contrast to the soluble secretion shown by cells on plastic (Rodriguez and Minguell 1992).

Several studies show that Sertoli cells have a receptor-mediated preference for a surface containing laminin. Rat Sertoli cells in culture attach preferentially to the lumenal surface of seminiferous tubule basement membranes, in contrast to a fibroblastic cell line (3T3) which binds to all surfaces. The binding of Sertoli cells could be inhibited by proteases but not by chondroitinase ABC, heparinase or hyaluronidase, and binding to trypsinized segments could be restored with laminin but not fibronectin (Enders *et al.* 1986). Both integrin and non-integrin-type laminin-binding proteins have been identified on immature rat Sertoli cells, particularly on the basolateral surface. Antisera against the non-integrin type partially prevented the spreading of Sertoli cells on a laminin-treated culture surface but those against the integrin-type did not, suggesting different roles for these receptors in cell–matrix interactions (Davis *et al.* 1991).

Sertoli cells grown within a thick matrigel matrix aggregated into highly organized cord arrangements but failed to do so in the presence of antisera to laminin or entactin. Peptide RGD, which represents the cell-binding domain of the laminin A chain as well as that of entactin and the integrins, prevented cord formation but did not prevent Sertoli cell attachment to the matrix. Peptide YIGSR, which is found in the laminin B1 chain, inhibited cord formation and prevented binding, as did its antiserum. These observations indicate that Sertoli cells have specific laminin receptors, probably involving both the YIGSR-laminin type and the RGD-integrin type domains (Hadley *et al*. 1990).

Culture substrate also influences the cellular response to hormones. Sertoli cells secrete transferrin and androgen binding protein; culture on matrigel not only increases the rates of basal secretion but also permits stimulation by a range of hormones including follicle-stimulating hormone (FSH), insulin, testosterone, and retinol (Hadley *et al*. 1985 and Skinner 1989). Cells grown on laminin show a significantly reduced response to FSH in the presence of peptides containing the RGD sequence (Dym *et al*. 1991). Evidence suggests that the concentration of G-protein is increased at the base of the Sertoli cells, that is to say, in the immediate vicinity of the substrate.

If, as suggested previously, laminin is exclusively the product of Sertoli cells, the evidence that Sertoli cells also respond specifically to laminin suggests that there is an autocrine promotion of cell-substrate interaction. Sertoli cells grown on collagen type I were initially immobile but after 4–6 days began to migrate into plaques following the gradual accumulation of laminin and collagen type IV in the culture (Tung and Fritz 1986*a*). In another study (Suárez-Quian *et al*. 1984), addition of laminin to a collagen type I substrate did not permit Sertoli cells to adopt a columnar shape, even though cells grown on laminin or on collagen type IV plus laminin were columnar. These experiments emphasize the specificity of the cell–substrate interaction but also suggest a level of surface recognition beyond that involving laminin alone.

The peritubular cells are also substrate-responsive, but studies on these cells are less extensive. In contrast to Sertoli cells, peritubular cells will migrate rapidly on a surface of collagen type I (Tung and Fritz 1986*a*). When cultured on glass, polystyrene, laminin, collagen types I, III, or IV, fibronectin, heparin, or liver biomatrix they assumed a flat, fibroblastic habit and were highly proliferative (Tung and Fritz 1986*b*). In contrast, cells cultured on seminiferous tubule biomatrix had a squamous polyhedral shape, with close or overlapping borders ('cobblestones') and low mitotic index, resembling the myoid cells of the tubule. Although further studies on peritubular cells are warranted, these results suggest that they have quite different matrix requirements from the laminin-dependent Sertoli cells.

5   Co-culture models of the seminiferous tubule

When Sertoli and myoid cells from 20-day rats were cultured together in serum-supplemented culture they formed 'limiting membranes' between the individual cells (Tung and Fritz 1980). This membrane closely resembled the lamina propria of the seminiferous tubule and its production was consistent with the hypothesis of cooperation between testicular cells.

In contrast to the soluble secretions of monocultures, Sertoli–peritubular co-cultures produced less soluble proteins which were deposited (Skinner *et al*. 1985; Anthony and Skinner 1989). In co-cultures on suboptimal substrata, peritubular cells in direct contact with Sertoli cells took on the polyhedral 'cobblestone' appearance exhibited by cells on biomatrix, while those resting some distance away retained the flat, proliferative, fibroblastic habit (Tung and Fritz 1986*b*). Sertoli cells migrated to form cords on top of the peritubular cells (Tung and Fritz 1986*a*); notwithstanding the laminin specificity of Sertoli cells, this mobility was shown to be independent of laminin and collagen and to require fibronectin produced by the peritubular cells. As already discussed in the context of seminiferous tubule embryology, fibronectin may serve to direct the otherwise random movements of Sertoli cells, perhaps guiding them along the top surface of the peritubular cells.

The deposition of ECM protein by Sertoli–peritubular co-cultures can be encouraged by hormonal treatment (FSH, insulin, testosterone, retinol; Skinner *et al*. 1985). Co-culture studies of pig Sertoli and Leydig cells suggest that a biomatrix substrate will increase the ability of each cell type to respond to gonadotrophins (Reventos *et al*. 1989). Similarly, the ECM secreted by a myoid-like cell line increased the FSH-response of Sertoli cells, an effect which could not be reproduced by collagen type I, fibronectin, or laminin (Mather *et al*. 1984).

6   Proteolytic enzymes

The testis contains a range of proteolytic enzymes and their regulators, which may be involved in both tissue remodelling and sperm development. Their substrates are presumably the ECM components of the seminiferous tubule wall and proteins secreted into the tubular lumen by cells of the germinal epithelium.

The activity of testicular plasminogen activator (PA), reported in the human by Albrechtsen (1957), has been studied in some detail in rodents. PA is the product of Sertoli rather than myoid cells. Its secretion, over an extended period of culture, can be stimulated by FSH, insulin, dibutyryl cyclic adenosine monophosphate (dbcAMP) and cholera toxin (Lacroix *et al*. 1977, 1981; Gore-Langton *et al*. 1981). Stimulation by testosterone is disputed (Marzowski *et al*. 1985; Lacroix and Fritz 1982). Some studies report that the FSH-stimulatable secretion is of the urokinase-type PA (uPA) rather than the tissue-type PA (tPA; Boitani *et al*. 1989; Nishimune *et al*. 1989), although others (Hettle *et al*. 1986; Nargolwalla *et al*. 1990) suggest that uPA is the basal secretion with most of the increase due to the tissue type. FSH and dbcAMP increase the expression of the tPA by rat Sertoli cells (Nargolwalla *et al*. 1990). Plasminogen is secreted within the tubule, indicating that testicular PA exploits a local rather than an extratesticular (liver) substrate (Saksela and Vihko 1986). Rat peritubular cells produce a PA inhibitor (PAI) type I which is active against tPA and the two-chain form of uPA but which does not inhibit plasmin (Hettle *et al.* 1986, 1988; Nargolwalla *et al.* 1990).

The seminiferous tubule also produces a plasminogen-independent protease in response to FSH (Hettle *et al.* 1986). Sertoli cells in the presence of laminin secrete several gelatinases (metalloproteinases) including a type IV collagenase, and secrete fibronectinolytic activity when stimulated by FSH (Sang *et al.* 1991). In co-culture experiments, the secretion of the type IV collagenase was related to the ratio of mesenchymal (peritubular) to epithelial (Sertoli) cells. Addition of laminin caused the enzyme to be secreted basally, while a peptide containing the RGD (fibronectin-integrin receptor) sequence caused it to be released into the peripheral plasma membrane region (Sang *et al.* 1991); in other words, there was a receptor-related redistribution of the enzyme. Myoid cells in culture produce metalloproteinases, predominantly a latent type IV procollagenase, while Sertoli and myoid cells both secrete a tissue inhibitor of metalloproteinase type 2 (TIMP 2; Ailenberg *et al.* 1991). Cheng *et al.* (1990) have demonstrated the synthesis of α_2-macroglobulin, a non-specific protease inhibitor, in the testis and identified it in rete testis fluid. They suggest that this is a Sertoli-derived inhibitor which 'prevents damaging effects of proteases released from degenerating late spermatids'.

7   Seminiferous tubule integrity: complex culture models

An appreciation of the laminar complexity of the seminiferous tubule, of the ECM specificity of its cells and of their interaction, has prompted several attempts to design more realistic experimental models. These usually consist of a two-compartment culture chamber in which Sertoli cells rest on a separating membrane, with or without a biomatrix or a layer of peritubular or other cells (Fig. 2.2). The tendency for optimally cultured Sertoli cells to secrete basement-membrane components and to form basal tight junctions can result in the formation of a semipermeable barrier between the chambers (Byers *et al.* 1986; Janecki and Steinberger 1986; Hadley *et al.* 1988). This is analogous to the so called 'blood–testis' barrier *in vivo* and can be used to investigate directional secretory activity or the transfer characteristics of the barrier (Hadley and Dym 1987*b*; Guillou 1990).

Ailenberg *et al.* (1988, 1990, 1991) and Ailenberg and Fritz (1988, 1989) studied the production of PA in a two-chamber culture system in which the interchamber barrier was formed by a Sertoli–matrigel–myoid sandwich, analogous to the seminiferous tubule membrane. The concentration of PA was highest in the lower compartment, indicating a basally directed secretion by the Sertoli cells, that is, across the matrix and into the 'peritubular' region. This was in contrast to transferrin which was secreted into the upper chamber, representing the 'adlumenal' direction. The degree of polarization reflected the integrity of the barrier; FSH (or dbcAMP) treatment, which stimulated PA secretion, decreased barrier integrity, while androgens had the opposite effect. Evidence from such models suggests that protease activity is responsible for modulating the resistance and function of the seminiferous tubule wall *in vivo*. uPA, released in the testis in response to human chorionic gonadotrophin (hCG), can increase vasopermeability and may therefore regulate the vascular transport of regulatory factors and secretions within the organ (Loukusa *et al.* 1990). Interestingly, the cryptorchid testis can be associated with a thickening of the lamina propria (Paniagua *et al.* 1990; Santamaria *et al.* 1990) and a markedly reduced secretion of PA (Okuyama *et al.* 1991).

8   The tubular ECM and sperm development

A role for enzymic alteration of the tubule wall has been suggested as a component of the spermatogenic process, responsible for permitting the translocation of early prophase spermatocytes from the basal to the adlumenal compartment (Hettle *et al.* 1986). PA activity was found to be higher in segments of tubule with spermatocytes at stages VII and VIII (i.e. just before release into the lumen) than in those at other stages of spermatogenesis (Lacroix *et al.* 1981; Fritz and Karmally 1983). During

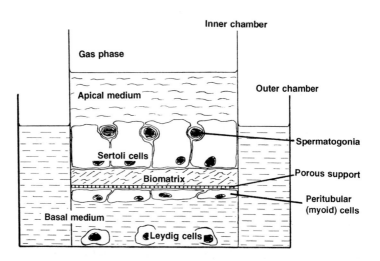

Fig. 2.2   Diagram of a typical dual-chamber culture system for testicular cells, designed to model the structure of the seminiferous tubule. (Composite from Ailenberg *et al* 1988; Hadley *et al.* 1988. Not to scale.)

postnatal development there is an increase in PA content at the time of rapid testis weight gain and rise in spermatid number, which does not occur if the testes are irradiated (Lacroix *et al.* 1982). The detachment of germ cells from the basal lamina at the pre-leptotene stage (Vihko *et al.* 1984) or Sertoli phagocytosis of residual bodies (Lacroix *et al.* 1982) may be signals for increased PA synthesis.

Mouse spermatogenic cells express laminin binding protein, irrespective of their contact with the basal lamina, and contain a putative ligand for such a protein (Fulcher *et al.* 1992). The role of a binding protein in spermatogenesis is uncertain but the presence of laminin has been shown to be critical to the elongation of gonocytes from the newborn mouse testis *in vitro* (Orth and McGuiness 1991) suggesting that laminin may be a Sertoli cell signal required for germ cell maturation.

Skinner (1990) summarizes the role of the testicular ECM as permitting the *environmental* interactions of testicular cells, either those between Sertoli and germ cells or those between Sertoli and peritubular cells. Sertoli cells and peritubular cells cooperate in the production of the ECM components which then provide the structural integrity of the tubule and permit its secretory and germinal functions.

IV   THE OVARY

1   Membranous ECM components of the stroma and follicle

Collagen is an abundant protein in most compartments of the ovary. The ovarian capsule contains collagen types I, III, and IV, while type V is located in the ovarian surface epithelium, the walls of atretic follicles and in capillary walls (Kaneko *et al.* 1984). The tunica albuginea and follicular theca externa contain abundant collagen fibrils of the interstitial types I and III (Christiane *et al.* 1988; Palotie *et al.* 1984); these pass from the interfollicular area and fan out around the follicle wall to provide mechanical support (Martin and Miller-Walker 1983). The meshwork becomes depleted in mature follicles, particularly around the apex as ovulation approaches (Okamura *et al.* 1978, 1980; Shiina 1990). Ovarian surface epithelial cells secrete collagen types I and III, keratin, and laminin; the autonomy of ovarian carcinoma cells may be due to the secretion of ECM by neoplastic epithelial cells, permitting them to form multilayers and to overcome the restrictions of crowding (Auersperg *et al.* 1991).

Within the follicle, a typical basal lamina, containing collagen type IV, laminin, fibronectin, and heparan sulphate proteoglycan, separates the theca and granulosa cell layers (Bagavandoss *et al.* 1983; Palotie *et al.* 1984; Christiane *et al.* 1988; Yoshimura *et al.* 1991; Leardkamolkarn and Abrahamson 1992). With the exception of fibronectin, these components are not found within the granulosa compartment itself. The quantity of laminin fraction P1, a marker of basal lamina turnover, increases as the follicle matures and correlates with progesterone concentrations (Christiane *et al*. 1988). The presence of laminin in this membrane, as well as in the thecal capillary membranes and the ovarian stroma, is consistent with the epithelial origin of follicles (Wordinger *et al*. 1983; Christiane *et al*. 1988). The developing follicle must be a very active site of membrane synthesis: the membrane surface area will increase as the square of the follicular radius and in some species the latter dimension may double on a daily basis as ovulation approaches (see Gosden *et al*. 1988). Both theca and granulosa cells may be sources of laminin (Leardkamolkarn and Abrahamson 1992).

Basement membranes define and separate cellular compartments but contact between cells and basement membrane directly influences cell function. Granulosa cells 'see' one of two membranes, depending on their location: the outermost layer of cells is in contact with the external (granulosa/theca) basal lamina, while towards the oocyte one layer of cumulus cells is in contact with the specialized membrane of the zona pellucida (Maresh *et al*. 1990). The 'antral' granulosa cells between these layers have no membrane contact, although the follicular fluid in which they are bathed represents a fluid matrix. Cells in the outer layer of the granulosa are cuboidal/columnar and tightly bound to the basal lamina, in contrast to those in the antral region which are loosely arranged and spherical/polyhedral (Bjersing and Cajander 1974*a*; Gosden *et al*. 1988; Zoller 1991). The more peripheral cells show greater differentiation in as much as they have a higher luteinizing hormone (LH) receptor density and a greater abundance of steroidogenic enzymes (Amsterdam *et al*. 1991; Telfer *et al*. 1992). If this is directly attributable to their membrane contact then the absence of a membranous matrix within the follicle could be seen as a means of maintaining antral cells in an undifferentiated state prior to ovulation.

The external basal lamina evidently prevents angiogenesis in the granulosa/germ cell compartment prior to follicular rupture (Espey 1991). The tissue changes of both follicular rupture and atresia are characterized by discontinuities in the membrane which permit the penetration of theca cells into the granulosa (Bagavandoss *et al*. 1983). Atretic follicles show internal staining for collagen types I, III, and V, suggesting an accumulation of interstitial collagens analogous to that which occurs during scarring or wound healing (Palotie *et al*. 1984; Christiane *et al*. 1988).

The zona pellucida membrane may influence its own production by cumulus cells: cells from early rabbit follicles produce specific zona proteins in culture but the degree of post-translational processing of these products is influenced by the nature of the substrate itself (Maresh *et al*. 1990) indicating an autocrine regulatory mechanism.

2   Fibronectin in the follicle

Fibronectin is present in the follicle as a component of the basal lamina and as a soluble fraction of follicular fluid. The quantity present in the fluid increases with follicular size (Hung *et al.* 1989) and probably originates from granulosa rather than theca cells (Lobb and Dorrington 1987; Yoshimura *et al.* 1991) since the follicular basal lamina would represent an exclusion barrier to molecules of the size of fibronectin (500 kD; Savion and Gospodarowicz 1980). Fibronectin accounts for about 20% of the protein produced by rat granulosa cells in culture; secretion can be stimulated by gonadotrophin-releasing hormone (GnRH), which reduces steroidogenesis, and reduced or inhibited by FSH, insulin or dbcAMP, which enhance steroidogenesis (Skinner and Dorrington 1984; Skinner *et al.* 1984; Dorrington and Skinner 1986; Lobb and Dorrington 1987; Bernath *et al.* 1990). Fibronectin production and steroidogenesis therefore represent two distinct states of granulosa differentiation and it is proposed that fibronectin production is high in those tissues (primordial follicles, atretic follicles, early corpus luteum) where ECM support is needed.

Several studies suggest that fibronectin secretion is influenced by the cellular environment, particularly in relation to follicle development and cell proliferation. Its secretion by rat granulosa cells was greatest during early follicular development and was associated with a tendency for the cells to spread in culture (Carnegie 1990). With bovine cells, secretion was greater in confluent cultures of cells from large (15 mm) follicles than in sparse cultures of cells from small (4–5 mm) follicles, and could be increased by epidermal growth factor (Savion and Gospodarowicz 1980). Expression of fibronectin was found to be low in granulosa cells from pre-ovulatory bovine follicles but greatly elevated in the same cells following serum-free culture (Luck *et al.* 1991). Since such culture is also associated with the expression of oxytocin and 'luteal' changes in steroidogenesis, it was concluded that fibronectin might be a marker for the postovulatory differentiation of granulosa cells.

3   Follicular cell culture models

Several culture models have been devised to investigate the control of granulosa cell function by ECM proteins. Though they have yielded much useful information, they are generally less complex than those used to study seminiferous tubule function and have been rather less successful in reproducing the tissue arrangements which pertain *in vivo*. Some models have arisen from the need to achieve cell attachment as well as significant hormone secretion under serum-free culture conditions. Bearing in mind the rounded morphology and loose attachment of the majority of (antral) granulosa cells in the follicle *in vivo*, this has led to some uncertainty about the state of differentiation of the cultured cells and the physiological state which they most nearly represent.

Human granulosa cells grown on a plastic surface formed a monolayer but tended to stay as isolated cells (Ben-Rafael *et al*. 1988). In contrast, cells grown on floating collagen (type I) gels formed multilayers with intercellular junctions, abundant mitochondria, and lipid droplets. They also secreted more oestradiol and progesterone than cells on plastic. Rat and human granulosa cells grown on basal lamina-type biomatrices, including Matrigel, showed better attachment, viability, steroid secretion, and gonadotrophin responsiveness than those on an uncoated plastic surface (Amsterdam *et al*. 1989; Richardson *et al*. 1992). They developed cytoplasmic processes and formed tight junctions between cells. The expression and secretion of fibronectin and inhibin by primary and early secondary rabbit follicles increased when they were cultured on Matrigel, in analogy perhaps with the effect of basal lamina contact during early follicular development (Timmons *et al*. 1990).

The endocrine and morphological characteristics of cultured cells suggest, either that they resemble the mural rather than the antral cells of the follicle or, that they are expressing a luteal rather than a follicular phenotype (Amsterdam *et al*. 1989; Richardson *et al*. 1992). Studies in the author's laboratory with bovine granulosa cells have led him to the opinion that a transition to the luteal phenotype may be an inevitable consequence of 'successful', culture, particularly under serum-free conditions. In other words anchorage in culture involves an obligate differentiation. Granulosa cells grown on a surface treated with collagen type I show good attachment and luteinize 'spontaneously' insofar as they begin to express and secrete oxytocin, secrete abundant progesterone, and undergo a marked decline in secretion of the follicular hormones oestradiol and inhibin (Luck *et al*. 1990). They acquire a flat, spreading habit with a limited amount of division. Such cells can recognize different types of collagen on the culture surface, the intensity of the luteinization response (degree of change in hormone secretion/cell) being observed as follows: type I>>type V≥type III>type IV (Luck *et al*. 1991). In these and other studies using a wide range of culture supplements and surface treatments, cells which show poor oxytocin and progesterone secretion are invariably those which retain a rounded habit and exhibit poor attachment to the culture surface.

An agar gel culture has been used to investigate the anchorage-dependence of bovine granulosa cells directly (Bartholomeusz *et al*. 1988). This model selects a subset of cells which can undergo clonal proliferation whilst suspended in the gel (supplemented with serum and erythrocytes) and which are therefore anchorage-independent. The proportion of cells represented is not clear, but they can be subcloned and remain steroidogenic and hormone-responsive. An inverse relationship between follicle size and the number of anchorage-independent cells suggested that follicular maturity is associated with a loss of anchorage independence; these cells may therefore represent a relatively immature 'stem cell' population.

Rat granulosa cells in conventional culture make use of endogenous fibronectin to achieve attachment (Morley et al. 1987a). A non-transformed rat cell line has been described which requires fibronectin to grow and divide under serum-free conditions (Orly and Sato 1979). Trypsin treatment inhibited growth but this could be reversed by the addition of fibronectin during the attachment phase of the culture. In other studies, rat cells cultured on fibronectin showed much greater spreading, protein synthesis, and proliferation than those cultured on plastic. However, fibronectin had no effect on endocrine markers of differentiation (FSH stimulation of cAMP, steroidogenic enzyme activity, acquisition of LH receptors). Treatment with FSH tended to cause cell rounding and aggregation and was therefore antagonistic to fibronectin with respect to morphological parameters (Morley et al. 1987b). Saumande (1991) found that fibronectin enhanced the survival of bovine granulosa cells in serum-free culture. It also reduced progesterone secretion and enhanced the oestradiol response to FSH (Saumande 1991). Somewhat in contrast, fibronectin caused a substantial increase in the expression of oxytocin by cultured bovine cells (Luck et al. 1991).

4   Fluid ECM components: proteoglycans

Follicular fluid has long been known to be a reservoir of hyaluronidase-digestible mucopolysaccharide material whose quantity and composition vary with follicular development and between species (Wislocki et al. 1947; Ax and Bellin 1988; Varner et al. 1991). This material is in close contact with the granulosa cells and is heavily sulphated as evidenced by the accumulation of injected ^{35}S both in the fluid and within the granulosa layer (Boström and Oldeblad 1952). Follicular fluid contains 0.2–0.3% glycosaminoglycans (Jensen and Zachariae 1958). These are sulphated, repeating unit polysaccharide chains making up 95% of the proteoglycan material (Alberts et al. 1989). When incubated with rat ovarian slices, an ^{35}S label became incorporated into glycosaminoglycans of the granulosa and theca of antral follicles, with 60% associated with heparan sulphate, 25% in dermatan sulphate and 15% in the chondroitin −4 and −6 sulphates (Lindner et al. 1977; Gebauer et al. 1978). Labelling was poor in pre-antral follicles, atretic follicles, corpora lutea, interstitial tissue, and oocytes (Rondell 1970; Mueller et al. 1978). The heparan sulphate of bovine follicular fluid is not protein-bound while chondroitin sulphate is covalently linked to protein (Grimek et al. 1984). Tadano and Yamada (1978) compared different areas of the follicle on the basis of their glycosaminoglycan content. All areas contained hyaluronic acid (a non-sulphated glycosaminoglycan), certain sulphated glycosaminoglycans, and neutral glycoproteins. Follicular fluid and the intercellular matrix contained isomeric chondroitin sulphates (A, B and C), while the zona pellucida of the oocyte contained complex acidic carbohydrates such as sialic acid. The higher sialic acid content of the outer region of the zona as compared with its inner region suggested the presence of two distinct layers.

Changes in proteoglycan composition occur during folliculogenesis, with an overall reduction in content as the follicle grows (Wislocki *et al.* 1947; Bushmeyer *et al.* 1984 Bellin and Ax 1987). The molecular weights of mucopolysaccharides in follicular fluid also fall as follicles mature, and this is associated with an increase in the ratio of chondroitin sulphate to hyaluronic acid (Jensen and Zachariae 1958). In pig follicular fluid the concentrations of chondroitin sulphate and heparan sulphate decreased with follicular maturation, but the degree of sulphation increased (Ax and Ryan 1979). In the cow, the heparan sulphate content of small follicles is similar to that of large follicles but the ratio of chondroitin sulphate to heparan sulphate increases with follicular maturity. Atretic follicles contain large amounts of chondroitin sulphate (Bushmeyer *et al.* 1984; Grimek *et al.* 1984). Dermatan sulphate has also been identified in ruminant follicular fluid (Bellin and Ax 1987) and tunica albuginea (Shiina 1990). Staining for chondroitin-4,6-sulphate, heparan sulphate and hyaluronic acid in the ovine follicular capsule increased with follicular development whilst the number of collagen fibrils fell (Shiina 1990).

Glycosminoglycan synthesis occurs in granulosa cells (Yanagishita and Hascall 1979) and can be stimulated by FSH unless high concentrations of hCG are present (Bellin *et al.* 1983). Stimulation by FSH could be blocked using inhibitors of protein and RNA synthesis, and cells from large follicles were found to be less responsive than those from small follicles (Schweitzer *et al.* 1981). Similarly, the incorporation of ^{35}S into ovarian proteoglycans was stimulated by FSH, but inhibited by LH and progesterone (Rondell 1970; Mueller *et al.* 1978). These data suggest that changes in proteoglycan production during follicle maturation are under close gonadotrophic regulation. The high level of chondroitin sulphate in atretic follicles may be related to the inhibition of gonadotropin binding (Nimrod and Lindner 1980; Salomon *et al.* 1978; Bushmeyer *et al.* 1984). A transformed rat ovarian cell line, DC-3, which was highly proliferative but undifferentiated, produced a different range of glycosaminoglycans from normal antral granulosa cells (Bellin *et al.* 1990).

Several functions have been proposed for these complex follicular carbohydrates. Tadano and Yamada (1978) speculated that they may create specific local regulatory environments within the follicle to control, for example, coagulation of fluid, transport of solutes and lubrication. They also proposed that sialic acid is present as a protectant for other carbohydrates and that the proteoglycans of the oocyte play a role in fertilization. Proteoglycans form highly hydrated gels (Alberts *et al.* 1989) and would therefore influence the fluid dynamics of follicular growth, although evidence on this topic is lacking (Gosden *et al.* 1988). They are certainly the principle reason for the relatively high viscosity of follicular fluid, a characteristic which is lost when fluid is treated with proteases (Yanagishita *et al.* 1979). Heparin may be present to counteract the increased tendency of follicular fluid to coagulate as ovulation approaches; the maintenance of fluidity is presumably a pre-requisite for oocyte release at ovulation (Lindner *et al.* 1977; Gebauer *et al.* 1978, Yanagishita *et al.* 1979). Partially purified glycosaminoglycans from mouse ovaries potentiated the angiogenic activity of epidermal growth factor by forming a complex with the growth factor (Sato *et al.* 1991).

Heparin binds strongly to granulosa cell membranes through one or more transmembrane receptors (Winer and Ax 1989; Yanagishita and McQuillan 1989). Binding is only semispecific and occurs to a greater extent in small follicles (Ax *et al.* 1984; Bushmeyer *et al.* 1985). The lower binding as the follicle matures may be due to a switch in competition from chondroitin sulphate to heparan sulphate rather than a change in membrane characteristics (Ax *et al.* 1986; Bellin *et al.* 1987*b*).

The physiological importance of granulosa-proteoglycan interactions is not entirely clear. Heparin can inhibit the LH-induced stimulation of adenylate cyclase activity in rat ovarian membranes by preventing the receptor binding of the gonadotrophin (Amsterdam *et al.* 1978; Salomon *et al.* 1978). Heparin and chondroitin sulphate C, but not chondroitin sulphate A or dextran sulphate, were able to overcome a serum-induced inhibition of LH receptor induction by FSH in rat granulosa cells despite having a negative effect of their own on gonadotrophin binding (Nimrod and Lindner 1980). Chondroitin sulphate and hyaluronic acid prevented or reduced FSH-stimulated progesterone accumulation by cultured rat granulosa cells (Campbell and Valiquett 1982). Other studies have suggested that proteoglycans influence the availability of lipoprotein as a steroidogenic substrate for granulosa cells. Chondroitin sulphate and heparan sulphate, at concentrations likely to be present in follicular fluid, inhibited low density lipoprotein (LDL) degradation and progesterone secretion by porcine granulosa cells (Bellin *et al.* 1987*a*). With pig granulosa cells, chondroitin sulphate inhibited progesterone production but had no effect on oestradiol production. Inhibition was lost in the presence of pregnenolone, demonstrating that it occurred at an early stage of steroidogenesis, perhaps by preventing uptake of LDL by the cell. The high concentrations of chondroitin sulphate in atretic follicles were therefore proposed as a contributory cause of atresia (Ledwitz-Rigby *et al.* 1987). Despite these speculations, other studies have demonstrated either low or insignificant binding of LDL to heparin or glycosaminoglycans from bovine follicular fluid (Brantmeier *et al.* 1988; Vanderboom *et al.* 1989).

5   Ovulation

The mechanism of ovulation depends upon a co-ordinated series of enzymic events which result in the breaching of the follicle wall (Cajander 1989). Espey (1967*a,b*) observed a thinning of collagen fibrils which occasioned a reduction in the tensile strength of the follicle wall during the preovulatory period. He suggested that this was an active degradation caused by enzymes released from fibroblasts in the follicle wall (Espey 1991). The histological events of follicle rupture have been described in some detail (mouse–Byskov 1969; Talbot *et al* 1987; rabbit–Bjersing and Cajander 1974*a-f*, 1975). Disintegration of the follicle wall takes place from the outside (tunica albuginea) inward, consistent with the view that at least some of the enzymes originate in the surface epithelium. Initially, the integrity of the surface epithelium is lost with disintegration of fibroblasts and epithelial cells; prominent dark bodies in germinal epithelial cells may be the source of proteolytic enzymes. In some species the basement membrane on which these cells rest undergoes oedema (Talbot *et al.* 1987) and its collagen bundles become fragmented. Subsequently, the membranes and endothelial structures of blood vessels within the theca also disintegrate, with resultant seepage of blood into the interstitium. In the last few hours before rupture the follicle expands and the basal lamina between the theca and granulosa breaks down; theca cells break into the granulosa layer and long processes from the granulosa cells begin to penetrate the theca.

These descriptive studies show that the proteolytic changes which follow the disintegration of the surface epithelium, occur in all regions of the follicle wall. The direction in which follicle contents are expelled will therefore depend on the thinning and loss of integrity in the thin apical 'stigma' in the tunica albuginea. Although the form of the stigma varies between species (Talbot *et al.* 1987), the collagen in the apical membrane undergoes considerable remodelling in the hours preceding ovulation (Dennefors *et al.* 1982). The location of this activity defines the site of the stigma and appears to be regulated by prostaglandins (Osman and Dullaart 1976; Dennefors *et al.* 1982) and steroid hormones (Tjugum *et al.* 1984).

6   Proteolytic enzymes

The proteolytic activity of porcine follicular fluid was demonstrated by Schochet (1916) and subsequently shown to be due to a complex of enzymes with a wide range of pH optima (Reichert 1962; Parr 1974; Beers 1975). The greatest amounts of enzyme activity were detected in large follicles and were thought to originate from the granulosa layer (Jung and Held 1959; Unbenhaun *et al.* 1965). Activity was also located in the follicle wall but as ovulation approached its quantity declined while that of the fluid increased, suggesting that it was being 'labilized' (Espey and Rondell 1968). A range of enzymes was investigated for their ability to affect the strength of the follicle wall and initiate ovulation; of these the collagenases were the most effective (Espey 1970). More recent studies indicate that the follicle secretes both metalloproteinases (collagenases) and the fibrinolytic serine proteases, together with their regulatory proteins (Fig. 2.3). These are probably released in a series of highly co-ordinated events which, with other evidence, have suggested comparisons between ovulation and inflammation (Espey 1980, 1991).

7   Metalloproteinases: collagenase

At least four different enzymes with collagenase activity have been identified in follicular tissues. These will cleave synthetic collagen substrates and have the typical biochemical characteristics of metalloproteinases (Espey and Coons 1976; Morales *et al.* 1978, 1983; Yajima *et al.* 1980; Fukumoto *et al.* 1981; Curry *et al.* 1985). Although the specific enzymes vary in their distribution within the follicle, activity can be located in the follicle wall, the

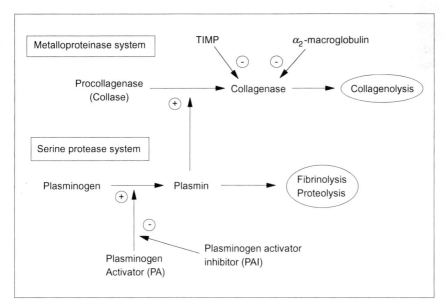

Fig. 2.3   Schematic summary of the ovarian protease systems and their interaction.

granulosa, and the fluid compartments. The activity in the wall occurs in all regions but is particularly intense at the apex (Murdoch and McCormick 1992). Activity generally falls as follicles mature, rises transiently in the granulosa in the immediate preovulatory period, and declines thereafter. One of the collagenases released into follicular fluid just before ovulation is specific for the basal lamina collagen type IV (Puistola *et al* 1986; Palotie *et al* 1987; Reich *et al* 1991). In the follicles of pregnant mare's serum gonadotrophin (PMSG)-treated immature rats, collagenase was detected immunohistologically at 8 h after hCG administration. It was located around cumulus cells and also in the cells and extracellular space of the theca and tunica albuginea at the apex of large follicles (Curry *et al.* 1986). Rat follicles show a decrease in hydroxyproline content on the afternoon of pro-oestrus, indicating collagen depletion, but this does not occur if nembutal is used to prevent the LH surge (Reich *et al.* 1985). Collagenase is considered to be secreted by granulosa cells (Fukumoto *et al.* 1981) and this has been confirmed by culture studies (Puistola *et al.* 1989).

The rise in collagenase immediately prior to ovulation is stimulated by LH but the mechanism by which activity is controlled is complex. Immature PMSG-treated rats given hCG showed an increase in mRNA for collase, an activatable collagenase, suggesting that potential activity was increased at the transcription level (Tsafriri *et al.* 1990). This latent form of the enzyme can be activated *in vitro* by trypsin (Morales *et al.* 1978; Reich *et al.* 1985). Ovulation of primed follicles can be blocked by collagenase inhibitors (Brännström *et al.* 1988) but also by a specific plasmin inhibitor, suggesting that an endogenous serine protease activates the procollagenase (Woessner *et al.* 1989). Several inhibitors of collagenase have been identified in follicles and follicular fluid. These increase in concentration just before ovulation and include serum-derived factors, such as α_2-macroglobulin, and classical TIMPs (Curry *et al.* 1988; Zhu and Woessner 1991). Granulosa cells of several species express the mRNA for TIMP (Rapp *et al.* 1990; Curry *et al.* 1990; Freudenstein *et al.* 1990; Reich *et al.* 1991) and expression is increased in response to preovulatory surge concentrations of gonadrotrophins (Moor and Crosby 1987; Mann *et al.* 1991; Smith and Moor 1991; Smith *et al.* 1992). The presence of TIMP mRNA in cultured bovine granulosa cells is closely associated with their spontaneous expression of the luteal phenotype (Freudenstein *et al.* 1990) and expression in the sheep ovary continues well into the luteal phase (Smith *et al.* 1992). Serum-derived inhibitors would gain access to the follicle at the time of basal lamina disintegration (Morales *et al.* 1983; Curry *et al.* 1990; Zhu and Woessner 1991); in addition, rat granulosa cells express α_2-macroglobulin after exposure to hCG (Gaddy-Kurten *et al.* 1989) indicative of a regulated endogenous follicular secretion. TIMP type 1, which preferentially inhibits interstitial collagenase, is expressed by theca and granulosa cells after exposure to hCG; expression also increases in the granulosa cells of non-ovulating antral follicles, suggesting a role for this inhibitor in preventing premature ovulation (Chun *et al.* 1992).

The gonadotrophin-induced rise in TIMP concentration at the time of ovulation appears to follow that of collagenase (Zhu and Woessner 1991) and may be viewed as a mechanism to prevent overshoot in the proteolytic activity which results in follicular rupture. In addition, both α_2-macroglobulin (Gaddy-Kurten *et al.* 1989) and TIMP (Freudenstein *et al.* 1990) are highly expressed well into the luteal phase, suggesting an important role in the remodelling and maintenance of luteal tissue (Mann *et al.* 1991).

8   Serine proteases: the plasmin cascade and its regulators

A large body of literature describes the characteristics of the plasminogen-plasmin system of fibrinolytic serine proteases which exists in the follicle (Cajander 1989). PA was first extracted from the human ovary by Albrechtsen (1957). Since then, follicular fluid has also been shown to contain plasminogen, plasmin-like proteolytic activity, and serine protease inhibitors such as plasminogen activator inhibitor (PAI). Plasmin, the fibrinolytic product of this enzyme cascade, decreases the tensile strength of the follicle wall (Beers 1975) and is presumed to play a role in follicle rupture. Fibrinolytic activity first appears in the rat follicle wall 12 h before ovulation, peaks at −2 h, and decreases after ovulation, but the activity is seen only in the presence of plasminogen, suggesting that fibrinolytic activity is dependent on PA (Akazawa *et al.* 1983*b*). Follicles treated with protease inhibitors before ovulation show markedly reduced rates of oocyte release (Ichikawa *et al.* 1983*a*, *b*; Pellicer *et al.* 1988).

Both the tPA and the uPA types of PA have been identified in the follicle (Ny *et al.* 1985), albeit with some dispute as to their origin and location. Studies in the rat suggested that tPA and uPA came from the granulosa and theca, respectively; since both are under gonadotrophic control this implied that ovulation requires the cooperation of both cell types (Canipari and Strickland 1985). Other investigations in the rat found that both PAs were produced by granulosa and theca cells with 80–90% of total activity coming from the granulosa cells; as with Sertoli cells, granulosa cells may switch to predominantly tPA production in the presence of FSH (Reich *et al.* 1986; O'Connell *et al.* 1987; Hettle *et al.* 1988). Oestrogenic follicle and those approaching ovulation express tPA, suggesting that this is the type involved in follicular rupture (Reich *et al.* 1986; Liu *et al.* 1991*b*). Immature rat ovaries were found to express uPA in the granulosa after PMSG treatment; subsequent hCG treatment caused the expression of tPA in the granulosa and uPA in the theca, but the granulosa of large follicles expressed uPA at the time of the LH surge (Park *et al.* 1991). Rats which were given intrabursal injections of α_2-antiplasmin or specific antibodies against tPA at the same time as hCG showed a suppression of ovulation, whereas injections given after the hCG were ineffective (Tsafriri *et al*. 1989). This and other data (Akazawa *et al*. 1983*a*; Palotie *et al*. 1987) suggest the involvement of plasminogen in early stages of the ovulatory process, for example, in the activation of collagenase.

In the mouse, uPA was expressed in both theca and granulosa, and granulosa cells produced uPA in response to gonadotrophins. In contrast to the rat, tPA could not be found in mouse follicular fluid but was expressed in the oocytes of preovulatory follicles (Canipari *et al*. 1987; Sappino *et al*. 1989). Primate follicles contain both types of PA but increasing oocyte maturation and the approach of ovulation are associated with an increase in tPA (Reinthaller *et al*. 1990; Liu *et al*. 1991*b*).

This somewhat confused picture implies species differences but also suggests a sequence of co-ordinated expression and activation events in the granulosa and theca of the peri-ovulatory follicle which may be difficult to study effectively. The concentrations of PA and plasmin in follicular fluid increase towards ovulation; concentrations of plasminogen remain constant and there is a decrease in the presence of PAI, indicating that the balance between PA and PAI regulate plasmin production (Beers *et al*. 1975; Akazawa *et al*. 1983*a*, *b*; Shimada *et al*. 1983; Espey *et al*. 1985; Politis *et al*. 1990; Reinthaller *et al*. 1990). PAI is expressed in both granulosa and theca, but at different times in relation to the impending follicular rupture; the activation of plasminogen is therefore a tissue- as well as time-coordinated process (Liu *et al*. 1991*a,b*). In one human study, PA activity in follicular fluid was closely correlated with oestradiol concentration (Weimer *et al*. (1984), but in another, very little free PA activity could be determined and there was an abundance of PAI (Jones *et al*. 1989).

It is likely that the plasminogen-plasmin system is only one component of a larger complex of protein-protease interactions, perhaps closely related to blood clotting mechanisms. Concentrations of α_1-antitrypsin, α_2-macroglobulin, antithrombin III, ceruloplasmin and fibrinogen are known to be higher in mature than in immature human follicles (Gulamali-Majid *et al*. 1987).

The loss of ovulation caused by intrabursal injections of α_2-antiplasmin or specific antibodies against tPA at the same time as hCG was associated with intraovarian release of oocytes; thus although the basal lamina had disintegrated in all cases, plasminogen activity may be essential to the early dissolution of the external follicular membranes (theca externa and tunica albuginea; Tsafriri *et al*. 1989) and the location of the stigma.

9   Hormonal regulation of proteases

The total PA activity secreted by granulosa cells is under gonadotrophic control (Beers *et al*. 1975; Strickland and Beers 1976; Wang and Leung

1983). As in the testis (Lacroix and Fritz 1982) it is particularly responsive to FSH (Dorrington *et al.* 1983), a property which has been exploited in developing highly sensitive FSH bioassays (Beers and Strickland 1978; Thakur *et al.* 1990). In the case of tPA, FSH produces an immediate and transient release whereas LH causes a delayed but sustained rise in concentration (O'Connell *et al.* 1987).

Prostaglandins play a role in the process of follicle rupture and have therefore been postulated as regulators of follicular proteases. Some of the experimental evidence is equivocal (Murdoch and McCormick 1991), but this may be related to a complex interaction between the serine proteases and collagenases. Regarding the serine proteases, Canipari and Strickland (1986) observed that indomethacin would prevent increases in uPA as well as LH-induced increases in tPA. They suggested the existence of two populations of granulosa cells: some would produce PA under the influence of FSH and a small fraction of these, responsive to LH, would produce prostaglandin which then stimulates all cells. In other words, prostaglandins mediate the effect of LH in cell-cell communication, and indomethacin inhibits ovulation by preventing the build up of PA activity within the cell population. Such a system would be consistent with an observed delay in the PA response to LH (O'Connell *et al.* 1987). Other studies found that inhibitors of the lipoxygenase and cyclooxygenase pathways reduced the preovulatory depletion of collagen from the follicle wall and prevented ovulation itself, but did so without preventing the rise in PA activity (Shimada *et al.* 1983; Espey *et al.* 1985; Reich *et al.* 1985; Grant *et al.* 1992). This suggested an effect on collagenase, but in other studies follicular collagenase activity was unaffected by prostaglandin (PG)E_2, $PGF_{2\alpha}$ (Norström and Tjugum 1986) or indomethacin (Curry *et al.* 1986). The preovulatory rise in TIMP mRNA could be blocked by inhibitors of eicosanoids (Reich *et al.* 1991).

Regarding other potential regulators, the distensibility of rabbit follicles can be increased by progesterone (Rondell 1970), and rat PA activity may be progesterone-dependent (Tanaka *et al* 1992a) although there appears to be no correlation between PA and progesterone concentrations in human follicular fluid (Weimer *et al.* 1984). The antiprogesterone RU486 prevented the hCG-induced rise in collagenolytic activity in PMSG-primed rat ovaries (Shibata 1990). Prolactin reduced the PA content of rabbit follicles and inhibited the breakdown of connective tissue at the apex of the follicle (Yoshimura *et al.* 1990). Relaxin can stimulate PA, collagenase and proteoglycanase activities in PMSG-primed rats and may act by increasing protease activation (Too *et al.* 1982, 1984). The expression of tPA by rat granulosa cells can be selectively stimulated by GnRH (which also inhibits uPA; Ny *et al.* 1987) and by catecholamines (Oikawa and Hsueh 1989). The PAI activity of cumulus and non-cumulus granulosa cells in the human ovary can be stimulated through mechanisms involving protein kinase-C (Kessel *et al*. 1992).

10   The ECM in oocyte maturation

The maturation of the oocyte involves an expansion of the cumulus oophorus as a result of gonadotrophic action. This is associated with the synthesis of hyaluronic acid and the appearance of a hyaluronidase-sensitive mucus between the cumulus cells in the last few hours before ovulation. The associated loss of cell-cell contacts can be modelled in culture using FSH (Dekel and Kraicer 1978; Eppig *et al*. 1982). Expansion can be prevented by an absence of substrate for hyaluronic acid synthesis or by the presence of specific synthesis inhibitors (Larsen *et al*. 1991). Sulphated glycosaminoglycans inhibited FSH-stimulated cumulus expansion and hyaluronic acid synthesis, with the following order of potency: heparin > heparan sulphate ≈ chondroitin sulphate B > chondroitin sulphate C > chondroitin sulphate A. The effect was not related to the binding of FSH to its receptor and probably involved an intracellular action subsequent to the generation of cAMP. These substances did not affect spontaneous oocyte maturation (Eppig *et al*. 1982).

The fibronectin content of human follicular fluid was found to correlate well with follicular size and oocyte maturation. Peptide GRGDS, which inhibits fibronectin by competing for cellular binding sites, prevented both spontaneous and induced meiosis and maturation of mouse oocytes *in vitro* (Hung *et al*. 1989).

The cumulus-oocyte complex of the hypophysectomized, immature, oestrogen-treated rat contains tPA. Its quantity rises after hCG injection and is correlated with the extent of cumulus cell expansion and dispersion. Although granulosa cells probably contain tPA and uPA, the denuded rat oocyte contains only tPA; this is located in the cytoplasm and does not, by itself, respond to FSH. In contrast, the cumulus-oocyte complex responds to FSH with an increase in tPA, suggesting that cumulus cells mediate the effects of FSH on the oocyte during ovulation or upon cumulus expansion/dispersal (Huarte *et al*. 1985; Liu *et al*. 1986, 1987; Liu and Hsueh 1987).

11   ECM components as predictors of success in assisted reproduction

Several attempts have been made to identify correlations between the results of assisted reproduction procedures, such as *in vitro* fertilization (IVF) and gamete intrafallopian tube transfer. The amounts of ECM components in follicular fluid and the relatively straightforward assays for many components might provide a basis for predicting the outcome of individual procedures and for improving success rates.

Bellin *et al*. (1986) report an association between low chondroitin sulphate concentrations and higher rates of oocyte fertilization in hMG-treated women. High heparan sulphate concentrations were significantly correlated with visual indices of oocyte-corona-cumulus complex maturation but not with the rate of fertilization. Tsuiki et al. (1988) found that fertilized oocytes tended to come from follicles with higher fibronectin/protein and fibronectin/glycosaminoglycan ratios and lower glycosaminoglycan/protein ratios.

Low collagenolytic activity and α_2-macroglobulin concentrations in follicular fluid were associated with pregnant cycles rather than non-pregnant cycles after IVF-embryo transfer procedures (Rom et al. 1987). Another study was unable to relate collagenase type IV activity in follicular fluid with the degree of oocyte maturity, perhaps because the normal selection mechanisms for dominant follicles were absent in stimulated cycles (Rönnberg et al. 1988).

Studies of PA in follicular fluid have given equivocal results. Rom et al. (1987) failed to observe a difference in total PA activity between successful and unsuccessful pregnancies, whereas Milwidsky et al. (1989) found higher levels in fertile than non-fertile patients. They detected latent activity in fluid from the non-fertile group and suggested that the difference was due to the presence of an endogenous inhibitor, perhaps related to lower follicular maturity. In another study, fertilized oocytes tended to originate from follicles with higher tPA, oestradiol and progesterone concentrations and lower testosterone concentrations. No differences in uPA or PAI were detected between fertile and non-fertile groups (Deutinger et al. 1988).

12   Luteinization and the luteal ECM

Compared with the follicle, our knowledge of the ECM of the corpus luteum is very poor and in need of systematic study. It is apparent from the foregoing that developments in the matrix are intimately involved in the later stages of follicular function and with ovulation itself, and that the follicular cells which will give rise to the corpus luteum respond dramatically to the matrix components to which they are exposed. An understanding of cell-matrix interactions during luteinization and the subsequent luteal phase will help in determining the factors upon which a functional corpus luteum depends.

Morphometric analysis of bovine tissue indicates that 'extracellular space' (that is, excluding luteal cells, erythrocytes, and blood vessels) comprises 30% of the cross-sectional area of corpus luteum on day 6 of the cycle and 15% on day 15 (Parry et al. 1980). In perfusion-fixed rat tissue, intercellular space accounted for 5% of the sectional area. The components of these regions have yet to be described although electron micrographs of sheep corpus luteum reveal an extensive network of fibrillar collagen, in close association with both large and small endocrine cells (Fig. 2.4). Luteal cells

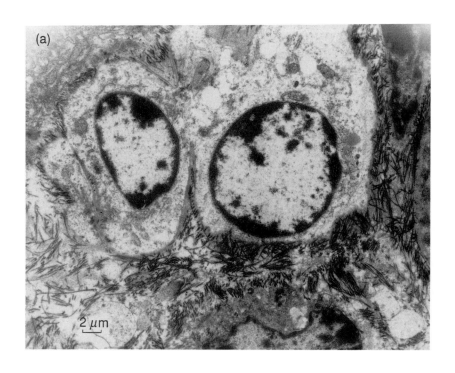

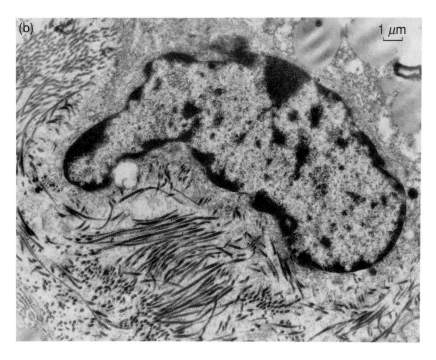

Fig. 2.4   Electron micrographs of sheep corpora lutea (mid-luteal phase) to illustrate collagen fibrils in close association with (a) large and (b) small luteal cells. Scale bars: (a) ×5000, 1cm = 2 μm; (b) ×10000, 1 cm = 1 μm. (Courtesy of Dr J.H Payne).

lack a basal lamina except where they are in close apposition to capillaries (capillary space comprised 20% of the total; Meyer 1991). Laminin has been located in the periphery of the corpus luteum and around individual luteal cells (Wordinger *et al.* 1983) and is particularly associated with perisiusoidal (vascular) areas (Leardkamolkarn and Abrahamson 1992). The bovine luteal tunica albuginea contains collagen fibres and hyaluronic acid which increase in amount during its formation (Shiina 1990).

The remodelling of ruptured follicular tissues to form the corpus luteum presumably involves considerable enzymic activity. A single report describes the presence of tPA in the outer layers of the newly formed corpus luteum (Park *et al.* 1991), and luteinizing primate granulosa cells express uPA (Liu *et al.* 1991*b*). In view of the role of collagenase in the destructive events of ovulation, and the requirement for tissue reconstruction during luteinisation, it is not surprising that TIMP is strongly expressed in luteinizing granulosa cells and luteal tissues, particularly those of the early luteal phase (Freudenstein *et al.* 1990; Tanaka *et al.* 1992*b*). TIMP occurs in ovine luteal tissue throughout the luteal phase independently of hormonal support, and accounts for approximately one-quarter of all secreted protein (Smith and Moor 1991). It is particularly associated with the large luteal cells (Smith *et al.* 1992).

The activities of two enzymes associated with collagen synthesis, prolyl and lysyl hydroxylase, were found to be maximal in rabbit ovaries in the period immediately following hCG-induced ovulation (Himeno *et al.* 1984). Chemical inhibitors of collagen biosynthesis (*cis*-hydroxyproline and 2,7,8,-trihydroxyanthraquinone) caused a dramatic but reversible inhibition of luteinization in cultured bovine granulosa cells (Luck *et al.* 1991) suggesting an autocrine role for collagen in the luteinization process.

13   Ascorbic acid in the ovary

The place of ascorbic acid in a review of this nature may not be apparent at first sight. However, evidence points to a role for this material in corpus luteum formation which can be considered in the context of the ECM. Ascorbic acid plays a well defined role in collagen biosynthesis; it promotes collagen gene expression and is an essential co-factor in the enzymatic hydroxylation of proline and lysine residues (Bornstein and Sage 1989; Chojkier *et al.* 1989 Padh 1991). As noted earlier, the hydroxylase activity of the postovulatory follicle is high. Ascorbate is also capable of promoting proteoglycan production (Kao *et al.* 1990; Fisher *et al.* 1991). The corpus luteum is an abundant storage site for ascorbate, achieving concentrations comparable with those in the adrenal cortex (Sheldrick and Flint 1989), although the reason for this has been somewhat obscure. This review has highlighted the turnover of collagen which accompanies follicular rupture and the remodelling of follicular tissues. The synthesis of collagen during remodelling will create an intense demand for ascorbate during the formation of luteal tissue. Such an hypothesis was proposed by Espey and Coons (1976) but has since received little attention.

Deane (1952) identified ascorbic acid in the rat ovary, located in the theca interna and interstitial cells but particularly in the luteal cells. It was also present in the granulosa cells of atretic follicles. The ability of LH to deplete ascorbic acid from the ovaries of PMSG and hCG-primed immature rats was used as an early bioassay for the gonadotrophin (Bell *et al.* 1965) based on independent proposals by Parlow (1958) and Karg (1957). High doses of ascorbate inhibited the collagenolytic activity of mature rabbit follicles (Espey and Coons 1976) and reduced the tensile strength of pig follicle wall *in vitro*, although at least the latter effect may have been the result of a very low pH (Espey 1970).

Ascorbate is one of a number of antioxidants in the ovary, whose uptake is energy-dependent and inhibited by LH and $PGF_{2\alpha}$ (Aten *et al.* 1992; Stanfield and Flint 1967). In the sheep, uptake by the corpus luteum is high during the first week of the luteal phase and substantially exceeds the amount calculated as necessary to support steroidogenesis and peptide hormone synthesis (Sheldrick and Flint 1989). Ascorbic acid excretion in women reaches a peak 3 days before ovulation, then declines and peaks again at ovulation (Loh and Wilson 1971) which would be consistent with significant changes in tissue uptake during this period. Ascorbate has been shown to promote the luteinization of bovine granulosa cells in serum-supplemented (Luck and Jungclas 1987) and serum-free (Luck *et al.* 1990) culture and its effects are associated with the initial response of the cells to culture conditions (Luck and Jungclas 1988).

V   CONCLUSIONS

The ECM evidently plays a crucial regulatory as well as structural role in all aspects of gonadal function. The location of extracellular structures not only directs the embryology of the gonad, but also maintains the segregation of tissues through later stages of remodelling. The process of tissue remodelling is itself involved in the release of germ cells, and the activities of somatic cells are profoundly influenced by the extracellular environment at all stages of the reproductive process.

A particularly striking feature which emerges is the high degree of similarity between the ovary and the testis in their ECM components, and the consistent association of these components with cells of similar embryological background. This emphasizes the parallels which can be drawn between Sertoli and granulosa cells on the one hand, and between Leydig and theca cells on the other. In each gonad there is a fluid environment, specific to the maturational requirements of the germ cells, which is separated by a basement membrane from the peripheral compartment. The membrane prevents angiogenesis in the germinal compartment but also emerges as a highly directional structure which can determine cell phenotype. Even within the limited confines of the follicular antrum, the degree of cellular differentiation can be related to distance from the basal lamina. In the case of the testis, the location of the membrane has been seen as a crucial factor in the structural evolution of the spermatogenic unit (Grier 1992), and it is likely that a similar interpretation could be applied to ovarian phylogeny.

It is evident that in both gonads the release of mature germ cells requires a degree of tissue remodelling, involving similar sets of proteolytic enzymes working in highly regulated sequences. One major difference is that the germ cells are released in opposite directions relative to the basement membrane: in the testis they remain on the lumenal side and remodelling is confined to the Sertoli barrier during sperm maturation, whereas in the ovary, release is to the peripheral side and involves a complete breach of the follicle wall. In the latter case, the repairs which follow are intimately concerned with the construction of a further ephemeral tissue and the provision of endocrine support for the conceptus. The somatic cells of both gonads remain in various stages of reversible differentiation, as dictated by the changing extracellular environment, but in the case of follicular cells may achieve a final stage of terminal differentiation during atresia or luteinization.

The teleological goal of gonadal function is the production of viable germ cells and, in the case of the ovary, the maintenance of pregnancy. This review demonstrates that, with the clear exception of the corpus luteum, we have a reasonably extensive knowledge of the gonadal ECM and its role in tissue dynamics, cellular differentiation, and the production of gametes.

ACKNOWLEDGEMENTS

I am indebted to many colleagues whose stimulating discussion has helped to clarify my ideas, especially Beate Jungclas, who contributed to the germinal stages of this review, Ray Rodgers and Malcolm Richardson. I am also grateful to Anne Grete Byskov for advice on the perplexing field of gonadal embryology and to Jo Payne for generously providing unpublished electron micrographs. Elke Rohrberg and Jan Barrowcliff provided the patient and efficient library services upon which a reviewer's work depends.

REFERENCES

Agelopoulou, R. and Magre, S. (1987). Expression of fibronectin and laminin in fetal male gonads *in vivo* and *in vitro* with and without testicular morphogenesis. *Cell Differentiation*, **21**, 31–6.

Ailenberg, M. and Fritz, I.B. (1988). Control of levels of plasminogen activator activity secreted by Sertoli cells maintained in a two-chambered assembly. *Endocrinology*, **122**, 2613–8.

Ailenberg, M. and Fritz, I.B. (1989). Influences of follicle-stimulating hormone, proteases, and antiproteases on permeability of the barrier generated by Sertoli cells in a two-chambered assembly. *Endocrinology*, **124**, 1399–407.

Ailenberg, M., Tung, P.S., Pelletier, M., and Fritz, I.B. (1988). Modulation of Sertoli cell functions in the two-chamber assembly by peritubular cells and by extra-cellular matrix. *Endocrinology*, **122**, 2604–12.

Ailenberg, M., McCabe, D., and Fritz, I.B. (1990). Androgens inhibit plasminogen activator activity secreted by Sertoli cells in culture in a two-chambered assembly. *Endocrinology*. **126**, 1561–8.

Ailenberg, M., Stetler-Stevenson, W.G., and Fritz, I.B. (1991). Secretion of latent type IV procollagenase and active type IV collagenase by testicular cells in culture. *Biochemical Journal*, **279**, 75–80.

Akazawa, K., Matsuo, O., Kosugi, T., Mihara, H., and Mori, N. (1983*a*). The role of plasminogen activator in ovulation. *Acta Physiologica Latinoamericana*, **33**, 105–10.

Akazawa, K., Mori, N., Kosugi, T., Matsuo, O., and Mihara, H. (1983*b*). Localization of fibrinolytic activity in ovulation of the rat follicle as determined by the fibrin slide method. *Japanese Journal of Physiology*, **33**, 1011–18.

Alberts, B., Bray, D., Lewis, J., Raff, M., Roberts, K., and Watson, J.D. (1989). *The molecular biology of the cell*, 2nd edn. Garland, New York.

Albrechtsen, O.K. (1957). The fibrinolytic activity of human tissues. *British Journal of Haematology*, **3**, 284–91.

Amsterdam, A., Reches, A., Amir, Y., Mintz, Y., and Salomon, Y. (1978). Modulation of adenylate cyclase activity by sulphated glycosaminoglycans. II. Effect of mucopolysaccharides and dextran sulfate on the activity of adenylate cyclase derived from various tissues. *Biochimica et Biophysica Acta*, **544**, 263–72.

Amsterdam, A., Rotmensch, S., Furman, A., Venter, E.A., and Vlodavsky, I. (1989). Synergistic effect of human chorionic gonadotrophin and extracellular matrix on *in vitro* differentiation of human granulosa cells: progesterone production and gap junction formation. *Endocrinology*, **124**, 1956–64.

Amsterdam, A., Suh, B.S., Himmelhoch, S., Baum, G., and Ben-Ze'ev, A. (1991). Modulation of granulosa cell ultrastructure during differentiation: the role of the cytoskeleton. In *Ultrastructure of the ovary*, Electron Microscopy in Biology and Medicine, **9**, (ed. G. Familiari, S. Makabe and P.M. Motta), pp. 101–12. Klewer Academic, Boston, USA.

Anthony, C.T. and Skinner, M.K. (1989). Actions of extracellular matrix on Sertoli cell morphology and function. *Biology of Reproduction*, **40**, 691–702.

Aten, R.F., Duarte, K.M., and Behrman, H.R. (1992). Regulation of ovarian antioxidant vitamins, reduced glutathione, and lipid peroxidation by luteinizing hormone and prostaglandin $F_{2\alpha}$. *Biology of Reproduction*, **46**, 401–7.

Auersperg, N., Maclaren, I.A., and Kruk, P.A. (1991). Ovarian surface epithelium: autonomous production of connective tissue-type extracellular matrix. *Biology of Reproduction*, **44**, 717–24.

Ax, R.L. and Bellin, M.E. (1988). Glycosaminoglycans and follicular development. *Journal of Animal Science*, **66**, (Suppl. 2), 32–49.

Ax, R.L. and Ryan, R.J. (1979). The porcine ovarian follicle. IV. Mucopolysaccharides at different stages of development. *Biology of Reproduction*, **20**, 1123–32.

Ax, R.L., Bushmeyer, S.M., Boehm, S.K., and Bellin, M.E. (1984). Binding of the glycosaminoglycan ^{3}H-heparin to bovine granulosa cells varies with size and oestrogen content of ovarian follicles. *Endocrine Research*, **10**, 63–72.

Ax, R.L., Stodd, C.M., Boehm, S.K., and Bellin, M.E. (1986). Removal of glycosaminoglycans from bovine granulosa cells contributes to increased binding of hydrogen-3 heparin. *Journal of Dairy Science*, **69**, 531–4.

Bagavandoss, P., Midgley, A.R., and Wicha, M. (1983). Developmental changes in the ovarian follicular basal lamina detected by immunofluorescence and electron microscopy. *Journal of Histochemistry and Cytochemistry*, **31**, 633–40.

Bartholomeusz, R.K., Bertoncello, I., and Chamley, W.A. (1988). A correlation between ovarian follicular maturity and anchorage-independent growth of bovine granulosa cells. *International Journal of Cell Cloning*, **6**, 106–15.

Beers, W.H. (1975). Follicular plasminogen and plasminogen activator and the effect of plasmin on follicle wall. *Cell*, **6**, 379–86.

Beers, W.H. and Strickland, S. (1978). A cell culture assay for follicle-stimulating hormone. *Journal of Biological Chemistry*, **253**, 3877–81.

Beers, W.H., Strickland, S., and Reich, E. (1975). Ovarian plasminogen activator: relationship to ovulation and hormonal regulation. *Cell*, **6**, 387–94.

Bell, E.T., Loraine, J.A., Mukerji, S., and Visutakul, P. (1965). Further observations on the ovarian ascorbic acid depletion test for luteinizing hormone. *Journal of Endocrinology*, **32**, 1–7.

Bellin, M.E. and Ax, R.L. (1987). Purification of glycosaminoglycans from bovine follicular fluid. *Journal of Dairy Science*, **70**, 1913–19.

Bellin, M.E., Lenz, R.W., Steadman, L.E., and Ax, R.L (1983). Proteoglycan production by bovine granulosa cells *in vitro* occurs in response to FSH. *Molecular and Cellular Endocrinology*, **29**, 51–65.

Bellin, M.E., Ax, R.L., Laufer, N., Tarlatzis, B.C., DeCherney, A.H., Feldberg, D., and Haseltine, F.P. (1986). Glycosaminoglycans in follicular fluid from women undergoing *in vitro* fertilization and their relationship to cumulus expansion, fertilization and development. *Fertility and Sterility*, **45**, 244–8.

Bellin, M.E., Veldhuis, J.D., and Ax, R.L (1987a). Follicular fluid glycosaminoglycans inhibit degradation of low density lipoproteins and progesterone production by porcine granulosa cells. *Biology of Reproduction*, **37**, 1179–84.

Bellin, M.E., Wentworth, B.C., and Ax, R.L (1987b). Comparisons of the ability of glycosaminoglycans and chemically desulphated heparin to compete for binding sites on granulosa cells. *Biology of Reproduction*, **37**, 293–300.

Bellin, M.E., Schmidt, W.A., and Ax, R.L. (1990). Glycosaminoglycan production by transformed ovarian cells. *Biology of Reproduction*, **42**, (Suppl. 1; Abstr. 349), 157.

Belt, W.D. and Cavazos, L.F. (1971). Fine structure of the interstitial cells of Leyding in the squirrel monkey during seasonal regression. *Anatomical Record*, **169**, 115–28.

Ben-Rafael, Z., Benadiva, C.A., Mastroianni, L., Garcia, C.J., Minda, J.M., Iozzo, R.V., and Flickinger, G.L. (1988). Collagen matrix influences the morphologic features and steroid secretion of human granulosa cells. *American Journal of Obstetrics and Gynecology*, **159**, 1570–4.

Bernath, V.A., Muro, A.F., Vitullo, A.D., Bley, M.A., Baranao, J.L., and Kornblihtt, A.R. (1990). Cyclic AMP inhibits fibronectin gene expression in a newly developed granulosa cells line by a mechanism that suppresses cAMP-response element-dependent transcriptional activation. *Journal of Biological Chemistry*, **265**, 18219–26.

Bjersing, L. and Cajander, S. (1974a). Ovulation and the mechanism of follicular rupture. I. Light microscopic changes in rabbit ovarian follicles prior to induced ovulation. *Cell and Tissue Research*, **149**, 287–300.

Bjersing, L. and Cajander, S. (1974b). Ovulation and the mechanism of follicular rupture. II. Scanning electron microscopy of rabbit germinal epithelium prior to induced ovulation. *Cell and Tissue Research*, **149**, 301–12.

Bjersing, L. and Cajander, S. (1974c). Ovulation and the mechanism of follicular rupture. III. Transmission electron microscopy of rabbit germinal epithelium prior to induced ovulation. *Cell and Tissue Research*, **149**, 313–27.

Bjersing, L. and Cajander, S. (1974d). Ovulation and the mechanism of follicular rupture. IV. Ultrastructure of membrana granulosa of rabbit Graafian follicles prior to induced ovulation. *Cell and Tissue Research*, **153**, 1–14.

Bjersing, L. and Cajander, S. (1974e). Ovulation and the mechanism of follicular rupture. V. Ultrastructure of tunica albuginea and theca externa of rabbit Graafian follicles prior to induced ovulation. *Cell and Tissue Research*, **153**, 15–30.

Bjersing, L. and Cajander, S. (1974f). Ovulation and the mechanism of follicular rupture. VI. Ultrastructure of theca externa and the inner vascular network surrounding rabbit Graafian follicles prior to induced ovulation. *Cell and Tissue Research*, **153**, 31–44.

Bjersing, L. and Cajander, S. (1975). Ovulation and the role of the ovarian surface epithelium. *Experientia*, **15**, 605–8.

Boitani, C., Farini, D., Canipari, R., and Bardin, C.W. (1989). Estradiol and plasminogen activator secretion by cultured rat Sertoli cells in response to melanocyte-stimulating hormones. *Journal of Andrology*, **10**, 202–9.

Borland, K., Muffly, K.E., and Hall, P.F. (1986). Production and components of extracellular matrix by cultured rat Sertoli cells. *Biology of Reproduction*, **35**, 997–1008.

Bornstein, P. and Sage, H. (1989). Regulation of collagen gene expression. *Progress in Nucleic Acid Research and Molecular Biology*, **37**, 67–106.

Boström, H. and Odeblad, E. (1952). Autoradiographic observations on the uptake of ^{35}S in the genital organs of the female rat and rabbit after the injection of labelled sodium sulfate. *Acta Endocrinologica*, **10**, 89–96.

Brännström, M., Woessner, J.F., Koos, R.D., Sear, C.H.J., and LeMaire, W.J.

(1988). Inhibitors of mammalian tissue collagenase and metalloproteinases suppress ovulation in the perfused rat ovary. *Endocrinology*, **122**, 1715–21.

Brantmeier, S.A., Grummer, R.R., and Ax, R.L (1988). High density lipoproteins from bovine plasma and follicular fluid do not possess a high affinity for glycosaminoglycans. *Lipids*, **23**, 269–74.

Bushmeyer, S.M., Bellin, M.E., Brantmeier, S.A., Boehm, S.K., Kubajak, C.L., and Ax, R.L. (1984). Relationships between bovine follicular fluid, glycosaminoglycans and steroids. *Endocrinology*, **117**, 879–85.

Bushmeyer, S.M., Bellin, M.E., and Ax, R.L. (1985). Specific binding of [^{3}H]heparin to bovine granulosa cell membranes. *Molecular and Cellular Endocrinology*, **42**, 135–44.

Byers, S.W., Hadley, M.A., Djakiew, D., and Dym, M. (1986). Growth and characterization of polarized monolayers of epididymal epithelial cells and Sertoli cells in dual environment culture chambers. *Journal of Andrology*, **7**, 59–69.

Byskov, A.G.S. (1969). Ultrastructural studies on the preovulatory follicle in the mouse ovary. *Zeitschrift der Zellforschung*, **100**, 285–99.

Byskov, A.G. (1978). The anatomy and ultrastructure of the rete system in the fetal mouse ovary. *Biology of Reproduction*, **19**, 720–35.

Byskov, A.G. and Høyer, P.E. (1988). Embryology of mammalian gonads and ducts. In *The physiology of reproduction* (ed. E. Knobil and J. Neill), pp. 265–302. Raven Press, New York.

Cajander, S.B. (1989). Periovulatory changes in the ovary. Morphology and expression of tissue-type plasminogen activator. *Progress in Clinical and Biological Research*, **296**, 91–101.

Campbell, K.L. and Valiquett, T.R. (1982). Do nonhormonal macromolecules cause correlated alterations of granulosa cell responses to FSH *in vitro*? *Biology of Reproduction*, **26**, (Suppl. 1), 102A (Abstr. no. 131).

Canipari, R. and Strickland, S. (1985). Plasminogen activator in the rat ovary. *Journal of Biological Chemistry*, **260**, 5121–5.

Canipari, R. and Strickland, S. (1986). Studies on the hormonal regulation of plasminogen activator production in rat ovary. *Endocrinology*, **118**, 1652–9.

Canipari, R., O'Connell, M.L., Meyers, G., and Strickland, S. (1987). Mouse ovarian granulosa cells produce urokinase-type plasminogen activator, whereas the corresponding rat cells produce tissue-type plasminogen activator. *Journal of Cell Biology*, **105**, 977–81.

Carnegie, J.A. (1990). Secretion of fibronectin by rat granulosa cells occurs primarily during early follicular development. *Journal of Reproduction and Fertility*, **89**, 579–89.

Cheng, C.Y., Grima, J., Stahler, M.S., Guglielmotti, A., Silvestrini, B., and Bardin, C.W. (1990). Sertoli cell synthesizes and secretes a protease inhibitor, α_2-macroglobulin. *Biochemistry*, **29**, 1063–8.

Chojkier, M., Houglum, K., Solis-Herruzo, J., and Brenner, D.A. (1989). Stimulation of collagen gene expression by ascorbic acid in cultured human fibroblasts. *Journal of Biological Chemistry*, **264**, 16957–62.

Christiane, Y., Demoulin, A., Gillain, D., Leroy, F., Lambotte, R., Lapiere, C.M., *et al.* (1988). Laminin and type III procollagen peptide in human preovulatory follicular fluid. *Fertility and Sterility*, **50**, 48–51.

Chun, S.Y., Popliker, M., Reich, R., and Tsafriri, A. (1992). Localization of preovulatory expression of plasminogen activator inhibitor type-1 and tissue inhibitor of metalloproteinase type-1 mRNAs in the rat ovary. *Biology of Reproduction*, **47**, 245–53.

Connell, C. and Christensen, A.K. (1975). The ultrastructure of the canine testicular interstitial tissue. *Biology of Reproduction*, **12**, 368–82.

Christl, H.W. (1990). The lamina propria of vertebrate seminiferous tubules: a comparative light and electron microscopic investigation. *Andrologia*, **22**, 85–94.

Curry, T.E., Dean, D.D., Woessner, J.F., and LeMaire, W.J. (1985) The extraction of a tissue collagenase associated with ovulation in the rat. *Biology of Reproduction*, **33**, 981–91.

Curry, T.E., Clark, M.R., Dean, D.D., Woessner, J.F., and LeMaire, W.G. (1986). The preovulatory increase in collagenase activity in the rat is independent of prostaglandin. *Endocrinology*, **118**, 1823–8.

Curry, T.E., Sanders, S.L., Pedigo, N.G., Estes, R.S., Wilson, E.A., and Vernon, M.W. (1988). Identification and characterization of metalloproteinase inhibitor activity in human ovarian follicular fluid. *Endocrinology*, **123**, 1611–18.

Curry, T.E., Mann, J.S., Estes, R.S., and Jones, P.B.C. (1990). α_2-macroglobulin and tissue inhibitor of metalloproteinases: collagenase inhibitors in human preovulatory ovaries. *Endocrinology*, **127**, 63–8.

Dang, D.C. and Fouquet, J.P. (1979). Differentiation of the fetal gonad of *Macaca fascicularis* with special reference to the testis. *Annales de Biologie Animale, Biochemie Biophysique*, **19**, 1197–1209.

Davis, C.M., Papadopoulos, V., Sommers, C.L., Kleinman, H.K., and Dym, M. (1990). Differential expression of extracellular matrix components in rat Sertoli cells. *Biology of Reproduction*, **43**, 860–9.

Davis, C.M., Papadopoulos, V. Jia, M.C., Yamada, Y., Kleinman, H.K., and Dym, M. (1991). Identification and partial characterization of laminin binding proteins in immature rat Sertoli cells. *Experimental Cell Research*, **193**, 262–73.

Deane, H.W. (1952). Histochemical observations on the ovary and oviduct of the albino rat during the estrous cycle. *American Journal of Anatomy*, **91**, 363–413.

Dekel, N. and Kraicer, P.F. (1978). Induction *in vitro* of mucification of rat cumulus oophorus by gonadotropins and adenosine 3', 5' monophosphate. *Endocrinology*, **102**, 1797–802.

Dennefors, B., Tjugum, J., Norström, A., Janson, P.O., Nilsson, L., Hamberger, L., and Wilhelmsson, L. (1982). Collagen synthesis inhibition by PGE_2 within the human follicular wall: one possible mechanism underlying ovulation. *Prostaglandins*, **24**, 295–302.

Deutinger, J., Kirchheimer, J., Reinthaller, A., Christ, G., Tatra, G., and Binder, B.R. (1988). Elevated tissue type plasminogen activator in human granulosa cells correlates with fertilizing capacity. *Human Reproduction*, **3**, 597–9.

Dorrington, J.H. and Skinner, M.K. (1986). Cytodifferentiation of granulosa cells induced by gonadotrophin-releasing hormone promotes fibronectin secretion. *Endocrinology*, **118**, 2065–71.

Dorrington, J.H., McKeracher, H.L., Chan, A.K., and Gore-Langton, R.E.

(1983). Hormonal interactions in the control of granulosa cells differentiation. *Journal of Steroid Biochemistry*, **19**, 17–32.

Dym, M. and Fawcett, D.W. (1970). The blood-testis barrier in the rat and the physiological compartmentation of the seminiferous epithelium. *Biology of Reproduction*, **3**, 308–26.

Dym, M., Lamsam-Casalotti, S., Jia, M.C., Kleinman, H.K., and Papadopoulos, V. (1991). Basement membrane increases G-protein levels and FSH responsiveness of Sertoli cell adenylyl cyclase activity. *Endocrinology*, **128**, 1167–76.

Elkington, J.S.H. and Fritz, I.B. (1980). Regulation of sulfoprotein synthesis by Sertoli cells in culture. *Endocrinology*, **107**, 970–6.

El Ouali, H., Leheup, B.P., Gelly, J.L., and Grignon, G. (1991). Laminin ultrastructural immunolocalization in rat testis during ontogenesis. *Histochemistry*, **95**, 241–6.

Enders, G.C., Henson, J.H., and Millette, C.F. (1986). Sertoli cell binding to isolated testicular basement membrane. *Journal of Cell Biology*, **103**, 1109–19.

Eppig, J.J., Ward-Bailey, P.F., Potter, J.E.R, and Schultz, R.M. (1982). Differential action of sulphated glycosaminoglycans on follicle-stimulating hormone-stimulated functions of cumuli-oophori isolated from mice. *Biology of Reproduction*, **27**, 399–406.

Espey, L.L. (1967*a*). Ultrastructure of the apex of the rabbit Graafian follicle during the ovulatory process. *Endocrinology*, **81**, 267–76.

Espey, L.L. (1967*b*). Tenacity of the porcine Graafian follicle as it approaches ovulation. *American Journal of Physiology*, **212**, 1397–401.

Espey, L.L. (1970). Effect of various substances on tensile strength of sow ovarian follicles. *American Journal of Physiology*, **219**, 230–3.

Espey, L.L. (1980). Ovulation as an inflammatory reaction: a hypothesis. *Biology of Reproduction*, **22**, 73–106.

Espey, L.L. (1991). Ultrastructure of the ovulatory process. In *Ultrastructure of the ovary*, (ed. G. Familiari, S. Makabe, and P.M. Motta), Electron Microscopy in Biology and Medicine, **9**, pp. 143–59. Klewer Academic, Boston, USA.

Espey, L.L. and Coons, P.J. (1976). Factors which influence ovulatory degradation of rabbit ovarian follicles. *Biology of Reproduction*, **14**, 233–45.

Espey, L.L. and Rondell, P. (1968). Collagenolytic activity in the rabbit and sow Graffian follicle during ovulation. *American Journal of Physiology*, **214**, 326–9.

Espey, L.L., Shimada, H., Okamura, H., and Mori, T. (1985). Effect of various agents on ovarian plasminogen activator activity during ovulation in pregnant mare's serum gonadotropin-primed immature rats. *Biology of Reproduction*, **32**, 1087–94.

Fisher, E., McLennan, S.V., Tada, H., Heffernan, S., Yue, D.K., and Turtle, J.R. (1991). Interaction of ascorbic acid and glucose on production of collagen and proteoglycan by fibroblasts. *Diabetes*, **40**, 371–6.

Ffrench-Constant, C., Hollingsworth, A., Heasman, J., and Wylie, C.C. (1991). Response to fibronectin of mouse primordial germ cells before, during and after migration. *Development*, **113**, 1365–73.

Freudenstein, J., Wagner, S., Luck, M.R., Einspanier, R., and Scheit, K.H. (1990). mRNA of bovine tissue inhibitor of metalloproteinase. Sequence and expression in bovine ovarian tissue. *Biochemical and Biophysical Research Communications*, **171**, 250–6.

Fritz, I.B. and Karmally, K. (1983). Hormonal influences on formation of plasminogen activator by cultured testis tubule segments at defined stages of the cycle of the seminiferous epithelium. *Canadian Journal of Biochemistry and Cell Biology*, **61**, 553–60.

Fukumoto, M., Yajimba, Y., Okamura, H., and Midorikawa, O. (1981). Collagenolytic enzyme activity in human ovary: an ovulatory enzyme system. *Fertility and Sterility*, **36**, 746–50.

Fulcher, K.D., Welch, J.E., O'Brien, D.A., and Eddy, E.M. (1992). Identification of the 44 and 70 kilodalton forms of the high affinity laminin receptor in mouse spermatogenic cells. *Biology of Reproduction*, **46**, (Suppl. 1; Abstr. 17), 55.

Gaddy-Kurten, D. Hickey, G.J., Fey, G.H., Gauldie, J., and Richards, J.S. (1989). Hormonal regulation and tissue-specific localization of α_2-macroglobulin in rat ovarian follicles and corpora lutea. *Endocrinology*, **125**, 2985–95.

Gebauer, H., Lindner, H.R., and Amsterdam, A. (1978). Synthesis of heparin-like glycosaminoglycans in rat ovarian slices. *Biology of Reproduction*, **18**, 350–8.

Gelly, J.L., Richoux, J.P., Leheup, B.P., and Grignon, G. (1989). Immunolocalization of type IV collagen and laminin during rat gonadal morphogenesis and postnatal development of the testis and epididymis. *Histochemistry*, **93**, 31–8.

Gore-Langton, R.E., Lacroix, M., and Dorrington, J.H. (1981). Differential effects of luteinizing hormone releasing hormone on follicle-stimulating hormone-dependent responses in rat granulosa cells and Sertoli cells *in vitro*. *Endocrinology*, **108**, 812–9.

Gosden, R.G., Hunter, R.F.H., Telfer, E., Torrance, C., and Brown, N. (1988). Physiological factors underlying the formation of ovarian follicular fluid. *Journal of Reproduction and Fertility*, **82**, 813–25.

Grant, G.F., Kwan, I., and Downey, B.R. (1992). The association between plasminogen activator and PGs in the periovulatory porcine follicle. *Biology of Reproduction*, **46**, (Suppl. 1; Abstr. 452), 163.

Grier, H.J. (1992). Chordate testis: the extracellular matrix hypothesis. *Journal of Experimental Zoology*, **261**, 151–60.

Grimek, H.J., Bellin, M.E., and Ax, R.L. (1984). Characteristics of proteoglycans from small and large ovarian follicles. *Biology of Reproduction*, **30**, 397–409.

Guillou, F. (1990). Characterization of a new superfusion, two-compartment culture system for Sertoli cells: influence of extracellular matrix on the cell permeability and dynamics of transferrin secretion. *Journal of Andrology*, **11**, 182–94.

Guillou, F., Monet-Kuntz, C., Fontaine, I., and Flechon, J.E. (1990). Expression of fetal-type intermediate filaments by 17-day-old rat Sertoli cells cultured on reconstituted basement membranes. *Cell and Tissue Research*, **260**, 395–401.

Gulamali-Majid, F., Ackerman, S., Veeck, L., Acost, A., and Pleban, P. (1987). Kinetic immunonephelometric determination of protein concentrations in follicular fluid. *Clinical Chemistry*, **33**, 1185–9.

Hadley, M.A. and Dym, M. (1987*a*). Immunocytochemistry of extracellular matrix in the lamina propria of the rat testis: electron microscopic localization. *Biology of Reproduction*, **37**, 1283–9.

Hadley, M.A. and Dym, M. (1987b). Role of the extracellular matrix in the culture of Sertoli cells and peritubular myoid cells in bicameral (dual compartment) chambers. *Anatomical Record*, **218**, 54A.

Hadley, M.A., Byers, S.W., Suarez-Quian, C.A., Kleinman, H.K., and Dym, M. (1985). Extracellular matrix regulates Sertoli cell differentiation, testicular cord formation and germ cell development *in vitro*. *Journal of Cell Biology*, **101**, 1511–22.

Hadley, M.A., Djakiew, D., Byers, S.W., and Dym, M. (1988). Factors involved in the maintenance of differentiated Sertoli cell structure and function in culture. In *Advances in andrology: Carl Schirren Symposium* (ed. A.F. Holstein, F. Leidenberger, K.H. Hölzer, and G. Bettendorf), pp. 59–66. Diesbach, Berlin.

Hadley, M.A., Weeks, B.S., Kleinman, H.K., and Dym, M. (1990). Laminin promotes formation of cord-like structures by Sertoli cells *in vitro*. *Developmental Biology*, **140**, 318–27.

Hatakeyama, S. (1965). A study of the interstitial cells of the human testis, especially on their fine-structural pathology. *Acta Pathologica Japonica*, **15**, 155–97.

Hettle, J.A., Walker, E.K., and Fritz, I.B. (1986). Hormonal stimulation alters the type of plasminogen activator produced by Sertoli cells. *Biology of Reproduction*, **34**, 895–904.

Hettle, J.A., Balekjian, E., Tung, P.S., and Fritz, I.B. (1988). Rat testicular peritubular cells in culture secrete an inhibitor of plasminogen activator activity. *Biology of Reproduction*, **38**, 359–71.

Himeno, N., Kawamura, N., Okamura, H., Mori, T., Fukumoto, M., and Midorikawa, O. (1984). Collagen synthetic activity in rabbit ovary during ovulation and its blockage by indomethacin. *Acta Obstetrica et Gynaecologica Japonica*, **36**, 1930–4.

Huarte, J., Belin, D., and Vassalli, J.D. (1985). Plasminogen activator in mouse and rat oocytes: induction during meiotic maturation. *Cell*, **43**, 551–8.

Hung, T.T., Tsuiki, A., and Yemini, M. (1989). Fibronectin in reproduction. *Steroids*, **54**, 575–82.

Hayashi, K., Hayashi, M., Jalkanen, M., Firestone, J.H., Trelstad, R.L., and Bernfield, M. (1987). Immunocytochemistry of cell surface heparan sulfate proteoglycan in mouse tissues. A light and electron microscope study. *Journal of Histochemistry and Cytochemistry*, **35**, 1079–88.

Ichikawa, S., Ohta, M., Morioka, H., and Murao, S. (1983a). Blockage of ovulation in the explanted hamster ovary by a collagenase inhibitor. *Journal of Reproduction and Fertility*, **68**, 17–19.

Ichikawa, S., Morioka, H., Ohta, M., Oda, K., and Murao, S. (1983b). Effect of various proteinase inhibitors on ovulation of explanted hamster ovaries. *Journal of Reproduction and Fertility*, **68**, 407–12.

Janecki, A. and Steinberger, A. (1986). Polarized Sertoli cell functions in a new two-compartment culture system. *Journal of Andrology*, **7**, 69–71.

Jensen, C.E. and Zachariae, F. (1958). Studies on the mechanism of ovulation. II.

Isolation and analysis of acid mucopolysaccharides in bovine follicular fluid. *Acta Endocrinologica*, **27**, 356–64.

Jeppesen, T.H. (1975). Surface epithelium of the fetal guinea-pig ovary: a light and electron microscopic study. *Anatomical Record*, **183**, 499–516.

Jones, P.B.C., Vernon, M.W., Muse, K.N., and Curry, T.E. (1989). Plasminogen activator and plasminogen activator inhibitor in human pre-ovulatory follicular fluid. *Journal of Clinical Endocrinology and Metabolism*, **68**, 1039–45.

Jost, A., Magre, S., and Agelopoulou, R. (1981). Early stages of testicular differentiation in the rat. *Human Genetics*, **58**, 59–63.

Jost, A., Valentino, O., Agelopoulou, R., and Magre, S. (1985). Action d'un analogue de la proline (acide-L-azétidine-2-carboxylique) sur la différentiation in vitro du testicule foetal de rat. *Compte Rendue de l'Academie des Sciences (Serie D)*, **301**, 225–32.

Jung, G. and Held, H. (1959). Über Fermente in der Follikelflussigkeit. *Archiv für Gynäkologie*, **192**, 146–50.

Kaneko, Y., Hirakawa, S., Momose, K., and Konomi, H. (1984). Immunochemical localization of Type I, III, IV and V collagens in the normal and polycystic ovarian capsules. *Acta Obstetrica et Gynaecologica Japonica*, **36**, 2473–4 (Abstr.).

Kao, J., Huey, G., Kao, R., and Stern, R. (1990). Ascorbic acid stimulates production of glycosaminoglycans in cultured fibroblasts. *Experimental and Molecular Pathology*, **53**, 1–10.

Karg, H. (1957). Ascortbinsäuredynamik im Ovar als Gonadotropinnachweis. *Kilinische Wochenschrift*, **35**, 643–4.

Kessel, B., Bicsak, T.A., and Garzo, V.G. (1992). Phorbol ester increases plasminogen activator inhibitor accumulation in cultures of human granulosa cells. *Journal of Clinical Endocrinology and Metabolism*, **74**, 33–8.

Kuopio, T. and Pelliniemi, L.J. (1989). Patch basement membrane of rat Leydig cells shown by ultrastructural immunolabelling. *Cell and Tissue Research*, **256**, 45–51.

Kuopio, T., Paranko, J., and Pelliniemi, L.J. (1989). Basement membrane and epithelial features of fetal-type Leydig cells in rat and human testis. *Differentiation*, **40**, 198–206.

Lacroix, M. and Fritz, I.B. (1982). The control of the synthesis and secretion of plasminogen activator by rat Sertoli cells in culture. *Molecular and Cellular Endocrinology*, **26**, 247–58.

Lacroix, M., Smith, F.E., and Fritz, I.B. (1977). Secretion of plasminogen activator by Sertoli cell enriched cultures. *Molecular and Cellular Endocrinology*, **9**, 227–36.

Lacroix, M., Parvinen, M., and Fritz, I.B. (1981). Localization of testicular plasminogen activator in the discrete portions (stages VII and VIII) of the seminiferous tubule. *Biology of Reproduction*, **25**, 143–6.

Lacroix, M., Smith, F.E., and Fritz, I.B. (1982). Changes in levels of plasminogen activator activity in normal and germ-cell-depleted testes during development. *Molecular and Cellular Endocrinology*, **26**, 259–67.

Larsen, W.J., Wert, S.E., Chen, L., Russel, P., and Hendrix, E.M. (1991). Expansion of the cumulus-oocyte complex during the preovulatory period: possible roles in oocyte maturation, ovulation, and fertilization. In *Ultrastructure of the ovary* (ed. G. Familiari, S. Makabe, and P.M. Motta), Electron Microscopy in Biology and Medicine, **9**, pp. 45–61. Klewer Academic, Boston, USA.

Leardkamolkarn, V. and Abrahamson, D.R. (1992). Immunoelectron microscopic localization of laminin in rat ovarian follicles. *Anatomical Record*, **233**, 41–52.

Ledwitz-Rigby, F., Gross, T.M., Shjeide, O.A., and Rigby, B.W. (1987). The glycosaminoglycan chondroitin-4-sulphate alters progesterone secretion by porcine granulosa cells. *Biology of Reproduction*, **36**, 320–7.

Leeson, C.R. and Leeson, T.S. (1963). The postnatal development and differentiation of the boundary tissue of the seminiferous tubule of the rat. *Anatomical Record*, **147**, 243–60.

Lindner, H.R., Amsterdam, A, Salomon, Y., Tsafriri, A, Nimrod, A., Lamprecht, S.A., *et al.* (1977). Intraovarian factors in ovulation: determinants of follicular response to gonadotrophins. *Journal of Reproduction and Fertility*, **51**, 215–35.

Liu, Y.X. and Hsueh, A.J.W. (1987). Plasminogen activator activity in cumulus-oocyte complexes of gonadotropin-treated rats during periovulatory period. *Biology of Reproduction*, **36**, 1055–62.

Liu, Y.X., Ny, T., Sarker, D., Loskutoff, D., and Hsueh, A.J.W. (1986). Identification and regulation of tissue plasminogen activator activity in rat cumulus-oocyte complexes. *Endocrinology*, **119**, 1578–87.

Liu, Y.X., Cajander, S.B., Ny, T., Kristensen, P., and Hsueh, A.J.W. (1987). Gonadotrophin regulation of tissue-type and urokinase-type plasminogen activators in rat granulosa and theca-interstitial cells during the periovulatory period. *Molecular and Cellular Endocrinology*, **54**, 221–9.

Liu, Y.X., Peng, X-R., and Ny, T. (1991*a*). Tissue-specific and time-coordinated hormone regulation of plasminogen-activator-inhibitor type I and tissue-type plasminogen activator in the rat ovary during gonadotropin-induced ovulation. *European Journal of Biochemistry*, **195**, 549–55.

Liu Y.X., Feng, Q., and Zou, R.J. (1991*b*). Changes of ovarian plasminogen activator and inhibitor during gonadotropin-induced ovulation in rhesus monkey. *Acta Physiologica Sinica (Sheng Li Hsueh Pao)*, **43**, 572–9.

Lobb, D.K. and Dorrington, J.H. (1987). Human granulosa and theca cells secrete distinct protein profiles. *Fertility and Sterility*, **48**, 243–8.

Loh, H.S. and Wilson, C.W.M. (1971). Relationship of human ascorbic acid metabolism to ovulation. *Lancet*, **i**, 110–3.

Loukusa, A.K., Veijola, M., and Rajaniemi, H. (1990). Plasminogen activator is involved in the hCG-induced neutrophil extravasation and vasopermeability increase in the rat testis. *International Journal of Andrology*, **13**, 306–14.

Luck, M.R. and Jungclas, B. (1987). Catecholamines and ascorbic acid as stimulators of bovine ovarian oxytocin secretion. *Journal of Endocrinology*, **114**, 423–30.

Luck, M.R. and Jungclas, B. (1988). The time course of oxytocin secretion from cultured bovine granulosa cells, stimulated by ascorbate and catecholamines. *Journal of Endocrinology*, **116**, 247–58.

Luck, M.R., Rodgers, R.J., and Findlay, J.K. (1990). Secretion and gene expression of inhibin, oxytocin and steroid hormones during the *in vitro* differentiation of bovine granulosa cells. *Reproduction, Fertility and Development*, **2**, 11–25.

76   Martin R. Luck

Luck, M.R., Münker, M., and Praetorius, C. (1991). Autocrine control of phenotype by extracellular matrix proteins in luteinising granulosa cells. *Journal of Reproduction and Fertility*, Suppl. 43, 101–2.

Makabe, S., Naguro, T., Nottola, S.A., Pereda, J., and Motta, P.M. (1991).Migration of germ cells, development of the ovary, and folliculogenesis. In *Ultrastructure of the ovary* (ed. G. Familiari, S. Makabe, and P.M. Motta), Electron Microscopy in Biology and Medicine, **9**, pp. 1–27. Klewer Academic, Boston, USA.

Mann, J.S., Kindy, M.S. Edwards, D.R., and Curry, T.E. (1991). Hormonal regulation of matrix metalloproteinase inhibitors in rat granulosa cells and ovaries. *Endocrinology*, **128**, 1825–32.

Maresh, G.A., Thimmons, T.M., and Dunbar, B.S. (1990). Effects of extracellular matrix proteins on the expression of specific ovarian proteins. *Biology of Reproduction*, **43**, 965–76.

Martin, G.G. and Miller-Walker, C. (1983). Visualization of the three dimensional distribution of collagen fibrils over preovulatory follicles in the hamster. *Journal of Experimental Zoology*, **225**, 311–9.

Marzowski, J., Sylvester, S.R., Gilmont, R.R., and Griswold, M.D. (1985). Isolation and characterization of Sertoli cell plasma membranes and associated plasminogen activator activity. *Biology of Reproduction*, **32**, 1237–45.

Mather, J.P., Wolpe, S.D., Gunsalus, G.L., Bardin, C.W., and Phillips, D.M. (1984). Effect of purified and cell-produced extracellular matrix components on Sertoli cell function. *Annals of the New York Academy of Sciences*, **438**, 572–5.

Merchant, H. (1975). Rat gonadal and ovarian organogenesis with and without germ cells. An ultrastructural study. *Developmental Biology*, **44**, 1–21.

Merchant-Larios, H. (1976). The onset of testicular differentiation in the rat: an ultrastructural study. *American Journal of Anatomy*, **145**, 319–30.

Merchant-Larios, H. (1979a). Origin of the somatic cells in the rat gonad: an autoradiographic approach. *Annales de Biologie Animale, Biochemie Biophysique*, **19**, 1219–29.

Merchant-Larios, H. (1979b). Ultrastructural events in horse gonadal morphogenesis. *Journal of Reproduction and Fertility*, Suppl. **27**, 479–85.

Meyer, G.T (1991). Ultrastructural dynamics during corpus luteum development and growth. In *Ultrastructure of the ovary*. (ed. G. Familiari, S. Makabe, and P.M. Motta), Electron Microscopy in Biology and Medicine, **9**, pp.161–76. Klewer Academic, Boston, USA.

Milwidsky, A., Kaneti, H., Finci, Z., Laufer, N., Tsafriri, A., and Mayer, M. (1989). Human follicular fluid protease and antiprotease activities: a suggested correlation with ability of oocytes to undergo *in vitro* fertilization. *Fertility and Sterility*, **52**, 274–80.

Moor, R.M. and Crosby, I.M. (1987). Cellular origin, hormonal regulation, and biochemical characteristics of polypeptides secreted by Graffian follicles of sheep. *Journal of Reproduction and Fertility*, **79**, 469–83.

Morales, T.I., Woessner, J.F., Howell, D.S., Marsh, J.M., and LeMaire, W.J. (1978). A microassay for the direct demonstration of collagenolytic activity in Graafian follicles of the rat. *Biochimica et Biophysica Acta*, **524**, 428–34.

Morales, T.I., Woessner, J.F., Marsh, J.M., and LeMaire, W.J. (1983). Collagen, collagenase and collagenolytic activity in rat Graafian follicles during follicular growth and ovulation. *Biochimica et Biophysica Acta*, **756**, 119–22.

Morley, P, Armstrong, D.T., and Gore-Langton, R.E. (1987*a*). Adhesion and differentiation of cultured rat granulosa cells: role of fibronectin. *American Journal of Physiology*, **253**, C625–32.

Morley, P., Armstrong, D.T., and Gore-Langton, R.E. (1987*b*). Fibronectin stimulates growth but not follicle stimulating hormone-dependent differentiation of rat granulosa cells *in vitro*. *Journal of Cellular Physiology*, **132**, 226–36.

Mueller, P.L., Schreiber, J.R., Lucky, A.W., Schulman, J.D., Rodbard, D., and Ross, G.T (1978). Follicle-stimulating hormone stimulates ovarian synthesis of proteoglycans in the estrogen-stimulated hypophysectomised immature female rat. *Endocrinology*, **102**, 824–31.

Murdoch, W.J. and McCormick, R.J. (1991). Dose-dependent effects of indomethacin on ovulation in the sheep. Relationship to follicular PG production, steroidogenesis, collagenolysis, and leukocyte chemotaxis. *Biology of Reproduction*, **45**, 907–11.

Murdoch, W.J. and McCormick, R.J. (1992). Enhanced degradation of collagen within apical vs. basal wall of ovulatory ovine follicle. *American Journal of Physiology*, **263**, E221–5.

Nargolwalla, C., McCabe, D., and Fritz, I.B. (1990). Modulation of levels of messenger RNA for tissue-type plasminogen activator in rat Sertoli cells, and levels of messenger RNA for plasminogen activator inhibin in testis peritubular cells. *Molecular and Cellular Endocrinology*, **70**, 73–80.

Nimrod, A. and Lindner, H.R. (1980). Heparin facilitates the induction of LH receptors by FSH in granulosa cells cultured in serum-enriched medium. *FEBS Letters*, **119**, 155–7.

Nishimune, Y., Sawada, K., Tajima, Y., Watanabe, D., Maekawa, M., Sakamaki, K., *et al.* (1989). Secretion of plasminogen activator in response to FSH in culture medium of human testicular cells from biopsy specimens. *Journal of Andrology*, **10**, 283–8.

Norström, A. and Tjugum, J. (1986). Hormonal effects on collagenolytic activity in the isolated human ovarian follicular wall. *Gynecologic and Obstetric Investigation*, **22**, 12–16.

Ny, T., Bjersing, L., Hsueh, A.J.W., and Loskutoff, D.J. (1985). Cultured granulosa cells produce two plasminogen activators and an antiactivator, each regulated differently by gonadotrophins. *Endocrinology*, **116**, 1666–8.

Ny, T., Liu, Y.-X., Ohlsson, M., Jones, P.B.C., and Hsueh, A.J.W. (1987). Regulation of tissue-type plasminogen activator activity and messenger RNA levels by gonadotrophin-releasing hormone in cultured rat granulosa cells and cumulo-oocyte complexes. *Journal of Biological Chemistry*, **262**, 11790–3.

O'Connell, M.L., Canipari, R., and Strickland, S. (1987). Hormonal regulation of tissue plasminogen activator secretion and mRNA levels in rat granulosa cells. *Journal of Biological Chemistry*, **262**, 2339–44.

Odor, D.L. and Blandau, R.J. (1969). Ultrastructural studies on fetal and early postnatal mouse ovaries. I. Histogenesis and organogenesis. *American Journal of Anatomy*, **124**, 163–86.

Oikawa, M. and Hsueh, A.J.W. (1989). β-adrenergic agents stimulate tissue plasminogen activator activity and messenger ribonucleic acid levels in cultured rat granulosa cells. *Endocrinology*, **125**, 2550–7.

Okamura, H., Takenaka, A., Yajima, Y., and Nishimura, T. (1978). Ultrastructural study of the ovarian follicular apex in human tissue. *Fertility and Sterility*, **30**, 729–30 (Abstr.).

Okamura, H., Takenaka, A., Yajima, Y., and Nishimura, T. (1980). Ovulatory changes in the wall at the apex of the human Graafian follicle. *Journal of Reproduction and Fertility*, **58**, 153–5.

Okuyama, A., Koh, E., Kondoh, N., Nakamura, H., Namiki, M., Kiyohara, H., *et al.* (1991). Plasminogen activator in cultured cells of human undescended testis. *Urology International*, **46**, 324–8.

Orly, J. and Sato, G. (1979). Fibronectin mediates cytokinesis and growth of rat follicular cells in serum-free medium. *Cell*, **17**, 295–305.

Orth, J.M. and McGuinness, M.P. (1991). Neonatal gonocytes co-cultured with Sertoli cells on a laminin-containing matrix resume mitosis and elongate. *Endocrinology*, **129**, 1119–21.

Osman, P. and Dullaart, J. (1976). Intraovarian release of eggs from the rat after indomethacin treatment at pro-estrus. *Journal of Reproduction and Fertility*, **47**, 101–103.

Padh, H. (1991). Vitamin C: Newer insights into its biochemical functions. *Nutrition Reviews*, **49**, 65–70.

Page, K.C., Killian, G.J., and Nyquist, S.E. (1990). Sertoli cell glycosylation patterns as affected by culture age and extracellular matrix. *Biology of Reproduction*, **43**, 659–64.

Palotie, A., Peltonen, L., Foidart, J.M., and Rajaniemi, H. (1984). Immuno-histochemical localization of basement membrane components and interstitial collagen types in preovulatory rat ovarian follicles. *Collagen Related Research*, **4**, 279–87.

Palotie, A. Salo, T., Vihko, K.K., Peltonen, L., and Rajaniemi, H. (1987). Types I and IV collagenolytic and plasminogen activator activities in preovulatory ovarian follicles. *Journal of Cellular Biochemistry*, **34**, 101–12.

Paniagua, R., Martinez-Onsurbe, P., Santamaria, L., Saez, F.J., Amat, P., and Nistal, M. (1990). Quantitative and ultrastructural alterations in the lamina propria and Sertoli cells in human cryptorchid testes. *International Journal of Andrology*, **13**, 470–87.

Paranko, J. (1987). Expression of type I and III collagen during morphogenesis of fetal rat testis and ovary. *Anatomical Record*, **219**, 91–101.

Paranko, J., Pelliniemi, L.J., Vaheri, A., Foidart, J-M., and Lakkala-Paranko, T. (1983). Morphogenesis and fibronectin in sexual differentiation of rat embryonic gonads. *Differentiation*, **23**, S72–81.

Park, O.K., Burtea, E.D., Leslie, N.D., Kessler, C.A., Degens, J.L., and Mayo, K.E. (1991). Differential expression and regulation of urokinase- and tissue-type plasminogen activator mRNAs in the rat ovary. *Biology of Reproduction*, **44**, (Suppl. 1; Abstr. 403), 153.

Parlow, A.F. (1958). A rapid bioassay method for LH and factors stimulating LH secretion. *Federation Proceedings*, **85**, 501–20.

Parr, E.L. (1974). Absence of neutral proteinase activity in rat ovarian follicle wall at ovulation. *Biology of Reproduction*, **11**, 509–12.

Parry, D.M., Willcox, D.L., and Thorburn, G.D. (1980). Ultrastructural and cytochemical study of the bovine corpus luteum. *Journal of Reproduction and Fertility*, **60**, 349–57.

Patek, C.E., Kerr, J.B., Gosden, R.G., Jones, K.W., Hardy, K., Muggleton-Harris, A.L., *et al*. (1991). Sex chimaerism, fertility and sex determination in the mouse. *Development*, **113**, 311–25.

Pellicer, A., Lightman, A, Ariza, A., DeCherney, A.H., Naftolin, F., and Littlefield, B.A. (1988). Follicular development is impaired by inhibitors of serine proteases in the rat. *American Journal of Obstetrics and Gynecology*, **158**, 670–6.

Pelliniemi, L.J. (1976). Ultrastructure of the indifferent gonad in male and female pig embryos. *Tissue and Cell*, **8**, 163–74.

Pelliniemi, L.J., Kellokumpu-Lehtinen, P., and Lauteala, L. (1979). Development of sexual differences in the embryonic genitals. *Annales de Biologie Animale, Biochemie Biophysique*, **19**, 1211–17.

Pelliniemi, L.J., Paranko, J., Grund, S., Fröjdman, K., Foidart, J-M., and Lakkala-Paranko, T. (1984). Extracellular matrix in testicular differentiation. *Annals of the New York Academy of Sciences*, **438**, 405–16.

Politis, I., Srikandakumar, A., Turner, J.D., Tsang, B.K., Ainsworth, L., and Downey, B.R. (1990). Changes in and partial identification of the plasminogen activator and plasminogen activator inhibitor systems during ovarian follicular maturation in the pig. *Biology of Reproduction*, **43**, 636–42.

Puistola, U., Salo, T., Martikainen, H., and Rönnberg, L. (1986). Type IV collagenolytic activity in human preovulatory follicular fluid. *Fertility and Sterility*, **45**, 578–80.

Puistola, U., Rönnberg, L., Martikainen, H., and Turpeenniemi-Hujanen, T. (1989). The human embryo produces basement membrane collagen (type IV collagen)-degrading protease activity. *Human Reproduction*, **4**, 309–11.

Rapp, G., Freudenstein, J., Klaudiny, J., Mucha, J., Wempe, F., Zimmer, M., and Scheit, K.H. (1990). Characterization of three abundant mRNAs from human ovarian granulosa cells. *DNA and Cell Biology*, **9**, 479–85.

Reich, R., Tsafriri, A., and Mechanic, G.L. (1985). The involvement of collagenolysis in ovulation in the rat. *Endocrinology*, **116**, 522–7.

Reich, R., Miskin, R., and Tsafriri, A. (1986). Intrafollicular distribution of plasminogen activators and their hormonal regulation *in vitro*. *Endocrinology*, **119**, 1588–1601.

Reich, R., Daphna-Iken, D., Chun, S.Y., Popliker, M., Slager, R., Adelmann-Grill, B.C., and Tsafriri, A. (1991). Preovulatory changes in ovarian expression of collagenases and tissue metalloproteinase inhibitor mRNA role of eicosanoids. *Endocrinology*, **129**, 1869–75.

Reichert, L.E. (1962). Endocrine influences on rat ovarian proteinase activity. *Endocrinology*, **70**, 697–700.

Reinthaller, A., Kirchheimer, J.C., Deutinger, J., Bieglmayer, C., Christ, G., and Binder, B.R. (1990). Plasminogen activators, plasminogen activator inhibitor, and fibronectin in human granulosa cells and follicular fluid related to oocyte maturation and intrafollicular gonadotrophin levels. *Fertility and Sterility*, **54**, 1045–51.

Reventos, J., Perrard-Sapori, M.H., Chatelain, P.G., and Saez, J.M. (1989). Leydig cell and extracellular matrix effects on Sertoli cell function: biochemical and morphologic studies. *Journal of Andrology*, **10**, 359–65.

Richardson, M.C., Davies, D.W., Watson, R.H., Dunsford, M.L., Inman, C.B., and Masson, G.M. (1992). Cultured human granulosa cells as a model for corpus luteum function: relative roles of gonadotrophin and low density lipoprotein studied under defined culture conditions. *Human Reproduction*, **7**, 12–18.

Rodriguez, J.P. and Minguell, J.J. (1989). Synthesis of proteoglycans and hyaluronic acid by long-term cultures of testicular cells from immature and pubertal rats. *Cell Biochemistry and Function*, **7**, 293–300.

Rodriguez, J.P. and Minguell, J.J. (1992). Collagen increases the synthesis of membrane-associated proteoglycans produced by Sertoli cells. *Journal of Cellular Biochemistry*, **50**, 21–5.

Rodriguez, J.P., Fernández, M., and Minguell, J.J. (1991). Interstitial collagen synthesis by somatic testicular cells in culture. *Cell Biochemical Function*, **9**, 63–7.

Rönnberg, L., Puistola, U., Martikainen, H., and Turpeenniemi-Hujanen (1988). Follicular basement membrane collagen (type IV) degrading protease activity and oocyte maturation. *Human Reproduction*, **3**, (Suppl. 1; Abstr. 361), page 114.

Rom, E., Reich, R., Laufer, N., Lewin, A., Rabinowitz, R., Pevsner, B., *et al.* (1987). Follicular fluid contents as predictors of success in *in-vitro* fertilization-embryo transfer. *Human Reproduction*, **2**, 505–10.

Rondell, P. (1970). Follicular processes in ovulation. *Federation Proceedings*, **29**, 1875–9.

Rosselli, M. and Skinner, M.K. (1992). Developmental regulation of Sertoli cell aromatase activity and plasminogen activator production by hormones, retinoids and the testicular paracrine factor PModS. *Biology of Reproduction*, **46**, 586–94.

Saksela, O. and Vihko, K.K. (1986). Local synthesis of plasminogen by the seminiferous tubules of the testis. *FEBS Letters*, **204**, 193–7.

Salomon, Y., Amir, Y., Azulai, R., and Amsterdam, A. (1978). Modulation of adenylate cyclase activity by sulphated glycosaminoglycans. 1. Inhibition by heparin of gonadotropin-stimulated ovarian adenylate cyclase. *Biochimica et Biophysica Acta*, **544**, 273–82.

Sang, Q-X., Thompson, E.W., Grant, D., Stetler-Stevenson, W.G., and Byers, S.W. (1991). Soluble laminin and arginine-glycine-aspartic acid containing peptides differentially regulate type IV collagenase messenger RNA, activation, and localization in testicular culture. *Biology of Reproduction*, **45**, 387–94.

Santamaria, L., Martinez-Onsurbe, P., Paniagua, R., and Nistal, M. (1990). Laminin, type IV collagen, and fibronectin in normal and cryptorchid human testes. An immunohistochemical study. *International Journal of Andrology*, **13**, 135–46.

Sappino, A.P., Huarte, J., Belin, D. and Vassalli, J.D. (1989). Plasminogen activators in tissue remodelling and invasion: mRNA localization in mouse ovaries and implanting embryos. *Journal of Cell Biology*, **109**, 2471–9.

Sato, E., Tanaka, T., Takeya, T., Miyamoto, H., and Koide, S.S. (1991). Ovarian glycosaminoglycans potentiate angiogenic activity of epidermal growth factor in mice. *Endocrinology*, **128**, 2402–6.

Saumande, J. (1991). Culture of bovine granulosa cells in a chemically defined serum-free medium: the effect of insulin and fibronectin on the response to FSH. *Journal of Steroid Biochemistry and Molecular Biology*, **38**, 189–96.

Savion, N. and Gospodarowicz, D. (1980). Patterns of cellular peptide synthesis by cultured bovine granulosa cells. *Endocrinology*, **107**, 1798–807.

Schochet, S.S. (1916) A suggestion as to the process of ovulation and ovarian cyst formation. *Anatomical Record*, **10**, 447–57.

Schweitzer, M., Jackson, J.C., and Ryan, R.J. (1981). The porcine ovarian follicle. VIII. FSH stimulation of *in vitro* [^{3}H]-glucosamine incorporation into mucopolysaccharides. *Biology of Reproduction*, **24**, 332–40.

Sheldrick, E.L. and Flint, A.P.F. (1989). Post-translational processing of oxytocin-neurophysis prohormone in the ovine corpus luteum: activity of peptidyl glycine α-amidating mono-oxygenase and concentrations of its co-factor, ascorbic acid. *Journal of Endocrinology*, **122**, 313–22.

Shibata, S. (1990). Effect of RU486 on collagenolytic enzyme activities in immature rat ovary. *Acta Obstetrica et Gynaecologica Japonica*, **42**, 136–42.

Shiina, K. (1990). Collagen and acid glycosaminoglycans in bovine tunica albuginea during ovulatory cycle. *Acta Obstetrica et Gynaecologica Japonica*, **42**, 1510–17.

Shimada, H., Okamura, H., Noda, Y., Suzuki, A., Tojo. S., and Takada, A. (1983). Plasminogen activator in rat ovary during the ovulatory process: independence of prostaglandin mediation. *Journal of Endocrinology*, **97**, 201–5.

Skinner, M.K. (1990). Mesenchymal (stromal) — epithelial interactions in the testis and ovary which regulate gonadal function. *Reproduction Fertility and Development*, **2**, 237–43.

Skinner, M.K. and Dorrington, J.H. (1984). Control of fibronectin synthesis by rat granulosa cells in culture. *Endocrinology*, **115**, 2029–31.

Skinner, M.K. and Fritz, I.B. (1985). Structural characterization of proteoglycans produced by testicular peritubular cells and Sertoli cells. *Journal of Biological Chemistry*, **260**, 11874–83.

Skinner, M.K. McKeracher, H.L., and Dorrington, J.H. (1984). Fibronectin as a marker of granulosa cell differentiation. *Endocrinology*, **117**, 886–92.

Skinner, M.K., Tung, P.S., and Fritz, I.B. (1985). Cooperativity between Sertoli cells and testicular peritubular cells in the production and deposition of extracellular matrix components. *Journal of Cell Biology*, **100**, 1941–7.

Skinner, M.K., Stallard, B., Anthony, C.T., and Griswold, M.D. (1989). Cellular localization of fibronectin gene expression in the seminiferous tubule. *Molecular and Cellular Endocrinology*, **66**, 45–52.

Smith, G.W., Anthony, R.V., and Smith, M.F. (1992). Molecular cloning and characterization of expression of an ovine ovarian tissue inhibitor of metalloproteinases. *Biology of Reproduction*, **46**, (Suppl. 1; Abstr. 496) 174,

Smith, M.F. and Moor, R.M. (1991). Secretion of putative metalloproteinase inhibitor by ovine granulosa cells and luteal tissue. *Journal of Reproduction and Fertility*, **91**, 627–35.

Stansfield, D.A. and Flint, A.P. (1967). The entry of ascorbic acid into the corpus luteum *in vivo* and *in vitro* and the effect of luteinizing hormone. *Journal of Endocrinology*, **39**, 27–35.

Strickland, S. and Beers, W.H. (1976). Studies on the role of plasminogen activation in ovulation. *Journal of Biological Chemistry*, **251**, 5694–702.

Strübing, C. (1989). Immunocytochemical detection of fibronectin in the testis of cattle. *Anatomica Histologia Embryologia*, **18**, 278.

Stryer, L (1988). *Biochemistry*, 3rd edn. W.H. Freeman, New York.

Suarez-Quian, C.A., Hadley, M.A., and Dym, M. (1984). Effect of substrate on the shape of Sertoli cells *in vitro*. *Annals of the New York Academy of Sciences*, **438**, 417–34.

Tadano, Y. and Yamada, K. (1978). The histochemistry of complex carbohydrates in the ovarian follicles of adult mice. *Histochemistry*, **57**, 203–15.

Takaba, H. (1990). A morphological study of the testes in patients with idiopathic male infertility. Immunohistochemical analysis of collagens and laminin in human testes. Hinyokika Kiyo *(Acta Urologica Japonica)*, **36**, 1173–80.

Takaba, H., Nagai, T., Hashimoto, J., Yanamoto, M., and Miyake, K. (1991). Identification of collagens in the human testis. *Urology International*, **46**, 180–3.

Talbot, P., Martin, G.G., and Ashby, H. (1987). Formation of the rupture site in preovulatory hamster and mouse follicles: loss of the surface epithelium. *Gamete Research*, **17**, 287–302.

Tanaka, N., Espey, L.L., Stacy, S., and Okamura, H. (1992*a*). Epostane and indomethacin actions on ovarian kallikrein and plasminogen activator activities during ovulation in the gonadotropin-primed immature rat. *Biology of Reproduction*, **46**, 665–70.

Tanaka, T., Andoh, N., Takeya, T., and Sato, E. (1992*b*). Differential screening of ovarian cDNA libraries detected the expression of the porcine collagenase inhibitor gene in functional corpora lutea. *Molecular and Cellular Endocrinology*, **83**, 65–72.

Telfer, E., Ansell, J.D., Taylor, H., and Gosden, R.G. (1988). The number of clonal precursors of the follicular epithelium in the mouse ovary. *Journal of Reproduction and Fertility*, **84**, 105–10.

Telfer, E.E., Johnson, K.A., and Eppig, J.J. (1992). Expression of cytochrome p-450 aromatase mRNA during murine follicular development *in vitro* and *in vivo*. *Journal of Reproduction and Fertility*, Abstr. Ser. No. 9, abstr. 19.

Thakur, A.N., Coles, R., Sesay, A., Earley, B., Jacobs, H.S., and Ekins, R.P. (1990). A rat granulosa cell plasminogen activator bioassay for FSH in human serum. *Journal of Endocrinology*, **126**, 159–68.

Timmons, T.M., Lee, V.H., and Dunbar, B.S. (1990). Differential expression of inhibin and fibronectin by granulosa cells cultured from primary follicles. *Biology of Reproduction*, **42**, Supplement 1 (Abstract 343) 155.

Tjugum, J., Dennefors, B., and Norström, A. (1984). Influence of progesterone, androstenedione and oestradiol-17β on the incorporation of (^{3}H) proline in the human follicular wall. *Acta endocrinologica*, **105**, 552–7.

Too, C.K.L., Weiss, T.J., and Bryant-Greenwood, G.D. (1982). Relaxin stimulates plasminogen activator by rat granulosa cells *in vitro*. *Endocrinology*, **115**, 1424–6.

Too, C.K.L., Bryant-Greenwood, G.D., and Greenwood, F.C. (1984). Relaxin increases the release of plasminogen activator, collagenase, and proteoglycans from rat granulosa cells *in vitro*. *Endocrinology*, **115**, 1043–50.

Tsafriri, A. Bicsak, T.A., Cajander, S.B., Ny, T., and Hsueh, A.J.W. (1989). Suppression of ovulation rate by antibodies to tissue-type plasminogen activator and α_2-antiplasmin. *Endocrinology*, **124**, 415–21.

Tsafriri, A., Daphna-Iken, D., Popliker, M., Goldberg, G.I., Adelmann-Grill, B.C., and Reich, R. (1990). Preovulatory stimulation of ovarian collagenase mRNA and its role in ovulation. *Biology of Reproduction*, **42**, (Suppl. 1; Abstract 239), 120.

Tsuiki, A., Preyer, J., and Hung, T.T. (1988). Fibronectin and glycosaminoglycans in human preovulatory follicular fluid and their correlation to follicular maturation. *Human Reproduction*, **3**, 425–9.

Tung, P.S. and Fritz, I.B. (1980), Interactions of Sertoli cells with myoid cells *in vitro*. *Biology of Reproduction*, **23**, 207–17.

Tung, P.S. and Fritz, I.B. (1984). Extracellular matrix proteins promote rat Sertoli cell histotype expression *in vitro*. *Biology of Reproduction*, **30**, 213–29.

Tung, P.S. and Fritz, I.B. (1986*a*). Extracellular matrix components and testicular peritubular cells influence the rate and pattern of Sertoli cell migration *in vitro*. *Developmental Biology*, **113**, 163–34.

Tung, P.S. and Fritz, I.B. (1986*b*). Cell-substratum and cell–cell interactions promote testicular peritubular myoid cell histotypic expression *in vitro*. *Developmental Biology*, **115**, 155–70.

Tung, P.S., Skinner, M.K., and Fritz, I.B. (1984*a*). Fibronectin synthesis is a marker for peritubular cell contaminants in Sertoli-enriched cultures. *Biology of Reproduction*, **30**, 199–211.

Tung, P.S., Skinner, M.K., and Fritz, I.B. (1984*b*). Cooperativity between Sertoli cells and the peritubular myoid cells in the formation of the basal lamina in the seminiferous tubule. *Annals of the New York Academy of Sciences*, **438**, 435–46.

Unbenhaun, V., Jung, G., and Kidess, E. (1965). Enzymuntersuchungen im Liquor Folliculi. *Archiv für Gynäkologie*, **202**, 225–8.

Upadhyay, S., Luciani, J.M., and Zamboni, L. (1979). The role of the mesonephros in the development of indifferent gonads and ovaries in the mouse. *Annales de Biologie Animale, Biochemie Biophysique*, **19**, 1179–96.

van der Rest, M. and Garrone, R. (1991). Collagen family of proteins. *FASEB Journal*, **5**, 2814–23.

Vanderboom, R.J., Carroll, D.J., Bellin, M.E., Schneider, D.K., Miller, D.J., Grummer, R.R., and Ax, R.L. (1989). Binding of bovine follicular fluid glycosaminoglycans to fibronectin, laminin and low-density lipoproteins. *Journal of Reproduction and Fertility*, **87**, 81–7.

Varner, D.D., Forrest, D.W., Fuentes, F., Taylor, T.S., Hooper, R.N., Brinsko, S.P., and Blanshard, T.L. (1991). Measurements of glycosaminoglycans in follicular, oviductal and uterine fluids of mares. *Journal of Reproduction and Fertility*, Suppl. **44**, 297–306.

Vernon, R.B., Lane, T.F., Angello, J.C., and Sage, H. (1991). Adhesion, shape, proliferation, and gene expression of mouse Leydig cells are influenced by extracellular matrix *in vitro*. *Biology of Reproduction*, **44**, 157–70.

Vihko, K.K., Souminen, J.J.O., and Parvinen, M. (1984). Cellular regulation of plasminogen activator secretion during spermatogenesis. *Biology of Reproduction*, **31**, 383–9.

Wang, C. and Leung, A. (1983). Gonadotropins regulate plasminogen activator production by rat granulosa cells. *Endocrinology*, **111**, 1201–7.

Wartenberg, H., Kinsky, I., Viebahn, C., and Schmolke, C. (1991). Fine structural characteristics of testicular cord formation in the developing rabbit gonad. *Journal of Electron Microscopy Technique*, **19**, 133–57.

Weimer, S.L., Campeau, J.D., Marrs, R.P., and DiZerega, G.S. (1984). Alteration of human follicular fluid plasminogen activator activity by ovarian hyperstimulation. *Journal of In Vitro Fertilization and Embryo Transfer*, **1**, 263–6.

Winer, M.A. and Ax, R.L. (1989). Properties of heparin binding to purified plasma membranes from bovine granulosa cells. *Journal of Reproduction and Fertility*, **87**, 337–48.

Wislocki, G.B., Bunting, H., and Dempsey, E.W. (1947). Metachromasia in mammalian tissues and its relationship to mucopolysaccharides. *American Journal of Anatomy*, **81**, 1–38.

Woessner, J.F., Morioka, N., Zhu, C., Mukaida, T., Butler, T., and LeMaire, W.J. (1989). Connective tissue breakdown in ovulation. *Steroids*, **54**, 491–9.

Wordinger, R.J., Rudick, V.L., and Rudick, M.J., (1983). Immunohistochemical localization of laminin within the mouse ovary. *Journal of Experimental Zoology*, **228**, 141–3.

Yajima, Y., Motohashi, T., Takenaka, A., Okamura, H., and Nishimura, T. (1980). Activities of collagenolytic enzymes in the human ovary. *Acta Obstetrica et Gynaecologica Japonica*, **32**, 1–5.

Yanagishita, M. and Hascall, V.C. (1979). Biosynthesis of proteoglycans by rat granulosa cells cultured *in vitro*. *Journal of Biological Chemistry*, **254**, 12355–64.

Yanagishita, M. and McQuillan, D.J. (1989). Two forms of plasma membrane-intercalated heparan sulfate proteoglycan in rat ovarian granulosa cells. *Journal of Biological Chemistry*, **264**, 17551–8.

Yanagishita, M., Rodbard, D., and Hascall, V.C. (1979). Isolation and characterization of proteoglycans from porcine ovarian follicular fluid. *Journal of Biological Chemistry*, **254**, 911–20.

Yoshimura, Y., Maruyama, K., Shiraki, M., Kawakami, S., Fukushima, M., and Nakamura, Y. (1990). Prolactin inhibits plasminogen activator activity in the preovulatory follicles. *Endocrinology*, **126**, 631–6.

Yoshimura, Y., Okamoto, T., and Tamura, T. (1991). Localization of fibronectin in the bovine ovary. *Animal Science and Technology*, **62**, 529–32.

Yoshinaga, K., Hess, D.L., Hendrickx, A.G., and Zamboni, L. (1988). The development of the sexually indifferent gonad in the prosimian *Galago crassicaudatus crassicaudatus*. *American Journal of Anatomy*, **181**, 89–105.

Zamboni, L. and Upadhyay, S. (1982). The contribution of the mesonephros to the development of the sheep fetal testis. *American Journal of Anatomy*, **165**, 339–56.

Zamboni, L., Bézard, J., and Mauléon, P. (1979). The role of the mesonephros in the development of the sheep fetal ovary. *Annales de Biologie Animale, Biochemie Biophysique*, **19**, 1153–78.

Zoller, L.C. (1991). Quantitative analysis of the membrana granulosa in developing and ovulatory follicles. In *Ultrastructure of the ovary*, (ed. G. Familiari, S. Makabe, and P.M. Motta), Electron Microscopy in Biology and Medicine, 9, pp. 73–87. Klewer Academic, Boston, USA.

Zhu, C. and Woessner, J.F. (1991). A tissue inhibitor of metalloproteinases and α-macroglobulins in the ovulating rat ovary: possible regulation of collagen matrix breakdown. *Biology of Reproduction*, **45**, 334–42.

3 Cell adhesion processes in implantation

CAROLE C. WEGNER and DANIEL D. CARSON

I   INTRODUCTION

The events that occur during early stages of pregnancy are fascinating from a variety of developmental, cellular, and molecular standpoints. Mammals have adopted the strategy of integrating their young at an extremely early stage of development with tissue of the maternal reproductive tract, i.e. the uterus. The embryo develops to the blastocyst stage within a glycoprotein coat called the zona pellucida and is not capable of attaching to uterine cells until shortly after it hatches from this coat. In parallel with the development of the embryo to an attachment competent state, the uterus is converted from a condition which will not support embryo attachment to one which will support this process. The initial attachment in all species occurs between the exterior surface of trophectoderm cells of the embryo and the apical surface of uterine epithelial cells. Embryos of some species, e.g., rodents and humans, also penetrate the epithelial layer and its subjacent basal lamina into the underlying stromal tissue. Consequently, the interactions of embryos with the uterus include multiple trophoblast cell–uterine cell as well as trophoblast cell–extracellular matrix binding and invasion events.

This review will emphasize the implantation reaction from the standpoint of cell adhesion. Since a large body of evidence is available in rodents as well as human models, most of the ideas presented will be based upon observations made in these species. Many other species, e.g., large animals, do not display an invasive type of implantation and so their strategies may differ from those described below. Nonetheless, the initial event of implantation, i.e., attachment of trophoblast to the apical surface of the uterine epithelium, is common to all and, therefore, may have some conserved aspects.

Embryo transfer experiments have demonstrated that blastocysts have the ability to bind to a variety of cell and tissue types (cf. Cowell 1969; Glass et al. 1979; Van Blerkom and Chavez 1981) as well as extracellular matrix components (reviewed in Carson et al. 1990). Likewise many extracellular matrix components have multiple adhesion supporting domains and/or can complex with other extracellular matrix components (Yamada 1985; Martin and Timpl 1987). Thus, it is not surprising that an extensive list of cell adhesion systems have been reported to function during embryo implantation. Multiple adhesion systems may be required at each step of implantation. In this case, interruption of any one of these systems might perturb the implantation reaction. It also must be considered that some adhesion systems may be present in either trophoblast or uterine cells or extracellular matrices which are redundant or at least not essential for all implantation reactions. For example, molecules used to support initial trophoblast binding to epithelial cells may persist on the cell surface of invading trophoblast, but may not be required for the subsequent interactions with extracellular matrices or decidual cells.

I   THE ROLE OF OVARIAN STEROIDS IN CONTROLLING UTERINE RECEPTIVITY

The condition in which the uterus will support the attachment of the embryo is referred to as the receptive state. This state of receptivity is transient and is followed by a period in which the uterus not only fails to support embryo attachment, but is also hostile to non-implanted embryos (Dickmann and Noyes 1960; Finn 1986; Psychoyos 1986). These observations indicate that there is a defined period or 'window' during which embryo implantation may occur. Furthermore, three functionally distinct uterine states exist with regard to embryo implantation: pre-receptive, receptive, and refractory (see Fig. 3.1). Neither the pre-receptive not the refractory state of the uterus supports embryo attachment, perhaps for different reasons. In any case, both the pre-receptive and refractory states constitute functionally non-receptive uterine conditions.

Given the transient nature of uterine receptivity, it is critical that development of the embryo to an attachment-competent state be co-ordinated with the conversion of the uterus from a non-receptive to a receptive state. This co-ordination is accomplished by the actions of ovarian steroids (Psychoyos 1986). Levels of these hormones generally rise during the pre-receptive phase and are maintained if successful implantation occurs. The 'window' of receptivity is transient and a refractory period follows during which embryo attachment does not occur. If implantation does not occur, corpora lutea degenerate and the hormone levels fall. This decline in steroid hormone levels is required for the uterus to leave the refractory state and re-enter a pre-receptive period. The pre-receptive state of the uterus and the implantation potential of blastocysts can be maintained and implantation delayed for an extended period if an appropriate hormonal milieu is maintained. 'Delayed' implantation occurs naturally (Given and Enders 1989) and can be induced in various species experimentally (Bergstrom 1978).

Although it is clear that ovarian steroids regulate uterine receptivity, it is unclear if the responses are direct or indirect. Steroid hormones may only directly regulate responses in a subset of uterine cells, e.g. stroma. In fact, changes in the cellular distribution of uterine steroid hormone receptors have been reported during the peri-implantation period (Glasser and McCormack 1981; De Hertogh et al. 1986). The responses in these target cells may include modulated secretion of growth factors, cytokines,

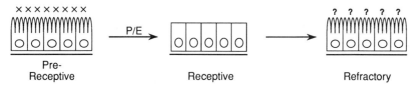

Pre-
Receptive   Receptive   Refractory

Fig. 3.1   Receptivity of the uterine epithelium. In the pre-receptive state, the uterine lumenal epithelium displays abundant apical microvilli and apical glycocalyx (indicated by X). Components of this apical glycocalyx may act as inhibitors of blastocyst and microbial attachment. In rodents, the influences of progesterone (P) and oestrogen (E) convert these cells transiently to a receptive state. The receptive state is characterized by diminished apical microvilli and loss of the apical glycocalyx. A third state of the uterine epithelium, the refractory state, prevails after the receptive state until steroid hormone levels fall. Although apical microvilli are shown in the figure, it is not clear if these structures or the apical glycocalyx return to the same extent as observed in the pre-receptive state. The decline of steroid hormone levels that occur during the cycle in non-pregnant animals permits the uterine epithelium to recycle back to the pre-receptive state. For further details, see text.

or lipid hormones which, in turn, act in a paracrine or autocrine fashion to produce a given response. For example, extensive remodelling of the stromal extracellular matrix occurs during early pregnancy in a steroid hormone-dependent fashion. Transforming growth factor type-β is a potent modulator of extracellular matrix expression in various cell types (Massagué 1990) and transforming growth factor type-β isoform expression also is modulated during early pregnancy (Tamada *et al.* 1990; Das *et al.* 1992). If paracrine loops occur for extracellular matrix remodelling in the uterus, it may be necessary to recombine appropriate uterine cell types *in vitro* along with appropriate steroid hormone or growth factor supplementation to replicate these responses. *In vitro* culture systems now exist that permit these types of experiments to be performed (Jacobs *et al.* 1992; Wegner and Carson 1992; Jacobs and Carson 1993).

III   CELLULAR ASPECTS OF IMPLANTATION

Upon hatching from the zona pellucida, the embryo not only develops the ability to bind to the uterine epithelium, but also can attach to a variety of substrates, cell types, and tissues, including those outside the reproductive tract (cf Cowell 1969). Owing to morphological, temporal, and biochemical distinctions among the processes involved, it is proposed that attachment and invasion of the uterine epithelial layer are separable events and may involve different adhesion systems (see Fig. 3.2). The trophectoderm and trophoblast cells surrounding the blastocyst mediate blastocyst attachment and invasive behaviour. In rodents, the mural trophectoderm, rather than the polar, mediates this attachment (Kirby *et al.* 1967). Therefore, regional specialization of the trophectoderm must exist. Trophoblast cells are among the most highly invasive cell types known and have been likened to metastatic tumour cells (Gurchot 1975; Liotta *et al.* 1983; Yagel *et al.* 1988; Turpeenniemi-Hujanen *et al.* 1992). High levels of various extracellular matrix-degrading proteinases are secreted by trophoblast cells, which facilitate this invasive activity (Fisher *et al.* 1985; Aplin 1989; Glass *et al.* 1983; Yagel *et al.* 1988; Lala and Graham 1990; Behrendtsen *et al.* 1992).

The multiplicity of cells and tissues which blastocysts can bind to and invade demonstrates that embryos can use adhesion systems found in adult tissues outside the uterus. Cell adhesion systems display cell-type, developmental or tissue-specific restriction (Ekblom *et al.* 1986); however, to date all of the cell adhesion molecules that have been described for either blastocysts or uterine cells also have been found in other adult tissues. These observations can largely account for the ability of embryos to bind, invade or outgrow on a wide variety of biological substrates. Inded, the remarkable feature of the uterus may not be its ability to support

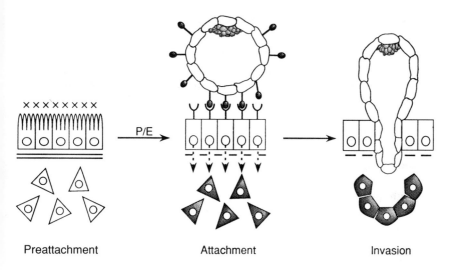

Preattachment   Attachment   Invasion

Fig. 3.2   Embryo attachment and invasion. Prior to embryo attachment, the uterine lumenal epithelium displays abundant apical microvilli and apical glycocalyx (X) as described in Fig. 3.1. Underlying stromal cells are displayed in the preattachment phase separated from the epithelium by an intact basal lamina. Under the appropriate steroid hormone influences, the uterus is converted to a receptive state. Hatched, attachment-competent blastocysts are capable of binding to the apical surface of the uterine epithelium which at this stage displays greatly reduced apical microvilli and glycocalyx. Cell surface components on the exterior surface of the trophectoderm (♀) and the apical surface of the epithelium (Ψ) are postulated to interact during the attachment phase. Embryo attachment in rodents is accompanied by the stimulation of differentiation of stromal cells at the implantation sites (decidual cell response). This local response is accompanied by partial disruption of the underlying basal lamina, and alterations in the pattern of gene expression of the underlying stroma (indicated by shading of the stromal cells). The decidual cell response is presumably initiated by release of appropriate biochemical signals from the basolateral aspect of the epithelial cells at the site of embryo attachment (indicated by the downwardly pointing arrows). By the time the embryo has started to penetrate the epithelial cell layer at the outset of the invasive phase, the basal lamina in partially compromised and the stromal cells in the immediate area (primary decidual zone) have differentiated both morphologically and biochemically. For example, tighter cell–cell apposition is apparent in this area and a molecular permeability barrier are established. For further details, see text.

embryo attachment, but rather its unique ability to regulate this process. In addition to controlling the attachment and invasion of embryos, the uterus also can control the invasion of various metastatic tumour cell types (Schlesinger 1962; Wilson and Potts 1970; Maharajan and Maharajan 1986). Consequently, the uterine cell types involved express activities generally capable of preventing cell attachment and invasion. Recently, isolated mouse uterine epithelial cells have been shown to display similar activities *in vitro* (Valdizan *et al.* 1992).

The initial regulated event in embryo implantation is attachment to the apical surface of the uterine epithelium. The externally disposed, presumably apical, surfaces of the trophoblast encounter the epithelium and mediate the initial binding between the embryo and uterus. Thus, at attachment, any 'barriers' which the uterus otherwise presents must be lost or compromised. In various species, apical microvilli of the uterine epithelium become progressively shorter during the peri-implantation period (Potts 1966; Smith and Wilson 1974; Schlafke and Enders 1975). The resultant surface is relatively flattened, a feature that has been suggested to facilitate interactions between the trophectoderm and epithelium. It also has been suggested that the uterine epithelium may become at least transiently less polarized during the receptive phase (Denker 1990). A decrease or loss of polarity may permit movement of cell surface components normally present at the lateral or basal cell surface to the apical domain. This would permit apical expression of molecules normally functioning in epithelial attachment to the basal lamina. Since certain basal lamina components are known to be expressed at the exterior surface of the mural trophectoderm in mice (Leivo *et al.* 1980; Wu *et al.* 1983; Carson *et al.* 1993), membrane 'mixing' could provide apical molecules functioning as embryo 'receptors'. Increased membrane trafficking (Tung *et al.* 1988), cytoskeletal rearrangement (Luxford and Murphy 1989; Luxford and Murphy 1992), and enhanced transepithelial permeability (Tung *et al.* 1988) occur during conversion to a receptive uterine state; however, there is no evidence to date indicating increased expression of basal cell surface components at the apical surface of uterine epithelium. As a result, the depolarization theory still remains speculative, albeit intriguing.

Another general feature of the apical surface of the uterine epithelium during the transition to a receptive state is the loss of the normally dense glycoprotein coat and reduction in negative charge character of this cell surface (reviewed in Schlafke and Enders 1975; Aplin 1991). These observations suggest that the molecules normally expressed in abundance at this surface do not function to promote embryo attachment. Rather, such glycoconjugates are more likely to perform a barrier or anti-adhesive function at this surface. The side of the uterus into which the vasculature enters is defined as mesometrial. In rodents, embryo attachment occurs at the antimesometrial aspect of the uterine lumen (Kirby *et al.* 1967). Thus, molecular changes occurring at the apical surface of the uterine epithelium also appear to be regionalized, perhaps involving multiple molecular species. For example, molecules that function to prevent embryo attachment (anti-adhesins) may be lost throughout the uterine lumen; however, embryo 'receptors' may only be expressed at the antimesometrial surface. The net result would be antimesometrial attachment.

After attachment, different species take different strategies for subsequent penetration of the trophoblast into the uterine wall (reviewed in Schlafke and Enders 1975). In some species, e.g. pigs and cows, the attachment reaction remains peripheral at the apical surface of the epithelium. In other species, e.g. rabbits, the trophoblast and epithelium fuse, but no further penetration occurs. Invasive species, e.g. rodents and humans, display two basic strategies for penetration of the uterine epithelium. In rodents, epithelial cells are displaced while the trophoblast intrudes between them. In contrast, human trophoblasts phagocytose the epithelial cells during the process. In either case, the embryo then encounters and must breach the underlying basal lamina. It seems likely that the embryo actively penetrates the basal lamina; however, there is evidence that the underlying stromal cells of the uterus contribute to, and may play a major role in, compromising the basal lamina (Schlafke *et al.* 1985; Blankenship and Given 1992). Consequently, extracellular matrix-degrading activities of uterine stroma may play an important role at early stages of implantation in some species. While a few studies of these uterine activities have been reported (Brenner *et al.* 1989; Sappino *et al.* 1989), more work needs to be performed to define the temporal and spatial pattern of their expression during early stages of the implantation process.

Even before trophoblasts of invasive species penetrate the uterine epithelium, the uterine stroma in the immediate vicinity of the implantation site initiate a differentiation process referred to as the decidual cell reaction (reviewed in Bell 1983 and Parr and Parr 1989). This differentiation involves changes in cellular morphology and expression of various extracellular matrix components, cytoskeletal proteins, and enzymes as well as lipid and polypeptide growth factors. Decidualization initially involves only the stromal cells immediately subjecent to the basal lamina at the implantation site and extends a few cells deep from this point. This area is called the primary decidual zone. This is a transient structure and is penetrated by the trophoblast within 2–3 days after formation. However, during its existence, the primary decidual zone provides an effective permeability barrier to soluble proteins of 68 kD or more (Parr and Parr 1989). It also has been suggested that the primary decidual zone provides a barrier to embryo invasion once the epithelium has been breached and also may exclude certain cells of the uterus, e.g. macrophages (Tachi *et al.* 1981; De *et al.* 1991). These suggestions are consistent with the known patterns of distribution of such cells during the peri-implantation period. Nonetheless, it is not immediately clear how this cellular barrier is provided since both trophoblast and macrophages are extremely capable of penetrating many tissues, including tissues containing tightly apposed cell junctions (see above). It is possible that a chemical barrier to penetration is provided by the primary decidual zone involving secreted soluble factors, protease inhibitors or extracellular matrix components. More work needs to be done to define how the primary decidual zone limits cellular migration.

Following initiation of the formation of the primary decidual zone, another area of stromal specialization is generated called the secondary decidual zone. Formation of this zone involves both local hyperplasia and hypertrophy of stromal cells of the antimesometrial region surrounding the primary decidual zone. In addition to altered morphology, these cells also express distinct patterns of extracellular matrix and cell surface components (Wewer et al. 1986; Nose and Takeichi 1986; Glasser et al. 1987; Farrar and Carson 1992), cytoskeletal proteins (Glasser et al. 1987), and various enzymes (reviewed in Bell 1983). Extensive remodelling of the extracellular matrix occurs in this area. A number of extracellular matrix components not normally expressed to an appreciable extent by fibroblasts, e.g. collagen type IV and laminin (Wewer et al. 1986; Glasser et al. 1987; Farrar and Carson 1992), are elaborated by secondary decidual cells. The trophoblast does not penetrate substantially into the secondary decidual zone which, in the form of the decidua capsularis, persists in apposition to the parietal yolk sac through at least day 16 of pregnancy in the mouse (Parr and Parr 1989). While a number of biochemical markers of the secondary decidual zone have been described (Bell 1983; Parr and Parr 1989), relatively little is known about the function of this tissue.

A third wave of uterine differentiation occurs at the mesometrial aspect and temporally follows the initiation of the primary and secondary decidual zones. The mesometrial decidua forms the decidua basalis, lies adjacent to the placenta, and persists throughout pregnancy. It has been suggested that cells in the mesometrial decidua may modulate immunological interactions between the mother and fetus (Kearns and Lala 1983; Parr and Parr 1989). Metrial gland cells found in this region have been shown to possess some properties of natural killer cells (Stewart and Mukhtar 1988; Linnemeyer and Pollack 1991). The potentially important role played by cells of the mesometrial decidua in immunological tolerance warrants more investigation of their biochemical and biological properties.

IV   IMPLANTATION MODELS

The most convincing way of studying implantation is to perform experiments in vivo. Unfortunately, this approach has been problematic. One reason for this may be the complexity of the system: multiple interactions occur among both embryonic and uterine cell types via both adhesion-related and growth factor/hormone- dependent processes. Due to the small size of rodent uteri, it is difficult to introduce agents into the uterine lumen, e.g. antibodies or polysaccharides, during the peri-implantation period without causing a non-specific, mechanical interference with implantation. None the less, Chavez (1990) has reported success with this approach by introducing these agents through the cervix. Mini-osmotic pumps (Alzet, Alza Corporation, PALO ALTO, California, USA) may be used to introduce small volumes of substances in a steady, long-term release mode; however, even very small catheters are difficult to stably insert and usually block the uterine lumen, preventing embryo passage from the oviduct. Skilled microsurgical techniques are required to perform such experiments *in vivo*. Consequently, studies of implantation *in vivo* have relied, for the most part, upon the use of specific probes to establish correlations in patterns of expression with developmental events at fixed points in time. Such approaches are essential to support any observations obtained using *in vitro* models. However, tissue staining does not test function and, therefore, provides only correlative information.

In vitro models of implantation are also limited. All models require that suitable attachment factors, e.g. extracellular matrix components coated on to plastic or present in serum or added to media, be present. Embryos do not attach to uncoated surfaces or surfaces coated with irrelevant proteins, e.g. serum albumin, IgG, or polysaccharides (Carson *et al.* 1990; Fisher *et al.* 1990). Most models also involve embryos developing in medium that probably does not contain the spectrum of growth factors and nutrients embryos would encounter *in vivo* (Pollard 1990; Tabibzadeh 1991). Furthermore, many systems examine development on two-dimensional matrices of particular extracellular matrix components. These systems do not require invasive, e.g. proteolytic, activities to the same degree as would be expected for embryo penetration of the endometrium. Some three-dimensional systems utilizing extracellular matrix preparations have been developed to study invasive activities (Yagel *et al.* 1988; Fisher *et al.* 1989; Kliman and Feinberg 1990), but lack the cellular influences of the uterus. Other models have introduced uterine cells as substrates to compensate for these influences (Van Blerkom and Chavez 1981; Farach *et al.* 1987; Valdizan *et al.* 1992; reviewed in Morris and Potter 1990). Uterine epithelial cells occur as polarized, non-receptive cells with regard to embryo attachment under most conditions *in vivo*. Thus, it has been a paradox that isolated non-polarized uterine epithelial cells readily support embryo attachment and outgrowth *in vitro* regardless of the state of the uterus from which they were derived (Sherman and Wudl 1976; Farach *et al.* 1987).

The development of procedures for culturing uterine epithelium in a polarized fashion (Glasser *et al.* 1988) has permitted re-examination of epithelial receptivity *in vitro*. Polarized uterine epithelium regain their non-receptive state *in vitro* and so more closely mimic the behaviour of these cells under most conditions *in vivo* (Julian *et al*. 1992; Valdizan *et al*. 1992). Development of systems that can be regulated *in vitro* to study embryo attachment to uterine epithelial cells would permit tremendous progress to be made in molecular aspects of implantation. Diverse influences from other uterine cell types, growth factors, and steroid hormones may be required to regulate this process. Moreover, components of serum-containing media may influence epithelial cells in complex ways. Therefore, a careful recombination of key uterine cell types along with media supplements may be required to reproduce control of epithelial receptivity *in vitro*.

V   ADHESION MOLECULES INFLUENCING EMBRYO–UTERINE ATTACHMENT

The following discussion will consider classes of molecules that have been implicated in performing functions relevant to embryo adhesion during implantation. The approach taken will be to divide the implantation reaction into the following phases: pre-receptive, receptive, refractory, basal lamina penetration, and stromal (decidual) penetration (see Figs. 3.1 and 3.2). Adhesion molecules that have been identified at each of these phases will be discussed in the context of that phase. In some cases, the same molecules have been implicated in more than one phase of the implantation process.

1   Pre-receptive phase

At the time the embryo initially enters the uterus, neither the embryo nor the uterine epithelium appear to be capable of binding to each other. The embryo makes its appearance while still sheathed in the glycoprotein coat of the zona pellucida. Zona-encased embryos cultured in the presence of various attachment-promoting substrata fail to bind (Sherman and Wudl 1976). Consequently, not only does the zona at this stage appear to be resistant to sperm binding (Wassarman 1987), but also seems to be generally non-adhesive. If the zona is removed from the preimplantation blastocyst, the embryos themselves also have been shown to be non-adhesive (Sherman and Wudl 1976). A caveat to these studies is that the procedures used to remove zona, e.g. chemical, mechanical, or enzymatic, may remove surface adhesion molecules either partially or totally. As a result, the lag in acquisition of adhesiveness may reflect replenishment of these surface components rather than their absence in the zona-encased embryo. None the less, *in situ* staining of sectioned embryos has demonstrated that substantial changes occur in the composition or distribution of surface components of blastocysts, e.g. heparan sulphate proteoglycans, during the hatching process (Sutherland *et al*. 1991; Carson *et al*. 1993). These observations support the contention that blastocysts are actually non-adhesive prior to hatching from the zona.

As discussed above, the apical surface of the uterine epithelium also is non-adhesive in the pre-receptive state. In addition to preventing embryo attachment, this surface must present a barrier to microbial invasion and has been shown to be resistant to invasion by various metastatic tumour cells both *in vivo* (Schlesinger 1962; Wilson and Potts 1970; Maharajan and Maharajan 1986) and *in vitro* (Valdizan *et al*. 1992). A general characteristic of the apical surface of uterine epithelial cells from a wide variety of species is the presence of an abundant glycocalyx (Schlafke and Enders 1975; Aplin 1991). Moreover, studies using various carbohydrate-specific staining techniques have demonstrated that this glycoprotein coat is greatly reduced during conversion to a receptive uterine state (reviewed in Schlafke and Enders 1975; Chavez and Anderson 1985; Anderson *et al*. 1986). Taken together, these observations suggest that the apically disposed surface glycoconjugates expressed by uterine epithelial cells under most conditions are non-adhesive and must be removed to generate a receptive surface.

Recent studies in mice indicate that a large fraction of the apically disposed glycoconjugates are integral mucin-type glycoproteins (Valdizan *et al*., 1992). An integral transmembrane mucin glycoprotein, muc-1, has been identified in uterine epithelium (Pemberton *et al*. 1992). Muc-1 expression appears to be regulated during the oestrous cycle and early pregnancy. Muc-1 expression declines to nearly undetectable levels at both the protein and mRNA level prior to generation of the receptive uterine state (Surveyor, Pemberton, Spicer, Gendler, and Carson, unpublished studies). In other systems, mucin glycoproteins have been implicated in performing anti-adhesive functions (Jentoft 1990). It is intriguing to speculate that these molecules perform similar functions at the apical cell surface of uterine epithelial cells. Experiments must be performed to determine whether selective alteration in the level of mucin glycoprotein expression also influences the ability of cells to support embryo attachment or penetration.

2   Receptive phase

Along with the removal of any anti-adhesive molecules that may exist at the exterior surface of embryos or the apical surface of uterine epithelial cells, adhesion-promoting molecules also must be expressed to support the initial phase of embryo–uterine attachment. Specific probes for a large number of cell surface and extracellular matrix components have been used to examine the distribution of such molecules at blastocyst and uterine epithelial cell surfaces, and implantation sites with largely negative results (reviewed in Carson *et al.* 1990). Interestingly, the few adhesion-promoting systems that have been implicated in early stages of embryo attachment generally seem to be carbohydrate-dependent. None the less, proof that any particular adhesion system is operative *in vivo* is not available. To obtain such data, it would be desirable to specifically perturb these systems during the implantation reaction, e.g. with antibodies. As discussed above, intrauterine injection experiments have been problematic; however, systemic introduction of antisera to growth factors has been reported to interfere with uterine processes (Nelson *et al.* 1991). In addition, both proteins and polysaccharides with molecular weights of up to 43 kD penetrate the primary decidual zone within 10 min after injection into the femoral vein (M.B. Parr and E.L. Parr 1986). Therefore, improved techniques of introduction of inhibitors of specific adhesion systems may permit a more direct examination of their function *in vivo*. In addition, application of homologous recombination technologies to genetically remove or alter these systems in embryos and uteri should be applicable.

A body of evidence has accumulated indicating that heparan sulphate proteoglycans support blastocyst interactions with a variety of substrates including uterine epithelial cells during early phases of attachment. Heparan sulphate proteoglycans support adhesive interactions of many cell types as well as extracellular matrix components (reviewed in Carson 1992). In addition, these molecules can complex with a number of growth factors and modulate a variety of cellular functions (Carson 1992). Heparan sulphate proteoglycans are expressed at the exterior surface of mouse blastocysts in a developmentally regulated fashion (Farach *et al.* 1987, 1988), are detected at sites of embryo–uterine epithelial cell attachment *in vivo* (Carson *et al.* 1993) and at the surface of human trophoblastic cell lines (Rohde and Carson 1993). While syndecan is expressed at the exterior surface of ·trophectoderm in preimplantation-stage blastocysts, this proteoglycan appears to be redistributed to the interior of the embryo prior to acquisition of attachment competence (Sutherland *et al.* 1991). In contrast, perlecan is expressed at low levels in preimplantation-stage embryos (Dziadek *et al.* 1985; Carson *et al.* 1993) and is expressed at the exterior surface of peri-implantation-stage embryos. Therefore, perlecan is a more likely candidate for supporting embryo attachment processes than syndecan.

Heparan sulphate proteoglycans also are expressed at the surface of mouse uterine epithelial cells (Tang *et al.* 1987; Morris *et al.* 1988) and human uterine epithelial cell lines (Rohde and Carson, 1993). Thus, it is possible that these molecules may act on both embryonic and uterine cell surfaces to support adhesive interactions. Heparan sulphate proteoglycans of mouse uterine epithelial cell surfaces are metabolically labile ($t_{1/2}$=30–60 min; Tang *et al.* 1987) and seem to be degraded more rapidly under conditions which produce a receptive uterine state (Morris *et al.* 1988). Moreover, probes which detect heparan sulphate polysaccharides, i.e. basic fibroblast growth factor, fail to stain the apical surface of uterine epithelial cells during the receptive phase (Carson *et al.* 1993). Therefore, rather than supporting embryo attachment, heparan sulphate proteoglycans present at apical surfaces of uterine epithelia appear to be removed and, perhaps, increase availability of endogenous heparan sulphate binding sites (see below).

Inhibition of proteoglycan synthesis by mouse embryos also inhibits attachment to various substrates, including uterine epithelial cells. Moreover, embryos acquire attachment competence following removal of such inhibitors from culture media (Farach *et al.* 1988). Inhibition of proteoglycan assembly in human trophoblastic cell lines also inhibits their adhesion to human uterine epithelial cell lines (Rohde and Carson 1993). Heparan sulphate, but not other highly sulphated polysaccharides, inhibits attachment of mouse embryos as well as human trophoblastic cell lines *in vitro* (Farach *et al.* 1987; Carson *et al.* 1988; Rohde and Carson 1993). Selective enzymatic removal of heparan sulphate from the surfaces of either mouse embryos of human trophoblastic cell lines results in inhibition of attachment to relevant substrates (Farach *et al.* 1987; Rohde and Carson 1993). Proteins with heparin-binding domains, e.g. laminin, fibronectin, vitronectin, uniformly support embryo attachment (Farach *et al.* 1987). Platelet factor IV is a low M_r protein whose only known binding activity is for heparin/heparan sulphate (Handin and Cohen 1976). While platelet factor IV effectively supports embryo attachment, it does not support embryo outgrowth (Farach *et al.* 1987). These observations indicate that functions other than heparan sulphate binding are required beyond the initial attachment reaction. Specific cell surface heparan sulphate binding sites have been described for both mouse uterine epithelial cells and human uterine epithelial cell lines (Wilson *et al.* 1990; Raboudi *et al.* 1992) and heparan sulphate binding peptides have been isolated from both sources (Carson *et al.* 1990; Raboudi *et al.* 1992). A 70 000 M_r heparan sulphate binding protein fraction isolated from primary cultures of mouse uterine epithelial cells supports mouse embryo attachment *in vitro* (Carson *et al.*, 1990); however, the identity of this protein or whether it functions in embryo-attachment related processes *in vivo* remains unknown. Molecular identification and generation of specific antibody and complementary DNA probes to these molecules is essential to determine if an heparan sulphate-dependant system is likely to operate at early stages of implantation.

A series of studies by Kimber and co-workers has implicated oligosaccharides containing lacto-*N*-fucopentaose-I (LNF-I) determinants on uterine epithelia in mediating embryo attachment. The LNF-I pentasaccharide, but not other closely related structures, substantially inhibits mouse embryo attachment to uterine cells *in vitro* (Lindenberg *et al.* 1988). Furthermore, expression of LNF-I-bearing molecules is regulated at the apical surface of uterine epithelial cells by steroid hormones (Kimber and Lindenberg 1990) and during early pregnancy (Kimber *et al.* 1988). Paradoxically, LNF-I expression at the apical surface of the lumenal epithelium decreases markedly during the receptive phase. The observation that the LNF-I-bearing molecules are largely extractable with organic solvents suggests an association with glycolipids. In this regard, glycolipids have been implicated in cell adhesion events in other systems (Roberts and Ginsburg 1988). The non-extractable LNF-I-bearing molecules are irregularly distributed and not confined to the antimesometrial side of the uterus. Thus, the patterns of expression of LNF-I-bearing molecules do not strictly match what one would expect of embryo adhesive components; however, a subset of these molecules may be involved in this process. Specific binding sites for LNF-I occur on the mural trophectoderm of mouse blastocysts (Lindenberg *et al.* 1990). The potential receptors for LNF-I have not been identified; however, recent progress has been made in the identification of carbohydrate-specific receptors which function in lymphocyte homing, i.e. LEC-CAMs or selections (Stoolman 1989). It is possible that related receptors may operate in the LNF-I-dependent system.

Cadherins (uvomorulin, L-CAM) are a family of calcium-dependent cell adhesion molecules (CAMs) that are most notably involved in maintenance of adhesion between epithelial cells (Takeichi 1991). Based on immunological differences, the cadherins have been placed into three groups: E-, N-, and P-cadherins. The E-cadherins were first identified in teratocarcinoma cells and are identical to uvomorulin and homologous to chicken L-CAM and human cell-CAM 120/80. Antibodies directed against cell-CAM 120/80 localized to precompaction eight-cell mouse embryos were able to block compaction (Damsky *et al.* 1983), suggesting E-cadherin is required for cell adhesion events in compaction. Since both the uterine epithelium and the trophectoderm are epithelia, it is reasonable to consider that cadherins might be involved in attachment occurring between these two cell types. It is clear that cadherins are distributed at the lateral surfaces of both uterine epithelia and trophectoderm (Vestweber *et al.* 1987; Kadokawa *et al.* 1989). Some evidence suggests that cadherins occur at points of trophectoderm–uterine epithelial interaction (Kadokawa *et al.* 1989); however, other studies do not find that substantial accumulations of cadherins occur at this area (Vestweber *et al.* 1987). No studies of cadherin function in embryo attachment processes have been published. Work in our lab indicates that various antibodies directed against E-cadherin do not inhibit embryo attachment to various substrates, including uterine epithelia, but do stain these cells in indirect immunofluorescence assays. Furthermore, we have not found E-cadherin localization at sites of trophectoderm–uterine epithelial cell contact Tang, Julian, and Carson, unpublished studies). As discussed below (see p. 103), a unique cadherin species, P-cadherin, was first detected in maternal decidua where it may support interactions occurring among decidual cells (Nose and Takeichi 1986). None the less, the role cadherins play in early stages of the implantation process remains uncertain. Expression of another putative adhesion molecule, CAM-105, has been described at the apical surfaces of both uterine epithelial cells and blastocysts (Svalander et al. 1987; Svalander et al. 1990); however, in both cases these molecules are lost from these areas prior to the time either surface becomes adhesive. Consequently, while CAM-105 may perform interesting functions in both cell types, its pattern of expression is inconsistent with a role in embryo attachment.

The abundant expression of glycoconjugates at the surfaces of both blastocysts and uterine epithelia has led some workers to examine the role of carbohydrate-binding proteins, i.e. lectins, in implantation-related events. Increased expression of epitopes bearing terminal galactosyl residues have been reported on peri-implantation stage blastocysts (Svalander et al. 1989; Chavez 1990). In addition, a galactose-binding lectin appears to be present in implanting embryos and decidua (Poirier et al. 1992). This lectin also binds lactosaminoglycans with high affinity (Leffler and Barondes 1986). Lactosaminoglycans are abundantly expressed at the surface of mouse uterine epithelial cells (Dutt and Carson 1990). Expression of uterine lactosaminoglycans also is regulated by steroid hormones (Carson et al. 1988; Dutt et al. 1988; Tang and Carson 1989) and during the menstrual cycle in humans (Hadley et al. 1990). Thus, interactions of the aforementioned lectins with lactosaminoglycan-bearing molecules have been proposed to occur during implantation. No functional studies of the galactose-binding lectin on blastocysts have been reported. Furthermore, close examination of the pattern of distribution of this lectin at implantation sites does not indicate coincidence with sites of embryo–uterine interaction. Therefore, a role for this lectin in embryo attachment is unclear. Galactosyltransferase also can function as a cell surface lectin that recognizes terminal N-acetylglucosaminyl residues as well as lactosaminoglycans and promotes cell adhesion, including sperm-egg binding and morula compaction (Shur 1984). Soluble galactosyltransferase recognizes mucin oligosaccharides which are abundantly expressed at the apical cell surface of polarized mouse uterine epithelial cells cultured in vitro (Valdizan et al. 1992). A galactosyltransferase-like activity may participate in adhesion among the uterine epithelial cells themselves (Dutt et al. 1987). Some evidence indicates that galactosyltransferase is present at the surfaces of blastocyst-stage embryos (Sato et al. 1984; Bayna et al. 1988) although it is not clear if this is at the exterior surface of trophectoderm. Importantly, the same conditions which perturb adhesion among uterine epithelial cells fail to interfere with blastocyst attachment and outgrowth on these cell layers (Dutt et al. 1987). A potential problem with the latter studies is that the epithelial cell layer is compromised by the treatment. As a result, embryos probably bind to the underlying extracellular matrix deposited by these cells *in vitro*, albeit in a galactosyltransferase-independent manner. We have found that high concentrations of the galactosyltransferase-perturbant, α-lactalbumin, inhibits mouse embryo outgrowth on laminin-coated, but not fibronectin-coated, surfaces; however, we were not able to demonstrate that other galactosyltransferase perturbing reagents, including specific antibodies (provided by Dr Barry Shur, M.D. Anderson Cancer Center) had this effect (Jacobs and Carson, unpublished studies). In contrast, Chavez (1990) reported that injection of galactosyltransferase perturbants specifically inhibited mouse blastocyst implantation. In addition, Babiarz's group has provided data implicating a role for galactosyltransferase in mouse trophoblast outgrowth (Romagnano and Babiarz 1990). In the latter case, the effects of the galactosyltransferase perturbants are transient and suggest that other adhesion systems can rapidly compensate. In addition, these studies utilized trophoblast obtained from post-implantation stage embryos and support the notion the galactosyltransferase may function during invasion, as opposed to attachment, *in vivo*. More work will be needed to clarify the potential role of galactosyltransferase in embryo attachment *in utero*.

Certain extracellular matrix components, e.g. laminin, entactin/nidogen, and perlecan, have been detected at the exterior surface of peri-implantation-stage blastocysts (Wu *et al.* 1983; Carson *et al.* 1993). Integrins are a large family of integral cell surface glycoproteins that mediate cellular attachment to various extracellular matrix components and certain cells (Hynes 1992). Only one extracellular matrix receptor, the α_v/β_3 integrin, has been described at the apical surface of human uterine epithelial cells at the receptive stage (Lessey *et al.* 1992). One potential ligand for the α_v/β_3 integrin, bone sialoprotein, has been detected in trophoblast cells in human placenta (Bianco *et al.* 1991); however, no function in trophoblast attachment has been assessed. It is possible that extracellular matrix receptors, e.g. integrins, may function at the apical cell surface of uterine epithelia to support embryo attachment. Limited studies of integrin expression in mouse uteri have been performed (Damjanov *et al.* 1986), but do not demonstrate apical localization of these molecules. Given the panorama of members of the integrin superfamily (Hynes 1992), it is quite possible that the antibody probes used to date have not been appropriate. Moreover, the finding that RGD (recognition sequence) containing peptides do not interfere with blastocyst attachment to cultured uterine epithelia is not necessarily damning since many integrins do not use this recognition sequence (Hynes 1992). Unfortunately, while many specific probes to integrin subtypes are available for human proteins, corresponding probes are not readily available for rodents where effects on implantation can be studied more rigorously. Hopefully, generation of appropriate rodent probes will provide a means to assess the role of integrins in embryo attachment.

3   Refractory phase

Very little is known about the expression of adhesion systems during the refractory phase of the uterus. Secretion of various embryotoxic substances, e.g. proteases (Moulton and Elangovan 1981), into the uterine lumen may obviate the need for a more specific regulation of cell adhesion systems during this period. It should be emphasized that the refractory period is quite different from the conditions which exist during implantation delay (Bergstrom 1978; Given and Enders 1989). In the latter case, the blastocyst hatches from the zona pellucida and can exist for extended periods of time in the uterine lumen intimately apposed to the epithelium (Given and Enders 1989). In the refractory phase, the embryos appear to be destroyed very quickly since it is difficult to recover embryos from refractory uteri following embryo transfer (Dickmann and Noyes 1960). None the less, in addition to release of embryotoxic agents into the uterine lumen, changes in the apical cell surface of uterine epithelial cell also may contribute to the refractory state.

VI   INVASION OF THE BASAL LAMINA AND DECIDUA BY TROPHOBLAST

Human and rodent trophoblasts are highly invasive and penetrate the uterine epithelium, associated basement membrane, the deep stroma, and finally the endothelium of maternal blood vessels to form the haemochorial placenta (Carlson 1981). Owing to their highly invasive nature, trophoblast cells are often compared to metastatic cells (Gurchot 1975; Liotta *et al.* 1983; Yagel *et al.* 1988; Turpeenniemi-Hujanen *et al.* 1992). Normally this invasion is limited; however, unlimited invasion of the trophoblast is observed in metastatic invasion by choriocarcinoma cells. Another pathological condition, pre-eclampsia, appears to be a consequence of insufficient invasion (Robertson *et al.* 1981). Therefore, invasion must be tightly regulated to achieve normal implantation and placentation and to sustain the pregnancy to term. Cell–cell and cell–extracellular matrix interactions must be regulated in a spatially and temporally specific manner for appropriate interactions between the trophoblast and, in turn, the epithelium, basement membrane, decidua, and endothelium of the uterus. To permit migration through these heterogeneous cell types and matrix, opposing requirements for cell adhesion and cell release must be met. This section will describe cell surface and extracellular matrix components which may play a role in mediating adhesion events in the invasive phase of implantation. Molecules such as SPARC (secreted protein, acidic and rich in cysteine), osteonectin, or tenascin may have anti-adhesive functions. Still other molecules such as laminin and thrombospondin may display either function, depending on the context in which the molecule is presented.

1   Cell adhesion molecules

As discussed above, cadherins are a class of CAMs which play a role in cell specific adhesion in adult and developing tissues. E-cadherin is masked or lost from the cell surface of the trophoblast during the transition from the trophectoderm to the invasive trophoblast (Damjanov *et al.* 1986; Fisher *et al.* 1989). P-cadherin, named for its localization to placenta, was first detected on day 5 of implantation on decidual tissue and persists throughout gestation (Nose and Takeichi 1986). Interestingly, whereas E-cadherin is present only on fetal cells, P-cadherin is localized to both fetal and maternal progenitor cells of the placenta. *In vitro* mixing of cells bearing either E-cadherin or P-cadherin molecules does not result in cell–cell adhesion between heterogeneous cell types. Regulated loss of some cadherins and up-regulation of other cadherins may be required for selective adhesion of the appropriate cell types for placentation as well as invasion (Takeichi 1987).

Anti-integrin antibodies can inhibit mouse embryo outgrowth on fibro-nectin, laminin, and collagen type IV in a ligand specific manner (Sutherland *et al.* 1988). These antibodies recognize a group of 140 kD glycoproteins on trophoblast similar to the 140 kD avian integrin complex. These 140 kD antigen sites appear coordinately with the invasive potential of these cells. The RGD recognition sequence of fibronectin can inhibit outgrowth on, but not attachment to, laminin and fibronectin by embryos (Armant *et al.* 1986*b*; Farach *et al.* 1987; Sutherland *et al.* 1988), suggesting that different adhesion molecules are required for attachment and outgrowth. *In vitro* attachment of ectoplacental cone trophoblast cells to decidual cells in unaffected by the addition of either the galactosyltransferase inhibitor α-lactalbumin, synthetic GRGDS or YIGSR peptides or heparan sulphate (Babiarz *et al.* 1992). In contrast, invasion, defined as the trophectoderm's ability to push aside decidual cells and spread, is disrupted by the presence of YIGSR peptides, heparan sulphate, and protease inhibitors (Babiarz *et al.* 1992). It appears that, at least *in vitro*, different adhesion molecules function in attachment and spreading.

In human endometrium, integrins are expressed in a menstrual cycle- and cell-specific manner (Lessey *et al.* 1992). Epithelial cells express α_2, α_3 and α_6 integrins which are collagen and laminin contrast, stromal cells express α_5 integrin. The α_1-integrin was expressed in the secretory phase whereas the α_v-integrin subunit increased throughout the cycle. Interestingly, β_3-integrin appears abruptly at the apical surface of the luminal epithelium on day 20, coincident with the receptive state of the uterus in fertile women; β_3-expression at this surface is delayed by 3 days in infertile women (Lessey *et al.* 1992). Osteopontin is one of the ligands for the α_v/β_3-integrin complex (Miyauci *et al.* 1991) and has been shown to support cell adhesion in other systems (Butler 1989). Osteopontin (2AR) expression is upregulated in mouse decidual tissue (Nomura *et al.* 1988), but has not been demonstrated on blastocyst surfaces. It also is unclear whether the α_v/β_3-complex is expressed in mouse uterine implantation sites or codistributes with osteopontin. The marked up-regulation of the β_3-subunits in uterine epithelia bears further investigation. Interestingly, cytokines such as interleukin-1α and tumour necrosis factor-α have been shown to regulate integrin expression in cell lines (Santala and Heino 1991). Since various uterine cell types including epithelia (Jacobs *et al.* 1992; Robertson *et al.* 1992; Jacobs and Carson 1993), stroma/decidua (Tabibzadeh *et al.* 1989; Jacobs *et al.* 1992; Robertson *et al.* 1992), and macrophages (Hunt 1989) are capable of producing cytokines, it would be interesting to determine what effect cytokines have on integrin expression in the uterus.

The distribution of extracellular matrix components and various integrins was determined in first trimester human cytotrophoblasts, cell columns, and uterine wall using a panel of antibodies (Damsky *et al.* 1992). Polarized non-migrating cytotrophoblast stained primarily for α_6/β_4 integrins and M and A chains of laminin. In contrast, non polarized migrating cytotrophoblasts expressed primarily α_5/β_1 integrins and a fibronectin rich matrix. Thus correlations exist between the expression of extracellular matrix components and their cognate receptors in trophoblast. These observations may have clinical implications in aetiology of pre-eclampsia (insufficient invasion) or malignancy (excess invasion).

2   Extracellular matrix molecules

As mentioned above, the extracellular matrix of the stromal tissue undergoes substantial remodelling during the implantation process. During this remodelling some components that are normally restricted to basal laminae begin to be synthesized and deposited, e.g. laminin (Wewer *et al.* 1986; Glasser *et al.* 1987; Farrar and Carson 1992), collagen type IV (Wewer *et al.* 1985; Farrar and Carson 1992), and heparan sulphate proteoglycans (Wewer *et al.* 1986; Kisalus and Herr 1988). Some matrix molecules, e.g. tenascin, are transiently expressed in stroma at implantation sites and actually disappear before the embryo has penetrated to this area (Julian *et al.*, 1994). In other cases, e.g. collagen type VI (Mulholland *et al.* 1992), matrix molecules are lost from the implantation site. There are also a number of matrix components such as fibronectin and chondroitin sulphate proteoglycans that appear to be constitutively expressed (Glasser *et al.* 1987; Jacobs and Carson 1991; Farrar and Carson 1992). It is common to speculate that these changes influence embryo interactions; however, it is not clear how many of these responses are directly related to embryo adhesion. Some matrix molecules may serve mainly as reservoirs for growth factors to promote embryo development or to propagate decidualization signals (Vlodavsky *et al*. 1990). Alternatively, some of the alterations in matrix expression may be epiphenomena induced by the rapidly changing influences of the myriad of growth factors and hormones during this time and irrelevant to embryo adhesion. This issue is further confounded by observations that some isolated matrix components facilitate embryo binding while others inhibit. Probably the most difficult prediction is the net effect of the multiple changes in stromal/decidual extracellular matrix on the trophoblast cells.

Human and mouse trophoblast cell invasion is not limited by the epithelium. Rather, these cells penetrate deep into the underlying decidualizing stroma. The extensive cellular and extracellular matrix remodelling that occurs ultimately results in formation of a decidual capsule and, finally, the maternal portion of the placenta (Carlson 1981). Penetration of the epithelial basement membrane during invasion of the trophoblast appears to be initiated by the decidual cells, not the trophoblasts, in rat (Schlafke *et al*. 1985) and mouse (Blankenship and Given 1992). However, extracellular matrix remodelling occurring during decidualization also may involve proteases produced by the trophoblast which act to locally degrade the extracellular matrix during trophoblast invasion (Glass *et al*. 1983; Fisher *et al*. 1985, Yagel *et al*. 1988; Aplin 1989; Lala and Graham 1990; Behrendtsen *et al*. 1992). These proteases (for review see Brenner *et al*. 1989) are apparently necessary for invasion since protease inhibitors have been used to block invasion (Yagel *et al*. 1988) and mouse blastocyst attachment and outgrowth *in vitro* (Kubo *et al*. 1981). Furthermore, disruption of the low density lipoprotein receptor-related protein (LRP) gene, whose product mediates uptake and degradation of urokinase-type plasminogen activator complexed with plasminogen activator inhibitor, results in implantation failure (Herz *et al*. 1992).* Human cytotrophoblasts derived from first trimester uterus are better able to digest extracellular matrices *in vitro* compared with non-invasive second or third trimester cytotrophoblasts. Moreover, first trimester cytotrophoblasts express several metalloproteinases which are not expressed by later stage cytotrophoblasts (Fisher *et al*. 1989). In addition to direct digestion of the intervening basement membrane, proteases may have other roles. Loss of E-cadherin is accompanied by appearance of the invasive phenotype (Fisher *et al*. 1989). Since cadherins and other adhesion molecules, e.g. integrins, are protease sensitive, it is tempting to speculate that one role of protease production by the trophoblast is to modify the complement of adhesion

* This observation was later retracted by these authors (Herz *et al*. (1993). Cell *73*, 428).

molecules presented on the surface. The extracellular matrix itself also may modulate protease activity. For example, laminin is able to induce type IV collagenase activity in malignant cells (Turpeenniemi-Hujanen *et al*. 1986; Emonard *et al*. 1990).

Another way in which basement membrane components may regulate invasiveness is by induction of differentiation. However, differentiation may act in some instances to facilitate and, in others, to inhibit invasion. Messenger RNA (mRNA) for collagenase, stromelysin, and tissue inhibitor of metalloproteinases is increased upon differentiation of embryonal carcinoma cells (Adler *et al*. 1990). In contrast, induction of differentiation in choriocarcinoma cells is believed to be the mechanism by which some chemotherapeutic agents are able to halt invasion (Speeg *et al*. 1976 Friedman and Skehan 1979). Upon differentiation, embryonal carcinoma cells exhibit a reduced adherence to laminin while adhesion to fibronectin is unchanged (Tienari *et al*. 1989). Thus, differentiation may modulate expression of a variety of adhesion regulating molecules with contrasting activities.

These observations demonstrate that a balance of degradative and synthetic events involving extracellular matrix occurs in normal implantation and placentation. How the uterus initially facilitates, then limits, blastocyst invasion is not understood. Comparison of embryo implantation with metastatic tumour invasion demonstrates that some of the mechanisms employed are similar. Furthermore, laminin is often the focal point of these mechanisms. Similar to embryo invasion, tumour invasion of the extracellular matrix has been described by Liotta *et al*. (1983) with a 'three step hypothesis' consisting of:

(1) tumour cell attachment to extracellular matrix;
(2) secretion of hydrolytic enzymes to degrade this matrix; and;
(3) locomotion of the tumour into the matrix.

The plasma membranes of highly invasive breast carcinoma cells have a 50-fold greater laminin binding capacity than normal or benign breast tumour cells (Liotta *et al*. 1983), suggesting that laminin binding of cells may facilitate invasion into the matrix. Laminin's ability to increase the release of type IV collagenase by malignant cells (Turpeenniemi- Hujanen *et al*. 1986) may facilitate their invasion. Furthermore, normal and malignant trophoblast cells have increased collagenolytic activities when grown on laminin or laminin-containing substrates such as Matrigel (Emonard *et al*. 1990), suggesting that laminin may regulate embryo invasiveness as well as provide a substrate for invasion. In addition, laminin interacts with both plasminogen and plasminogen activator (Salonen *et al*. 1984; Stephens *et al*. 1992), molecules which have previously been implicated in proteolytic extracellular matrix remodelling events associated with decidualization (Aplin 1991).

Laminin exists as a very large ($M_r = 1 \times 10^6$) heterocomplex of multiple isoforms of at least six different peptides (Kleinman *et al.* 1982; Leivo and Engvall 1988; Pikkarainen *et al.* 1988; Hunter *et al.* 1989; Ehrig *et al.* 1990; Kallunki *et al.* 1991, 1992). Each subunit is encoded by a different gene (Sasaki *et al.* 1987, 1988; Sasaki and Yamada 1987; Vuolteenaho *et al.* 1990). The function each subunit serves is unknown; however, various functional domains have been attributed to the A (Tashiro *et al.* 1989) and B1 (Graf *et al.* 1987) chains. Isoforms for the laminin B1 chain (S subunit) and for the A chain (M subunit) have been isolated from placental tissue (Engvall *et al.* 1990). Recently, a truncated form of the laminin B2 chain has been reported in human fetal tissues and in choriocarcinoma (JAR) cells (Kallunki *et al.* 1992). Antibodies against laminin B1, B2, S, M, and A chain of laminin demonstrate differential localization of these proteins in decidualizing human uterine tissue (Damsky *et al.* 1992). Commercially available preparations of laminin are usually derived from a murine tumour which produces one particular trimer of laminin, i.e. A-B1-B2 (Kleinman *et al.* 1982, 1986). Laminin complexes have multiple, distinct adhesion-supporting activities including peptide-dependent and carbohydrate-dependent functions (Martin and Timpl 1987). Laminin has numerous other biological activities including regulation of cell morphology, chemotaxis, migration, differentiation, and invasion (Kleinman *et al.* 1985; Graf *et al.* 1987; Iwamoto *et al.* 1987; Sephel *et al.* 1989; Tashiro *et al.* 1989; Timpl 1989; Dean *et al.* Emonard *et al.* 1990; Hadley *et al.* 1990; Vukicevic *et al.* 1990). Most studies of laminin expression have used. antibodies which recognize multiple components of the A-B1-B2 complex. Consequently, with the exception of the study by Damsky *et al.* (1992) little is known about the possible isoforms present at implantation sites or decidua.

Laminin synthesis appears to be hormonally regulated based on immuno-cytochemistry with laminin antibodies in human uterine tissue obtained at different hormonal stages (Loke *et al.* 1989). During the proliferative stage, very little laminin is observed in the stroma, but it is present in the basement membranes of endometrial glands and blood vessels. With progression into the secretory phase, discrete granules of laminin are observed on the plasma membrane of stromal cells. With decidualization, a marked increase in pericellular laminin in decidual cells is observed (Loke *et al.* 1989). Mouse uterine stromal cell mRNA for laminin B2 subunit increases five to six fold *in vitro* in response to exogenous progesterone (Jacobs, Wegner, and Carson, manuscript submitted), suggesting that at least in the mouse the laminin B2 subunit is under hormonal regulation.

Various extracellular matrix components are differentially regulated during decidualization both *in vivo* and *in vitro*. Northern analysis of uterine RNA demonstrated that the relative mRNA levels for entactin, fibronectin, glyceraldehyde-3-phosphate dehydrogenase, and laminin B1 mRNA remained relatively constant during days 4–7 of pregnancy; however, laminin B2 and to a lesser extent collagen IVα_2 and collagen IVα2 mRNA levels were elevated with progression of decidualization (Farrar and Carson 1992). Interestingly, the B1 and B2 chains of laminin are not co-ordinately regulated in the uterus. *In situ* hybridization using laminin B1 and B2 chain probes on sections of uteri obtained from days 4 to 8 of pregnancy demonstrates that B1 and B2 are already distributed differentially by day 7 in that laminin B1 mRNA is restricted to the primary decidual zone and B2 is found throughout the decidualizing uterus (Farrar and Carson 1992). Differential regulation of laminin subunit expression has been reported previously in kidney (Holm *et al.* 1988; Laurie *et al.* 1989), embryo (Cooper and MacQueen 1983), liver (Wewer *et al.* 1992), and neuromuscular junction (Sanes *et al.* 1990).

Electron microscopy utilizing antibodies against laminin have shown that laminin can be detected below the human trophoblast in the placental villi, in the basement membrane, and in the umbilical amniotic cells (Charpin *et al.* 1985). *In situ* hybridization studies have demonstrated that laminin mRNA is present in abundance in association with decidual cells and around the migrating trophoblast (Senior *et al.* 1988; Farrar and Carson 1992). Both human (Loke *et al.* 1989) and mouse embryos and trophoblast (see discussion above) attach to and outgrow on laminin. Its association with decidual and trophoblast cells and its increased production at the time of decidualization suggest that laminin plays a role in embryo implantation. Furthermore, embryo attachment to laminin appears to be mediated by the peptide, not carbohydrate, domain of laminin since binding activity could be reduced by boiling or trypsin treatment, but not by carbohydrate modification (Armant 1991). Outgrowth-promoting activity was localized to the E-8 fragment of laminin (Armant 1991). These observations are in apparent conflict with those implicating a role for galactosyltransferase in trophoblast spreading on laminin (Romagnano and Babiarz 1990).

Studies from several labs have demonstrated that blastocyst interactions with laminin are inhibitable by soluble RGD-containing peptides (Armant *et al.* 1986*b*; Farach *et al.* 1987; Sutherland *et al.* 1988). Moreover, specific antisera to integrins have been shown to inhibit blastocyst outgrowth on, but not attachment to laminin and certain other substrates. These same antisera have been used to detect integrins on trophoblast outgrowths by both immunofluorescence and immunoprecipitation assays (Sutherland *et al.* 1988). Consequently, trophoblast integrins appear to play an important role in interactions with extracellular matrix components including laminin. Sutherland *et al.* (1988) noted that integrin-perturbing reagents had a more severe effect on blastocyst outgrowth and relatively little effect on attachment. Through the use of proteolytic fragments of laminin, Armant (1991) recently reported that the domain containing the RGD-sequence, but not the major heparin/heparan sulphate binding regions, effectively supported blastocyst outgrowth *in vitro*. Collectively, these data indicate an important role for RGD sequence of laminin in trophoblast outgrowth processes. Other peptide sequences in laminin, e.g. YIGSR and IKVAV, play important functions in laminin-mediated adhesion processes (Martin and Timpl 1987). Studies emphasizing the role of these sequences in blastocyst interactions need to be performed as well.

Most *in vitro* studies of embryo attachment and outgrowth have used commercial preparations of laminin consisting of the A, B1, and B2 chains isolated from the EHS (mouse subcutaneous) tumour. It is well established that the patterns of protein glycosylation at a particular glycosylation site are regulated in a cell-type specific manner (Paulson and Colley 1989). Commercial laminin glycosylation patterns may not be similar to what is found in uterine laminin and, therefore, may not be the optimal source for implantation studies. For example, commercial preparations of laminin have an abundance of lactosaminoglycan structures (Fujiwara *et al*. 1988) as do placental forms of fibronection (Zhu *et al*. 1984). Studies of embryo-associated lectins, e.g. galactosyltransferase, are dependent upon the carbohydrate structures present on test substrates. Therefore, there may be bias due to differences in glycosylation patterns between uterine and commercially available forms of these proteins. Utilization of isolated attachment proteins from uteri would alleviate much of this concern, but have not been described. The form in which the extracellular matrix is presented, as well as its composition, may be critical in determining the function of an extracellular matrix component such as laminin. For example, a three-dimensional (gel) extracellular matrix is better able to induce markers of differentiation in choriocarcinoma cells than thin coatings of extracellular matrix on plastic (Hohn *et al*. 1992). There have been some studies examining embryo binding to complex extracellular matrix preparations resembling basal lamina. These studies have utilized matrix secreted by the EHS tumour cell line, a preparation commercially available as Matrigel. The composition of Matrigel has been examined and includes various growth factors in addition to basal lamina components including abundant amounts of laminin (Vukicevic *et al*. 1992). Thus, studies employing this or other tissue- or cell-derived matrix preparations may involve complex effects of both matrix and growth factor constituents. Mouse blastocysts readily attach and outgrow on Matrigel (Carson *et al*. 1990). Human trophoblasts also can bind to and invade Matrigel (Kliman and Feinberg 1990). In these studies, the trophoblasts cultured on various thicknesses of Matrigel varied in their ability to spread and digest the matrix. Trophoblasts produced flat aggregates on the thinnest substrate and rounded aggregates with intracellular processes on the intermediate thickness substrate. The intermediate thickness best supported invasion, and this process was inhibited by 8-bromo-cyclic adenosine monophosphate (cAMP). In contrast, the thickest substrate did not support invasion and trophoblasts grew unicellularly or in small aggregates. Interestingly, choriocarcinoma cells (JEG-3) exhibited the same culture characteristics with the exception that 8-bromo-cAMP was unable to inhibit invasiveness on the intermediate thickness. Whether the effect of extracellular matrix on invasion in these experiments is mediated by inducing differentiation is unknown. The basement membrane beneath the uterine epithelial cells is thin compared to the basement membrane-like pericellular matrix elaborated by decidual cells. In light of the above studies, it can be speculated that differences in the thickness of these matrices may modulate trophoblast invasive behaviour *in vivo*. Penetration of the thinner basal lamina may be facilitated by activation of trophoblast invasive functions. In contrast, the thicker matrix in the decidua may down-regulate such activities of the trophoblast and limit invasion of this tissue.

Finally, several studies are available that have examined interactions of blastocysts and trophoblast cells with complex matrices which may more closely resemble stromal extracellular matrix *in utero*. The studies using Matrigel described above may mimic interactions with the basal lamina components induced in this tissue; however, Matrigel lacks the interstitial matrix components, e.g. fibronectin, collagen types I and III, chondroitin sulphate proteoglycans, that also are abundantly expressed. Interstitial matrix preparations also have been found to support blastocyst attachment and outgrowth at rates similar to that observed for fibronectin (Carson *et al.* 1990); the adhesion molecules mediating these effects are unknown. Blastocyst attachment and outgrowth on matrices laid down by non-uterine cells also has been examined (Glass *et al.* 1979). These matrices are substantially more complex than purified matrix components and perhaps a better representation than Matrigel. None the less, in either case, the argument can be made that these matrices are not good reflections of the environment that the trophoblast encounters *in vivo*. One step closer to this environment is the use of matrices elaborated by isolated uterine stromal or decidual cells *in vitro*. These matrices are very complex and appear to contain most, if not all, of the components normally expressed by these cells *in vivo* (Julian and Carson, unpublished observations). As might be expected from the above discussion, blastocysts and trophoblast can use a variety of systems to interact with these complex matrices. Proteases appear to be involved along with peptide-dependent, galactosyltransferase- dependent, heparin-dependent, and hyaluronate-dependent systems (Romagnano and Babiarz 1990; Babiarz *et al.* 1992; Julian and Carson, unpublished observations). Thus, rather than one system predominating, it appears that multiple adhesion systems expressed by the embryo operate to support adhesion to these complex matrices.

In contrast to laminin, fibronectin appears to function predominately in promoting cell adhesion. However, oncofetal fibronectin is expressed specifically in extracellular matrix connecting extravillous trophoblast and trophoblastic cell columns to human decidua (Feinberg *et al.* 1991). Fibronectin adhesion may be mediated by two different binding sites acting cooperatively (Obara *et al.* 1988). *In vitro* studies have attributed cell attachment function of the 75 kD fragment of fibronectin to the RGDS sequence (McCarthy *et al.* 1986). The cell motility function of the 75 kD fragment is lost when a smaller 11.5 kD fragment containing RGDS is substituted suggesting this sequence cannot account for the cell mobility function of fibronectin. A second 33 kD tryptic/ catheptic carboxyl terminal heparin-binding fragment on the A chain also stimulated melanoma cell invasion. Therefore, multiple adhesion promoting domains are present on fibronectin. The formation of tumours in mice was inhibited in a dose-dependent manner by coinjecting the GRGDS sequence with melanoma cells (Humphries *et al.* 1986). Fibronectin has been demonstrated to support embryo attachment and outgrowth by many labs and is currently widely used for this purpose. Embryo interactions with fibronectin involve both integrin-dependent (Armant *et al.* 1986*a*; Farach *et al.* 1987; Sutherland *et al.* 1988) and heparan sulphate dependent (Farach *et al.* 1987) systems. Moreover, synthetic peptides designed from the primary amino acid sequence of fibronectin have been shown to support embryo attachment and outgrowth *in vitro* (Armant *et al.* 1986*b*). Fibronectin mRNA splice variants encoding different functional motifs appear to be expressed in distinct spatial patterns in non-pregnant and day-6 pregnant rat uteri (Rider *et al.* 1992). Therefore, fibronectin isoforms may perform different functions in these uterine locales. Fibronectin is abundantly expressed in the uterine stroma under all conditions and is very likely to facilitate embryo movement through stromal matrix *in vivo*.

Hyaluronate is a large ($M_r > 1 \times 10^6$ kD) pure polysaccharide synthesized by uterine endometrial explants and isolated uterine stromal cells *in vitro* (Jacobs and Carson 1991). Endometrial hyaluronate synthesis increases dramatically at the time of implantation in mice and purified hyaluronate has been shown to support blastocyst attachment and outgrowth *in vitro* at a similar rate as fibronectin (Carson *et al.* 1987; Carson *et al.* 1990). It is not clear what systems embryos use to bind to hyaluronate, but a family of proteins isolated initially as lymphocyte homing receptors (CD44) have been identified as transmembrane proteins that function as hyaluronate receptors at the surface of various cell types (Aruffo *et al.* 1990). Monoclonal antibodies directed against mouse CD44 specifically react with uterine tissue sections during the peri-implantation period; however, the pattern of expression and distribution in the stroma is not consistent with a direct role in mediating embryo-uterine adhesion. Furthermore, these same antibodies do not react with blastocyst cell surfaces either in sections of implantation sites or in whole mounts of blastocysts cultured *in vitro* (Julian and Carson, unpublished observations). Since the hyaluronate receptors are a family of proteins (Stamenkovic *et al*. 1989), it is possible that the available antibodies do not recognize the embryonic forms of these proteins. While it is apparent that blastocysts express functional hyaluronate binding activity on their surfaces, it remains uncertain whether CD44-related molecules are involved.

Collagen type IV is a component of uterine basal lamina and is produced by decidual cells. Blastocysts attach and outgrow on this substrate by systems which can be inhibited by RGD-containing peptides as well as antibodies directed against integrins (Carson *et al*. 1988 Sutherland *et al*. 1988). Heat-denatured collagen type IV also supports embryo attachment and outgrowth and is even more sensitive to inhibition by soluble RGD- containing peptides. Therefore, it appears that additional, functional RGD sequences are revealed by the denaturation process. Although the concentration of total collagen decreases at sites of rat uterine implantation sites compared with non-implantation sites (Myers *et al*. 1990), isoforms of collagen appear to be differentially regulated. Collagen type IV protein and mRNA for the α_1- and α_2-subunits increases substantially during decidualization in mouse (Wewer *et al*. 1986; Farrar and Carson 1992). In contrast, collagen type VI protein as detected by immunocytochemistry is lost from decidualizing rat stroma coincident with the appearance of desmin, a marker of decidualization (Mulholland *et al*. 1992). Interestingly, the reappearance of collagen type VI was observed with regression of decidual tissue and loss of desmin. Immunocytochemistry of first trimester decidua using collagen type-specific antibodies demonstrated the presence of collagen types I and III in the matrix in which stromal cells were embedded, whereas collagen type IV and laminin were present in the external basal lamina surrounding decidual cells. Collagen type V was detected in both areas (Kisalus *et al*. 1987).

The function of collagen matrix during implantation is unclear; however, collagen type I when used as an *in vitro* substrate for human chorionic villi explants can influence the hormone and protein production of these placental cells (Castellucci *et al*. 1990). In addition, collagen types I–VI support embryo attachment and outgrowth *in vitro* (Carson *et al*. 1988). Some of these functions appear to be mediated by the RGD sequence since this peptide can inhibit embryo outgrowth on collagen types II and IV. Remodelling of extracellular matrix appears to be hormonally regulated *in vitro*, since progesterone inhibits release of collagenase and gelatinase activity from human endometrial explants (Marbaix *et al*. 1992). Fibrillar collagens (types I and III) also are major components of the uterine stroma under all conditions. Both types of collagens support blastocyst attachment and outgrowth *in vitro* at a slower rate than fibronectin (Carson *et al*. 1988). Many cell types appear to attach to collagens via integrin-dependent systems (Hynes 1992) and, as discussed above, integrins appear to be involved in embryo attachment to certain collagens as well. Embryos can attach and outgrow on collagen types I–VI, albeit at different rates. Embryo attachment and outgrowth on collagen types II, V, and VI is heparan sulphate-independent (Carson *et al.* 1988). It is not clear what mechanisms embryos use to attach to all collagens of this series. Furthermore, it is unlikely that embryos encounter all of these collagens, e.g. collagen type II, during implantation. Thus, these observations provide an example of adhesion functions expressed by blastocysts.

Heparan sulphate proteoglycans are the major proteoglycan components of basal lamina (Yurchenco and Schittny 1990). While embryos can attach to certain polysaccharides used to coat plastic surfaces, e.g. hyaluronate, they do not attach to other glycosaminoglycans including heparin (Carson *et al.* 1987). It should be noted that in the cited studies the efficiency by which different glycosaminoglycans bind to plastic was not monitored for each glycosaminoglycan used. Therefore, it is possible that some differences may be due to suboptimal binding of the polysaccharides. Definitive information on the expression of heparan sulphate-binding sites on the surface of blastocysts is not available; however, we have found that human trophoblast cell lines display specific heparan sulphate-binding sites on their cell surfaces (Rohde and Carson, unpublished studies). Consequently, it is possible that embryos may be able to bind to heparan sulphate components of basal lamina. Entactin (nidogen), fibronectin, and some other extracellular matrix components occur in basal lamina; however, no data are available for blastocyst binding to entactin.

Chondroitin sulphate proteoglycans are the major class of proteoglycans synthesized by uterine endometrial explants and isolated uterine stromal cells *in vitro* (Jacobs and Carson 1991). These molecules are large (M_r > 1 ×10^6), heavily substituted with chondroitin sulphate chains and, thereby, highly negatively charged. The nature of the core protein(s) in uteri is unknown although several chondroitin sulphate proteoglycan core proteins have been cloned from other sources (see Carson 1992 for review). Uterine chondroitin sulphate proteoglycans bind selectively to collagen type I fibrils with high affinity (Jacobs and Carson 1991), and this property may contribute to their retention in the stromal matrix. In contrast to heparan sulphate proteoglycans, chondroitin sulphate proteoglycans appear to inhibit blastocyst adhesive interactions with collagen type I *et al.* 1992). Similar anti-adhesive properties of chondroitin sulphate proteoglycans have been reported in other systems (Perris and and Johansson 1987; Yamagata *et al.* 1989). By analogy with these other systems, uterine chondroitin sulphate proteoglycans may not act by binding to specific receptors on the embryo cell surface. Rather, due to their large hydrodynamic radii and their retention in stromal extracellular matrix, they may interfere with blastocyst adhesion by steric hindrance. Physiologically, this may help attenuate the invasive behaviour of the trophoblast during its penetration of the decidual tissue.

Three secreted glycoproteins which demonstrate anti-adhesive properties are SPARC, tenascin, and thrombospondin (TSP; reviewed by Sage and Bornstein 1991). All three glycoproteins when used as substrates are able to reduce the number of focal adhesion-positive cells by 50% (Murphy-Ullrich *et al.* 1991) and have been localized to uterine cells and embryos (Wewer *et al.* 1988; Howe *et al.* 1988). SPARC, also known as osteonectin, is a secreted calcium-binding glycoprotein which is inducible at the genomic level by tumour promoters and growth factors (Wrana *et al.* 1991). SPARC is highly similar to the bovine matrix protein osteonectin and the bovine endothelial 43 kD protein from EHS (BM-40, Lankat-Buttgereit *et al.* 1988). BM-40 was originally isolated from EHS and was found to bind collagen IV, but not collagens I, III, V, and VI. Binding to collagen IV was abolished by the presence of chelating or chaotropic agents, high salt or reduction of disulphide bonds (Mayer *et al.* 1991). SPARC inhibits cell spreading *in vitro* and has a high calcium-dependent binding affinity for collagen types III and V, and TSP (Sage *et al.* 1989*a*). Both the carboxyl and amino domains of SPARC exhibit antispreading activity but only the carboxyl domain interacts with the extracellular matrix (Lane and Sage 1990). High levels of SPARC mRNA and protein as determined by Northern analysis and immunocytochemistry are detected in human decidual tissue (Wewer *et al.* 1988). Intense immunostaining for SPARC is detected in the extracellular matrix around large mature decidual cells. Transcripts and protein for SPARC have been detected in embryonic, extra-embryonic, and placental tissue (Holland *et al.* 1987; Howe *et al.* 1988; Sage *et al.* 1989*b*). Embryos and extra-embryonic tissue expressed SPARC transcript from day 9 whereas placental transcripts were first detected on day 11. Interestingly, embryonal carcinoma cells, upon differentiation, also produce SPARC (Howe *et al.* 1988).

Another developmental process critical to embryo implantation which requires extensive extracellular remodelling and cell adhesion events is angiogenesis of the decidual compartment. Angiogenesis is a critical step in placentation (Welsh and Enders 1991). SPARC has been associated with the differentiation of endothelial cells during angiogenesis. For instance, increased expression of SPARC as well as collagen type I mRNA is correlated with spontaneous tube formation in angiogenesis (Iruela-Arispe *et al.* 1991). SPARC and collagen type I are both detected in large amounts in the cytoplasm of cells located near actively developing tubes. Collagen type III and TSP are localized by immunocytochemistry around cells organizing into tubular structures. In contrast, laminin and collagen type IV are present in the cytoplasm of monolayer cells but were absent from cells present in the developing tube. SPARC has been shown to increase plasminogen activator-1 production (Hasselaar *et al.* 1991) and modulate platelet-derived growth factor activity (Raines *et al.* 1992). SPARC added to cultured endothelial cells decreases synthesis of TSP1 and fibronectin and increases the synthesis of plasminogen activator-1 inhibitor (Lane *et al.*

1992). In contrast, collagen type I mRNA levels and secreted gelatinases are unaffected by the addition of SPARC. The ability of SPARC to regulate these genes suggests it may play a general role in co-ordinating these genes to facilitate extracellular matrix remodelling events.

TSP is a large trimeric adhesive glycoprotein encoded by three separate genes (TSP1, TSP2, TSP3; reviewed by Bornstein 1992). TSP1 is inducible by both serum and growth factors. In contrast, TSP2 is less inducible by serum and lacks a serum response element present in TSP1. Both TSP1 and TSP2 share homologies at the protein level including type 1 (TSP or properdin), type II (epidermal growth factor-like) and type III (Ca^{2+} binding repeats. However, their amino acid sequences differ. TSP1 and TSP2 can be expressed in both homo- and heterotrimeric forms which differ in their affinity for heparin (O'Rourke et al. 1992). TSP can interact with heparan sulphate on the surface of epithelial cells (Sun et al. 1989). Melanoma cells are capable of utilizing two cooperative receptor systems (glycoprotein IV and heparin) to modulate their attachment and spreading on a TSP substrate. Both receptor systems must be inhibited to prevent attachment (Asch et al. 1991). Suramin, a heparin analogue, can inhibit laminin and TSP-mediated melanoma cell adhesion and chemotaxis (Zabrenetzky et al. 1990).

TSP is able to support the adhesion of various cell types including platelets, melanoma cells, muscle cells, endothelial cells, fibroblasts, and epithelial cells (Tuszynski et al. 1987). In contrast, TSP has been shown to reduce focal adhesions in endothelial cells and fibroblasts (review Sage and Bornstein 1991). The molecular basis for these apparently conflicting activities of TSP are poorly understood, but may reflect differences in TSP isoform activities, receptor expression or a combination of both. TSP has been localized using immunocytochemistry to trophectoderm, endoderm and inner cell mass, and has been shown to support trophoblast outgrowth in vitro to a greater extent than laminin (O'Shea et al. 1990). The function of TSP in implantation is unclear. However, TSP has been implicated in regulating angiogenesis (Taraboletti et al. 1990) due to its ability to induce adhesion, spreading and migration in endothelial cells and to inhibit basic fibroblast growth factor action in a dose-dependent manner. Consequently, TSP may promote angiogenesis during placentation.

Tenascin is an oligomeric glycoprotein commonly found in tissues undergoing invasive/ remodelling processes such as during embryogenesis and metastasis (reviewed by Erickson 1989). Tenascin is uncommon in adult tissues but has been localized in mouse small intestine villus where it is believed to play a role in epithelial cell shedding (Probstmeier et al. 1990). Tenascin provides a very poor substrate for cell attachment. In fact, tenascin is able to reduce binding of cells to fibronectin, but does not interfere with cell binding to laminin or collagen types I–IV (Chiquet-Ehrismann et al. 1988; Probstmeier et al. 1990). Monoclonal antibodies to tenascin neutralize these inhibitory effects. Tenascin specifically reduces the number of focal adhesion-positive bovine aortic endothelial cells in a dose–dependent manner (Murphy-Ullrich *et al.* 1991). Interestingly, the ability of tenascin to reduce focal adhesion integrity is neutralized by chondroitin-6-sulphate and to a lesser degree by dermatan sulphate and chondroitin-4-sulphate (Murphy-Ullrich *et al.* 1991). In contrast, heparan sulphate and heparin are unable to neutralize tenascin activity.

Tenascin has been localized to both normal and malignant uterine tissue. During the menstrual cycle, tenascin is localized around the endometrial glands and persists until the early postovulatory phase (Vollmer *et al.* 1990). In the late secretory phase, tenascin is not detected in either the glandular or stromal cells, but is found in endometrial arterioles. In contrast, endometrial adenocarcinoma tissue exhibits tenascin in the entire extracellular space, but not in the malignant cells themselves. Tenascin expression has also been used as a stromal marker for epithelial carcinogenesis in mammary tissue (Mackie *et al.* 1987). It is expressed transiently in stromal cells in the primary decidual zone and is rapidly induced by the presence of a hatched blastocyst or an artificial decidual stimulus (Julian *et al.* 1994). Importantly, tenascin disappears prior to the time embryos penetrate the epithelial cell layer. Consequently, it is unlikely that tenascin plays a direct role in blastocyst binding. Rather, tenascin expression may reflect transmission of decidualizing signals to underlying implantation sites. As an anti-adhesin, tenascin may promote local disruption of basal lamina or 'loosening' of epithelial cells at the implantation site. Consistent with this, we have found that purified tenascin interferes with adhesion- related processes of isolated mouse uterine epithelial cells cultured *in vitro* (Julian *et al.* 1994).

Tenascin has been found in the human uterus during the formation of the placenta in the first trimester at sites of cytotrophoblast differentiation (Damsky *et al.* 1992). For example, tenascin staining was found at sites of column formation, in decidua, in myometrium, and in maternal blood vessels. Epithelial cells are able to induce tenascin secretion in mesenchymal cells of the mammary gland (Inaguma *et al.* 1988) and uterine stroma *in vitro* (Julian *et al.* 1994). The identification of tenascin at sites of epithelial-stromal interactions in developing systems including the uterus has led to the suggestion that this matrix component plays a role in morphogenetic events (Chiquet-Ehrismann *et al.* 1986; Erickson 1989). Therefore, it is surprising that mice lacking a functional tenascin gene develop and reproduce normally (Saga *et al.* 1992). The possibility that other gene products can replace tenascin function remains.

VII   SUMMARY AND CONCLUDING REMARKS

In this chapter, we have reviewed studies of a variety of cell adhesion molecules which have been localized to either trophoblast or uterine cells during the peri-implantation period and, thus, may play a role in embryo attachment. It is possible to draw several general conclusions from the studies to date. In the attachment phase of implantation, carbohydrate-mediated interactions appear to be crucial. As the blastocyst becomes attachment competent, some adhesion molecules, e.g. cell CAM-105 and syndecan, are rearranged from the exterior to the interior cell surface. At the same time, others, e.g. perlecan, laminin, and integrins, begin to be expressed at the exterior surface where they may play a role in initial phases of embryo attachment (see Fig. 3.3). The profile of adhesion molecule expression also changes on the surface of uterine cells (see Fig. 3.4). During conversion to a receptive state, certain glycoproteins including mucins, LNF-I-bearing molecules, and CAM-105 are decreased on the surface of uterine epithelial cells prior to acquisition of attachment competence and may be requisite for this process. These glycoproteins and others that may be discovered

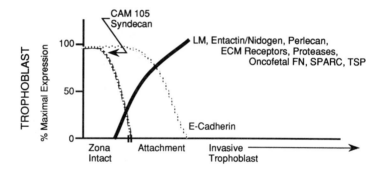

Fig. 3.3. Summary of changes in embryo adhesion molecule expression during the peri-implantation period. During the period in which the blastocyst hatches from the zona pellucida, some molecules, e.g. cell CAM-105 and syndecan, are lost from the exterior surface of the trophectoderm. Others are expressed either at the exterior of the attachment-competent blastocyst, i.e. laminin (LM), entactin/nidogen, perlecan, or on the cell surfaces of invasive trophoblast, e.g. extracellular matrix (ECM) receptors, proteases, SPARC and TSP. In contrast, E-cadherin expression appears to decrease on invasive trophoblast. The graph displays relative changes and, in most cases, is based on changes detected by immunohistochemistry. Consequently, no quantitative relationships among these various molecules should be interpreted from the figure. For further details, see text.

which exhibit this behaviour have been grouped as class I uterine adhesion molecules in Fig. 3.4. Once class I glycoproteins are removed, a second group of glycoproteins including heparan sulphate proteoglycans, and certain lectins are exposed or expressed which may support blastocyst attachment. Such molecules may function in ligand-receptor interactions between trophoblast and uterine epithelial cells, resulting in an initial tethering of trophectoderm and lumenal epithelia during the attachment phase. Other adhesion-modulators, e.g. hyaluronate and tenascin, are transiently expressed at implantation sites and are grouped as class II adhesion molecules in Fig. 3.4. In contrast to the carbohydrate-mediated events of attachment, the invasive phase of implantation appears to be characterized predominantly by protein-protein interactions. For instance, most interactions between the invading trophoblast occur between protease sensitive cell surface adhesion molecules like cadherins and integrins and the extracellular matrix of the basal lamina or that produced by decidual cells. In some cases, components of the stromal matrix appear to be removed from the decidua, e.g. collagen type VI, and constitute another class of adhesion-related proteins in this system (class III in Fig. 3.4). Others components of uterine cell surfaces and interstitial matrix appear to be constitutively expressed, e.g. E-cadherin, chondroitin sulphate proteoglycans and fibronectin, and have been grouped as class IV adhesion molecules in Fig. 3.4. Finally, a number of potential adhesion modulators markedly increase in decidual matrix and constitute a fifth class (class V) of uterine adhesion molecules indicated in Fig. 3.4.

Numerous cell adhesion molecules have been identified which may play a role in implantation. Some of these are relatively well characterized, others have simply been identified in the 'right place at the right time'. In most cases, very little conclusive data are available demonstrating a function and a mechanism of action for these molecules. Many adhesion molecules and extracellular matrix components display different activities in the presence of other adhesion molecules and matrix components. In most cases, including uteri, these molecules occur in complex combinations. Therefore, the uterine context in which these molecules function is crucial. For example, different laminin isoforms can provide different functional domains which can act alone or in concert with other subunits to produce a particular 'laminin' effect. Moreover, different isoforms may be expressed in different uterine regions during implantation, e.g. basal lamina, primary and secondary decidual zones, etc. Finally, other cell adhesion molecules in the vicinity can provide complementary or antagonistic interactions which can further modify function.

It seems likely that development of implantation models that can be regulated *in vitro* will substantially improve our understanding of embryo implantation. Presently, uterine epithelial cells can be efficiently isolated and studied alone or in combination with other uterine cell types or trophoblast cells. The ability to culture uterine epithelial cells in a polarized manner provides a system which retains many of the features of uterine epithelial cells *in utero*. Similar approaches are being used to study trophoblast function as well. Use of this type of *in vitro* culture system allows identification and determination of vectorial patterns of expression of molecules produced by uterine cells or trophoblasts under a variety of controlled conditions. Although many adhesion molecules permit the attachment of trophoblasts *in vitro*, it is unproven which of these may be utilized *in utero*. One major limitation of this system is that the receptive/non-receptive regulation of uterine epithelial cells presently cannot be duplicated *in vitro*. The development of a regulatable

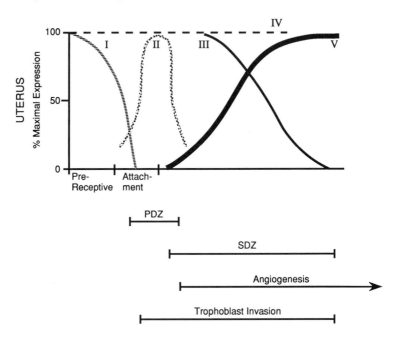

Class	Representative Members
I	CAM-105, LNF-1, Muc I
II	hyaluronate, tenascin
III	collagen type VI
IV	fibronectin, collagen types I & III
V	laminin B2, collagen types IV & V, P-cadherin, SPARC, osteopontin, HSPG, proteases and protease inhibitors

in vitro system in terms of receptivity would facilitate the identification and manipulation of molecules which function in the attachment phase of implantation. Another benefit of a regulatable system may be to facilitate identification of molecular signals produced by trophoblast, epithelial, or decidual cells involved in triggering of implantation-related responses, e.g. decidual cell reaction. Verification of such *in vitro* studies can be performed *in utero* using available *in situ* staining techniques. Function of candidate molecules should be examined *in vivo* by exploiting intrauterine injection of specific antagonists or use of gene disruption techniques.

Implantation research has identified numerous adhesion proteins which vary markedly in their temporal and/or spatial patterns of expression during the peri-implantation period. However, almost no information is available regarding the regulation of the genes encoding many of these proteins. Since these adhesion molecules are not uniquely expressed in the uterus, molecular mechanisms of regulation must exist to specifically activate, amplify, or down-regulate these genes in uterine cells during implantation. In addition, it seems likely that at least some of the multitude of genes which are up- or down-regulated co-ordinately during this interval share similar modes of regulation. Modern molecular biological approaches must be applied to studies of gene expression in trophoblast and uterine cells to develop this understanding. One approach may be to determine if *cis*-acting 'decidual response elements' or *trans*-acting decidual cell specific transcription factors are shared by genes regulated during implantation.

This review may provide a useful framework for extension of studies of cell adhesion systems operating during implantation. We have tried to summarize as many relevant areas and studies as possible. Undoubtedly,

Fig. 3.4 Summary of uterine adhesion molecule changes during the peri-implantation period. Several types of molecules displayed at the apical surface of the uterine epithelium are down-regulated during conversion to a receptive state (class I). At the time of embryo attachment, other adhesion modulators are transiently increased at implantation sites (class II). Timing of the expression of these molecules coincides approximately to the time at which formation of the primary decidual zone (PDZ) is initiated. As the secondary decidual zone (SDZ) forms, some extracellular matrix components become diminished in decidua (class III). Various extracellular matrix molecules do not appear to change their pattern of expression in decidualized versus non-decidualized uteri (class IV). In contrast, during this same interval a number of proteins begin to be expressed by decidual cells that are not expressed by stroma of non-pregnant uteri or outside of the implantation site of pregnant animals (class V). The timing of expression of class V molecules corresponds to the period in which trophoblast invasion and angiogenesis occurs. The graph displays relative changes and, in most cases, is based on changes detected by immunohistochemistry. Consequently, no quantitative relationships among these various molecules should be interpreted from the figure. For further details, see text. HSPG = heparan sulphate proteoglycans.

some important work has been missed or glossed over and we extend our regrets for this. We deeply appreciate the substantial contributions of all the scientists and laboratories who have lent their efforts to the study of this difficult and vital research effort.

ACKNOWLEDGEMENTS

The authors greatly appreciate the numerous discussions with Dr A.L. Jacobs and J. Julian, S. Liu, M. McIntosh, L. Rohde, and G. Surveyor as well as the critical reading of this manuscript by Drs J. Clemens and M.C. Farach-Carson. We are grateful to S. Jordan and H. Young for their typing. This work was supported by a National Research Service award (HD07568) awarded to C.C.W. and a National Institutes of Health grant (HD25235) awarded to D.D.C.

REFERENCES

Adler, R.R., Brenner, C.A., and Werb, Z. (1990). Expression of extracellular matrix-degrading metalloproteinases and metalloproteinase inhibitors is developmentally regulated during endoderm differentiation of embryonal carcinoma cells. *Development*, **110**, 211–20.

Anderson, T.L., Olson, G.E., and Hoffman, L. (1986). Stage-specific alterations in the apical membrane glycoproteins of endometrial epithelial cells related to implantation in rabbits. *Biology of Reproduction*, **34**, 701–20.

Aplin, J.D. (1989). Cellular biochemistry of the endometrium. In *Biology of the uterus* (ed. R.M. Wynn and W.P. Jollie), pp. 89–129. Plenum Press, New York.

Aplin, J.D. (1991). Review: Implantation, trophoblast differentiation and haemochorial placentation: mechanistic evidence *in vivo* and *in vitro*. Journal of Cell Science, **99**, 681–92.

Armant, D.R. (1991). Cell interactions with laminin and its proteolytic fragments during outgrowth of mouse primary trophoblast cells. *Biology of Reproduction*, **45**, 664–72.

Armant, D.R., Kaplan, H.A., and Lennarz, W.I. (1986a). Fibronectin and laminin promote *in vitro*-attachment and outgrowth of mouse blastocyst. *Developmental Biology*, **116**, 519–29.

Armant, R., Kaplan, H.A., Morer, H., and Lennarz, W.J. (1986b). The effect of hexapeptides on attachement and outgrowth of mouse blastocysts cultured *in vitro*: evidence for the involvement of the cell recognition tripeptide Arg-Gly-Asp. *Proceedings of the National Academy of Sciences, USA*, **83**, 6751–5.

Aruffo, A., Stamenkovic, I., Melnick, M., and Underhill, C.B. (1990). CD44 is the principle cell surface receptor for hyaluronate. *Cell*, **61**, 1303–13.

Asch, A.S., Teplar, J., Silloiger, S., Nachman, R.L. (1991). Cellular attachment to thrombospondin: cooperative interactions between receptor systems. *Journal of Biological Chemistry*, **266**, 1740–5.

Babiarz, B.S., Romagnano, L.C., and Kurilla, G.M. (1992). Interaction of mouse ectoplacental cone trophoblast and uterine decidua *in vitro*. *In Vitro Cellular and Developmental Biology*, **28A**, 500–508.

Bayna, E.M., Shaper, J.H., and Shur, B.D. (1988). Temporally specific involvement of cell surface β-1,4 galactosyltransferase during mouse embryo morula compaction. *Cell*, **53**, 145–57.

Behrendtsen, O., Alexander, C.M., and Werb, Z. (1992). Metalloproteinases mediate extracellular matrix degradation by cells from mouse blastocyst outgrowths. *Development*, **114**, 447–56.

Bell, S.C. (1983). Decidualization: regional differentiation and associated function. *Oxford Reviews of Reproductive Biology*, **5**, 220–71.

Bergstrom, S.I. (1978). Experimentally delayed implantation. In *Methods in mammalian reproduction* (ed. J.C. Daniel, pp. 419–35. Academic Press, New York.

Bianco, P., Fisher, L.W., Young, M.F., Termine, J.D., and Gehron Robey, P. (1991). Expression of bone sialoprotein (BSP) in developing human tissues. *Calcified Tissue International*, **49**, 421–6.

Blankenship, T.N., and Given, R.L. (1992). Penetration of the uterine epithelial basement membrane during blastocyst implantation in the mouse. *Anatomical Record*, **233**, 196–204.

Bornstein, P. (1992). Thrombospondins: structure and regulation of expression. *FASEB Journal*, **6**, 3290–9.

Brenner, C.A., Adler, R.R., Rappolee, D.A., Pedegwen, R.A., and Werb, Z. (1989). Genes for extracellular matrix-degrading metalloproteinases and their inhibitor, TIMP, are expressed during early embryo development. *Genes and Development*, **3**, 848–59.

Butler, W.T. (1989) The nature and significance of osteopontin. *Connective Tissue Research*, **23**, 123–36.

Carlson, B.M. (1981). Extraembryonic membranes and placenta In *Pattens foundations of embryology*, pp. 186–230. McGraw-Hill Inc. New York.

Carson, D.D. (1992). Proteoglycans in development. In *Cell surface carbohydrates and cell development* (ed. M. Fukuda), pp. 257–83 CRC Press, Inc., Boca Raton, FL.

Carson, D.D., Dutt, A., and Tang, J.P. (1987). Glycoconjugate synthesis during early pregnancy: hyaluronate synthesis and function. *Developmental Biology*, **120**, 228–35.

Carson, D.D., Tang, JP., and Gay, S. (1988). Collagens support embryo attachment and outgrowth *in vitro*: effects of the Arg-Gly-Asp sequence. *Developmental Biology*, **127**, 368–75.

Carson, D.D., Wilson, O.F., and Dutt, A. (1990). Glycoconjugate expression and interactions at the cell surface of mouse uterine epithelial cells and peri-implantation stage embryos. *Trophoblast Research*, **4**, 211–41.

Carson, D.D., Julian, J., and Jacobs, A.L. (1992). Uterine stromal cell chondroitin sulfate proteoglycans bind to collagen type I and inhibit embryo outgrowth *in vitro*. *Developmental Biology*, **149**, 307–16.

Carson, D.D., Tang, J -P. and Julian, J. (1993). Heparan sulfate proteoglycan (perlecan) expression by mouse embryos during acquisition of attachment competence. *Developmental Biology*, **155**, 97–106.

Castellucci, M., Kaufmann, P., and Bischof, P. (1990). Extracellular matrix influences hormone and protein production by human chorionic villi. *Cell Tissue Research*, **262**, 135–42.

Charpin, C., Kopp, F., Pourreau-Schneider, N., Lissitzky, J.C., Lavaut, M.N., Martin, P.N., and Toga, M. (1985). Laminin distribution in human decidua and immature placenta. *American Journal of Obstetrics and Gynecology*, **151**, 822–6.

Chavez, D.J. (1990). Possible involvement of D-galactose in the implantation process. *Trophoblast Research*, **4**, 259–72.

Chavez, D.J. and Anderson, T.L. (1985). The glycocalyx of the mouse uterine epithelium during estrus, early pregnancy, the peri-implantation period and delayed implantation. I. The acquisition of *Ricinus communis* I binding sites during pregnancy. *Biology of Reproduction*, **32**, 1135–45.

Chiquet-Ehrismann, R., Mackie, E.U., Pearson, C.A., and Sakakura, T. (1986). Tenascin: an extracellular matrix protein involved in tissue interactions during fetal development and oncogenesis. *Cell*, **47**, 131–9.

Chiquet-Ehrismann, R., Kalla, P., Pearson, C.A., Beck, K., and Chiquet, M. (1988). Tenascin interferes with fibronectin action. *Cell*, **53**, 383–90.

Cooper, A.R. and MacQueen, H.A. (1983). Subunits of laminin are differentially synthesized in mouse eggs and early embryos. *Developmental Biology*, **96**, 467–71.

Cowell, T.P. (1969). Implantation and development of mouse eggs transferred to the uterus of non- progestational mice. *Journal of Reproduction and Fertility*, **19**, 239–45.

Damjanov, I., Damjanov, A., and Damsky, C.H. (1986). Developmentally regulated expression of the cell-cell adhesion glycoprotein cell-CAM 120/80 in periimplantation mouse embryos and extraembryonic membranes. *Developmental Biology*, **116**, 194–202.

Damsky, C.H., Richa, J., Solter, D., Knudsen, K., and Buck, C.A. (1983). Identification and purification of a cell surface glycoprotein mediating intercellular adhesion in embryonic and adult tissue. *Cell*, **34**, 455–66.

Damsky, C.H., Fitzgerald, M.L., and Fisher, S.J. (1992). Distribution patterns of extracellular matrix components and adhesion receptors are intricately modulated during first trimester cytotrophoblast differentiation along the invasive pathway, *in vivo*. *Journal of Clinical Investigation*, **89**, 210–22.

Das, S.K., Flanders, K.C., Andrews, G.K., and Dey, S.K. (1992). Expression of transforming growth factor- β isoforms (β2 and β3) in the mouse uterus: analysis of the peri-implantation period and effects of ovarian steroids. *Endocrinology*, **130**, 3459–66.

De, M., Choudhuri, R., and Wood, G.W. (1991). Determination of the number and distribution of macrophages, lymphocytes, and granulocytes in the mouse uterus from mating through implantation. *Journal of Leukocyte Biology*, **50**, 252–62.

Dean, J.W., Chandrasekaran, S., and Tanzer, M.L. (1990). A biological role of carbohydrate moieties of laminin. *Journal of Biological Chemistry*, **265**, 12553–62.

DeHertogh, R., Ekka, E., Vanderheyden, I., and Glorieux, B. (1986). Estrogen and progesterone receptors in the implantation sites and interembryonic segments of rat uterus endometrium and myometrium. *Endocrinology*, **119**,, 680–4.

Denker, H–W. (1990). Trophoblast–endometrial interactions at embryo implantation: A cell biological paradox. *Trophoblast Research*, **4**, 3–29.

Dickmann, Z. and Noyes, R.W. (1960). The fate of ova transfered into the uterus of the rat. *Journal of Reproduction and Fertility*, **12**, 197–206.

Dutt, A. and Carson, D.D. (1990). Lactosaminoglycan assembly, cell surface expression, and release by mouse uterine epithelial cells. *Journal of Biological Chemistry*, **265**, 430–8.

Dutt, A., Tang, J-P., and Carson, D.D. (1987). Lactosaminoglycans are involved in uterine epithelial cell adhesion *in vitro*. *Developmental Biology*, **119**, 27–37.

Dutt, A., Tang, J–P., and Carson, D.D. (1988). Estrogen preferentially stimulates lactosaminoglycan- containing oligosaccharide synthesis in mouse uteri. *Journal of Biological Chemistry*, **263**, 2270–9.

Dziadek, M., Fujiwara, S., Paulsson, M., and Timpl, R. (1985). Immunological characterization of basement membrane types of heparan sulfate proteoglycan. *EMBO Journal*, **4**, 905–12.

Ehrig, K., Leivo, I., Argraves, W.S., Ruoslahti, E., and Engvall, E. (1990). Merosin, a tissue specific basement membrane protein, is a laminin-like protein. *Proceedings of the National Academy of Sciences, USA*, **87**, 3264–8.

Ekblom, P., Vestweber, D., and Kemler, R. (1986). Cell-matrix interactions and cell adhesion during development. *Annual Review of Cell Biology*, **2**, 27–47.

Emonard, H., Christiane, Y., Smet, M., Grimaud, J.A., and Foidart, J.M. (1990). Type IV and interstitial collagenolytic activities in normal and malignant trophoblast cells are specifically regulated by the extracellular matrix. *Invasion Metastasis*, **10**, 170–7.

Engvall, E., Earwicker, D., Haaparanta, T., Ruoslathi, E., and Sanes, J.R. (1990). Distribution and isolation of four laminin variants, tissue restricted distribution of heterotrimers assembled from five different subunits. *Cell Regulation*, **1**, 731–40.

Erickson, H.P. (1989). Tenascin: and extracellular matrix protein prominent in specialized embryonic tissues and tumours. *Annual Review of Cell Biology*, **5**, 71–92.

Farach, M.C., Tang, JP., Decker, G.L., and Carson, D.D. (1987). Heparin/heparan sulfate is involved in the attachment and spreading of mouse embryos *in vitro*. *Developmental Biology*, **123**, 401–10.

Farach, M.C., Tang, J-P., Decker, G.L., and Carson, D.D., (1988). Differential effects of *p*-nitrophenyl-D- xylosides on mouse blastocysts and uterine epithelial cells. *Biology of Reproduction*, **39**, 443–55.

Farrar, J.D. and Carson, D.D. (1992). Differential temporal and spatial expression of mRNA encoding extracellular matrix components in decidua during the peri-implantation period. *Biology of Reproduction*, **46**, 1095–108.

Feinberg, R.F., Kliman, H.J., and Lockwood, C.J. (1991). Is oncofetal fibronectin a trophoblast glue for human implantation? *American Journal of Pathology*, **138**, 537–43.

Finn, C.A. (1986). Implantation, menstruation and inflammation. *Biological Reviews*, **61**, 313–28.

Fisher, S.J., Leitch, M.S., Kantor, M.S., Basbaum, C.B., and Kramer, R.H. (1985). Degradation of extracellular matrix by the trophoblastic cells of first-trimester human placentas. *Journal of Cellular Biochemistry*, **27**, 31–41.

Fisher, S.J., Cui, T., Zhang, L., Hartmann, L., Grahl, K., Guo-Yang, Z., et al. (1989). Adhesive and degradative properties of human placental cytotrophoblast cells *in vitro*. *Journal of Cell Biology*, **109**, 891–902.

Fisher, S.J., Sutherland, A., Moss, L., Hartman, L., Crowley, E., Bernfield, M., et al. (1990). Adhesive interactions of murine and human trophoblast cells. *Trophoblast Research*, **4**, 115–38.

Friedman, S.J., and Skehan, P. (1979). Morphological differentiation of human choriocarcinoma cells induced by methotrexate. *Cancer Research*, **39**, 1960–7.

Fujiwara, S., Shinkai, H., Deutzmann, R., Paulsson, M., and Timpl, R. (1988). Structure and distribution of N-linked oligosaccharide chains on various domains of mouse tumour laminin. *Biochemical Journal*, **252**, 453–61.

Given, R.L. and Enders, A.C. (1989). The endometrium of delayed and early implantation. In *Biology of the uterus*. ed. R.M. Wynn and W.P., Jolise, pp. 175–231. Plenum Press, New York.

Glass, R.H., Aggeler, J., Spindle, A., Pederson, R.A., and Werb, Z. (1983). Degradation of extracellular matrix by mouse trophoblast outgrowths: a model for implantation. *Journal of Cell Biology*, **96**, 1108–16.

Glass, R.H., Spindle, A.I., and Pedersen, R.A. (1979). Mouse embryo attachment to substratum and interaction of trophoblast with cultured cells. *Journal of Experimental Zoology*, **208**, 327–36.

Glasser, S.R. and Julian, J. (1986). Intermediate filament protein as a marker of uterine stromal cell decidualization. *Biology of Reproduction*, **35**, 463–74.

Glasser, S.R. and McCormack, S.A. (1981). Separated cell types as analytical tools in the study of decidualization and implantation. In *Cellular and molecular aspects of implantation* (ed. S.R., Glasser and D.W. Bullock), pp. 217–39. Plenum Press, New York.

Glasser, S.R., Lampelo, S., Munir M.I., and Julian, J. (1987). Expression of desmin, laminin and fibronectin during *in situ* differentiation (decidualization) of rat uterine stromal cells. *Differentiation*, **35**, 132–42.

Glasser, S.R., Julian, J., Decker, G.L., Tang, JP., and Carson, D.D. (1988). Development of morphological and functional polarity in primary cultures of immature rat uterine epithelial cells. *Journal of Cell Biology*, **107**, 2409–23.

Graf, J., Iwamoto, Y., Sasaki, M., Martin, G.R., Kleinman, H.K., Robey, F.A., and Yamada, Y. (1987). Identification of an amino acid sequence in laminin mediating cell attachment, chemotaxis, and receptor binding. *Cell*, **48**, 989–96.

Gurchot, C. (1975). The trophoblast theory of cancer (John Beard, 1857–1924) revisited. *Oncology*, **31**, 310–33.

Hadley, M.A., Weeks, B.S., Kleinman, H.K., and Dym, M. (1990). Laminin promotes formation of cord-like structures by Sertoli cells *in vitro*. *Developmental Biology*, **140**, 318–27.

Handin, R.I., and Cohen, H.J. (1976). Purification and binding properties of human platelet factor four. *Journal of Biological Chemistry*, **251**, 4272–82.

Hasselaar, P., Loskutoff, D.J., Sawdey, M., and Sage, E.H. (1991). SPARC induces the expression of type I plasminogen activator inhibitor in cultured bovine aortic endothelial cells. *Journal of Biological Chemistry*, **266**, 13178–84.

Herz, J., Clouthier, D.E., and Hammer, R.E. (1992). LDL receptor-related protein internalizes and degrades uPA-PAI-1 complexes and is essential for embryo implantation. *Cell*, **71**, 411–21.

Hohn, H-P., Parker, C.R., Boots, L.R., Denker, H-W., and Hook, M. (1992). Modulation of differentiation markers in human choriocarcinoma cells by extracellular matrix: on the role of a three-dimensional matrix structure. *Differentiation*, **51**, 61–70.

Holland, P., Harper, S., Mcvey, J., and Hogan, B.L.M. (1987). *In vivo* expression of mRNA for the Ca^{++} binding protein SPARC (osteonectin) revealed by *in situ* hybridization. *Journal of Cell Biology*, **105**, 473–82.

Holm, K., Risteli, L., and Sariola, H. (1988). Differential expression of the laminin A and B chains in chimeric kidneys. *Cell Differentiation*, **24**, 223–38.

Howe, C.C., Overton, C.G., Sawick, J., Solter, D., Stein, P., and Strickland, D. (1988). Expression of SPARC/osteonectin transcript in murine embryos and gonads. *Differentiation*, **37**, 20–5.

Humphries, M.J., Olden, K., and Yamada, K.M. (1986). A synthetic peptide from fibronectin inhibits experimental metastasis of murine melanoma cells. *Science*, **233**, 467–70.

Hunt, J.S. (1989). Cytokine networks in the uteroplacental unit: macrophages as pivotal regulatory cells. *Journal of Reproductive Immunology*, **16**, 1–17.

Hunter, D.D., Shah, V., Merlie, J.P., and Sanes, J.R. (1989). A laminin-like adhesive protein concentrated in the synaptic cleft of the neuromuscular junction. *Nature*, **338**, 229–34.

Hynes, R.O. (1992). Integrins: versatility, modulation, and signalling in cell adhesion. *Cell*, **69**, 11–25.

Inaguma, Y., Kusakabe, M., Mackie, E.J., Pearson, L.A., Chiquet-Ehrismann, R., and Sakakura, T. (1988). Epithelial induction of stromal cell tenascin in mouse mammary gland from embryogenesis and carcinogenesis. *Developmental Biology*, **128**, 245–55.

Iruela-Arispe, M.L., Hasselaar, P., and Sage, H. (1991). Differential expression of extracellular proteins is correlated with angiogenesis *in vitro*. *Laboratory Investigation*, **64**, 174–86.

Iwamoto, Y., Robey, F.A., Graf, J., Sasaki, M., Kleinman, H.K., Yamada, Y., and Martin, G.R. (1987). YIGSR, a synthetic laminin pentapeptide, inhibits experimental metastasis formation. *Science*, **238**, 1132–4.

Jacobs, A.L. and Carson, D.D. (1991). Proteoglycan synthesis and metabolism by mouse uterine stroma cultured *in vitro*. *Journal of Biological Chemistry*, **266**, 15464–73.

Jacobs, A.L. and Carson, D.D. (1993). Uterine epithelial cell secretion of interleukin-1α induces prostaglandin E_2 (PGF_2) and $PGF_{2\alpha}$ secretion by uterine stromal cells *in vitro*. *Endocrinology*, **132**, 300–8.

Jacobs, A.L., Sehgal, P.B., Julian, J., and Carson, D.D. (1992). Secretion and hormonal regulation of interleukin-6 production by mouse uterine stromal and polarized epithelial cells cultured *in vitro*. *Endocrinology*, **131**, 1037–46.

Jentoft, N. (1990). Why are proteins O-glycosylated? *Trends in Biochemical Sciences*, **15**, 291–4.

Julian, J., Carson, D.D., and Glasser, S.R. (1992). Polarized rat uterine epithelium *in vitro*: responses to estrogen in defined medium. *Endocrinology*, **130**, 68–78.

Julian, J., Chiquet-Ehrismann, R., Erickson, H.P., and Carson, D.D. (1994). Tenascin is induced at implantation sites in the mouse uterus and interferes with epithelial cell adhesion. *Development*, in press.

Kadokawa, Y., Fuketa, I., Nose, A., Takeichi, M., and Nakatsuji, N. (1989). Expression pattern of E- and P-cadherin in mouse embryos during the periimplantation period. *Development Growth and Differentiation*, **31**, 23–30.

Kallunki, T., Ikonen, J., Chow, L.T., Kallunki, P., and Tryggvason, K. (1991). Structure of the human laminin B2 chain gene reveals extensive divergence from the laminin B1 chain gene. *Journal of Biological Chemistry*, **266**, 221–8.

Kallunki, P., Sainio, K., Eddy, R., Byers, M., Kallunki, T., Sariola, H., *et al.* (1992). A truncated laminin chain homologous to the B2 chain: structure, spatial expression and chromosomal assignment. *Journal of Cell Biology*, **119**, 679–93.

Kearns, M. and Lala, P.K. (1983). Life history of decidual cells: a review. *American Journal of Reproductive Immunology*, **3**, 78–82.

Kimber, S.J. and Lindenberg, S. (1990). Hormonal control of a carbohydrate epitope involved in implantation in mice. *Journal of Reproduction and Fertility*, **89**, 13–21.

Kimber, S.J., Lindenberg, S., and Lundblad, A. (1988). Distribution of some Gal β1-3(4) GlcNAc related carbohydrate antigens on the mouse uterine epithelium in relation to the peri-implantational period. *Journal of Reproductive Immunology*, **12**, 297–313.

Kirby, D.R.S., Potts, D.M., and Wilson, I.B. (1967). On the orientation of the implanting blastocyst. *Journal of Embryology and Experimental Morphology*, **17**, 527–32.

Kisalus, L.L. and Herr, J.C. (1988). Immunocytochemical localization of heparan sulfate proteoglycan in human decidual cell secretory bodies and placental fibrinoid. *Biology of Reproduction*, **39**, 419–30.

Kisalus, L.L., Herr, J.C., and Little, C.D. (1987). Immunolocalization of extracellular matrix proteins and collagen synthesis in first-trimester human decidua. *Anatomical Record*, **218**, 402–15.

Kleinman, H.K., McGarvey, M.L., Liotta, L.A., Robey, P.G., Tryggvason, K., and Martin, G.R. (1982). Isolation and characterization of type IV procollagen, laminin and heparan sulfate proteoglycan from the EHS sarcoma. *Biochemistry*, **21**, 6188–93.

Kleinman, H.K., Cannon, F.B., Laurie, G.W., Hassell, J.R., Aumailley, M., Terranova, V.P. *et al.* (1985). Biological activities of laminin. *Journal of Cellular Biochemistry*, **27**, 317–25.

Kleinman, H.K., McGarvey, M.L., Hassel, J.R., Star, V.L., Cannon, F.B., Laurie, G.W., and Martin, G.R. (1986). Basement membrane complexes with biological activity. *Biochemistry*, **25**, 312–18.

Kliman, H.J. and Feinberg, R.F. (1990). Human trophoblast–extracellular matrix

(ECM) interactions *in vitro*: ECM thickness modulates morphology and proteolytic activity. *Proceedings of the National Academy of Sciences USA*, **87**, 3057–61.

Kubo, H., Spindle, A., and Pedersen, R.A. (1981). Inhibition of mouse blastocyst attachment and outgrowth by protease inhibitors. *Journal of Experimental Zoology*, **216**, 445–51.

Lala, P.K. and Graham, C.H. (1990). Mechanisms of trophoblast invasiveness and their control: the role of proteases and protease inhibitors. *Cancer Metastasis Reviews*, **9**, 369–79.

Lane, T.F. and Sage, E.H. (1990). Functional mapping of SPARC: peptides from two distinct Ca^{++} binding sites modulate cell shape. *Journal of Cell Biology*, **111**, 3065–76.

Lane, T.F., Iruela-Arispe, M.L., and Sage, E.H. (1992). Regulation of gene expression by SPARC during angiogenesis *in vitro*: changes in fibronectin, thrombospondin-1 and plasminogen activator-inhibitor-1. *Journal of Biological Chemistry*, **267**, 16736–45.

Lankat-Buttgereit, B., Mann, K., Deutzmann, R., Timpl, R., and Krieg, T. (1988). Cloning and complete amino acid sequences of human and murine basement protein BM-40 (SPARC, osteonectin). *FEBS Letters*, **236**, 352–6.

Laurie, G.W., Horikoshi, S., Killen, P.D., Sequi-Real, B., and Yamada, Y. (1989). *In situ* hybridization reveals temporal and spatial changes in cellular expression of mRNA for a laminin receptor, laminin, and basement membrane (Type IV) collagen in the developing kidney. *Journal of Cell Biology*, **109**, 1351–62.

Leffler, M. and Barondes, S.H. (1986). Specificity of binding of three soluble rat lung lectins to substituted and unsubstituted mammalian β-galactosides. *Journal of Biological Chemistry*, **261**, 10119–26.

Leivo, I. and Engvall, E. (1988). Merosin, a protein specific for basement membranes of Schwann cells, striated muscle and trophoblast, is expressed late in nerve and muscle development. *Proceedings of the National Academy of Sciences USA*, **85**, 1544–8.

Leivo, I., Vaheri, A., Templ, R., and Wartiovaara, J. (1980). Appearance and distribution of collagens and laminin in the early mouse embryo. *Developmental Biology*, **76**, 100–14.

Lessey, B.A., Damjanovich, L., Coutifaris, C., Castelbaum, A., Albeida, S.M., and Buck, C.A. (1992). Integrin adhesion molecules in the human endometrium: correlation with normal and abnormal menstrual cycle. *Journal of Clinical Investigation*, **90**, 188–95.

Lindenberg, S., Kimber, S.J., and Kallin, E. (1990). Carbohydrate binding properties of mouse embryos. *Journal of Reproduction and Fertility*, **89**, 431–9.

Lindenberg, S., Sundberg, K., Kimber, S.J., and Lundblad, A. (1988). The milk oligosaccharide, lacto-*N*-fucopentaose I, inhibits attachment of mouse blastocysts on endometrial monolayers. *Journal of Reproduction and Fertility*, **83**, 149–58.

Liotta, L.A., Rao, C.N., and Barsky, S.H. (1983). Tumour invasion and the extracellular matrix. *Laboratory Investigation*, **49**, 639–49.

Linnemeyer, P.A. and Pollack, S.B. (1991). Murine granulated metrial gland cells at uterine implantation sites are natural killer lineage cells. *Journal of Immunology*, **147**, 2530–5.

Loke, Y.W., Gardner, L., Burland, K., and King, A. (1989). Laminin in human trophoblast–decidua interaction. *Human Reproduction*, **4**, 457–63.

Luxford, K.A. and Murphy, C.R. (1989). Cytoskeletal alterations in the microvilli of uterine epithelial cells during early pregnancy. *Acta Histochimica*, **87**, 131–6.

Luxford, K.A. and Murphy, C.R. (1992). Changes in the apical microfilaments of rat uterine epithelial cells in response to estrogen and progesterone. *Anatomical Record*, **233**, 521–6.

Mackie, E.J., Chiquet-Ehrismann, R., Pearson, C.A., Inaguma, Y., Taya, K., Kawarada, Y., and Sakakura, T. (1987). Tenascin is a stromal marker for epithelial malignancy in the mammary gland. *Proceedings of the National Academy of Sciences, USA*, **84**, 4621–5.

Maharajan, P. and Maharajan, V. (1986). Behavioural pattern of tumour cells in the mouse uterus. *Gynecologic and Obstetric Investigation*, **21**, 32–9.

Marbaix, E., Donnez, J., Courtoy, P.J., and Eeckhout, Y. (1992). Progesterone regulates the activity of collagenase and related gelatinase A and B in human endometrial explants. *Proceedings of the National Academy of Sciences*, USA, **89**, 11789–93.

Martin, G.R. and Templ, R. (1987). Laminin and other basement membrane components. *Annual Review of Cell Biology*, **3**, 57–85.

Massagué, J. (1990). The transforming growth factor β family. *Annual Review of Cell Biology*, **6**, 597–641.

Mayer, I., Aumailley, M., Mann, K., Timpl, R., and Engel, J. (1991). Calcium-dependent binding of basement membrane protein BM-40 (osteonectin, SPARC) to basement membrane collagen type IV. *European Journal of Biochemistry*, **198**, 141–50.

McCarthy, J.B., Hagen, S.T., and Furcht, L.T. (1986). Human fibronectin contains distinct adhesion- and motility-promoting domains for metastatic melanoma cells. *Journal of Cell Biology*, **102**, 179–88.

Miyauchi, A., Alvarez, J., Greenfield, E.M., Teti, A., Grano, M., Calucci, S., et al. (1991). Recognition of osteopontin and related peptides by an alpha v, beta 3 integrin stimulates immediate cell signals in osteoclasts. *Journal of Biological Chemistry*, **266**, 20369–74.

Morris, J.E. and Potter, S.W. (1990). An *in vitro* model for studying interactions between mouse trophoblast and uterine epithelial cells. A brief review of *in vitro* systems and observations on cell surface changes during blastocyst attachment. *Trophoblast Research*, **4**, 51–69.

Morris, J.E., Potter, S.W., and Gaza-Bulseco, G. (1988). Estradiol-stimulated turnover of heparan sulfate proteoglycan in mouse uterine epithelium. *Journal of Biological Chemistry*, **263**, 4712–18.

Moulton, B.C. and Elangovan, S. (1981). Lysosomal mechanisms in blastocyst implantation and early decidualization. In *Cellular and molecular aspects of implantation*, (ed. S.R. Glasser and D.W., Bullock), pp.335–44. Plenum Press, New York.

Mulholland, J., Aplin, J.D., Ayad, S., Hong, L., and Glasser, S.R. (1992). Loss of collagen type VI from rat endometrial stroma during decidualization. *Biology of Reproduction*, **46**, 1136–43.

Murphy-Ullrich, J.E., Lightner, V.A., Aukhil, I., Yan, Y.Z., Erickson, H.P., and Hook, M. (1991). Focal adhesion integrity is down regulated by the alternatively spliced domain of human tenascin. *Journal of Cell Biology*, **115**, 1127–36.

Myers, D.B., Clark, D.E., and Hurst, P.R. (1990). Decreased collagen concentration in rat uterine implantation sites compared to non-implantation tissue at days 6–11 of pregnancy. *Reproduction and Fertility and Development*, **2**, 607–12.

Nelson, K.G., Takahashi, T., Bossert, N.L., Walmer, D.K., and Mchachlan, J.A. (1991). Epidermal growth factor replaces estrogen in the stimulation of female genital-tract growth and differentiation. *Proceedings of the National Academy of Sciences, USA*, **88**, 21–5.

Nomura, S., Wills, A.J., Edwards, D.R., Heath, J.K., and Hogan, B.L.M. (1988). Developmental expression of 2ar (osteopontin) and SPARC (osteonectin) RNA as revealed by *in situ* hybridization. *Journal of Cell Biology*, **106**, 441–50.

Nose, A. and Takeichi, M. (1986). A novel cadherin cell adhesion molecule: its expression patterns associated with implantation and organogenesis of mouse embryos. *Journal of Cell Biology*, **103**, 2649–58.

Obara, M., Kang, M.S., and Yamada, K.M. (1988). Site directed mutagenesis of the cell-binding domain of human fibronectin: separable, synergistic sites mediate adhesive function. *Cell*, **53**, 649–57.

O'Rourke, K.M., Laherty, C.D., and Dixit, V.M. (1992). Thrombospondin 1 and 2 are expressed as both homo- and heterotrimers. *Journal of Biological Chemistry*, **267**, 24921–4.

O'Shea, K.S., Liu, L.-H.J., Kinnunen, L.H., and Dixit, V.M. (1990). Role of the extracellular matrix protein thrombospondin in the early development of the mouse embryo. *Journal of Cell Biology*, **111**, 2713–23.

Parr, M.B. and Parr, E.L. (1986). Permeability of the primary decidual zone in the rat uterus: Studies using fluorescein-labeled proteins and dextrans. *Biology of Reproduction*, **34**, 393–403.

Parr, M.B. and Parr, E.L. (1989). The implantation reaction. In *Biology of the uterus* (ed. R.M. Wynn, and W.P., Jollie), pp. 233–77. Plenum Press, New York.

Paulson, J.C. and Colley, K.J. (1989). Glycosyltransferases: structure, localization, and control of cell-type-specific glycosylation. *Journal of Biological Chemistry*, **264**, 17615–18.

Paulsson, M. (1992). Basement membrane proteins: structures, assembly, and cellular interactions. *Critical Reviews in Biochemistry and Molecular Biology*, **27**, 93–127.

Pemberton, L., Taylor-Papadimitrou, J., and Gendler, S.J. (1992). Antibodies to the cytoplasmic domain of MUC1 mucin show conservation throughout mammals. *Biochemical and Biophysical Research Communications*, **185**, 167–75.

Perris, R. and Johansson, S. (1987). Amphibian neural crest cell migration on purified extracellular matrix components: a chondroitin sulfate proteoglycan inhibits locomotion on fibronectin substrates. *Journal of Cell Biology*, **105**, 2511–21.

Pikkarainen, T., Kallunki, T., and Tryggvason, K. (1988). Human laminin B2 chain. *Journal of Biological Chemistry*, **263**, 6751–8.

Poirier, F., Timmons, P.M., Chan, G-T.J., Gvenet, J-L., and Rigby, P.W.J. (1992). Expression of the L14 lectin during mouse embryogenesis suggests multiple roles during pre- and post-implantation development. *Development*, **115**, 143–55.

Pollard, J.W. (1990). Regulation of polypeptide growth factor synthesis and growth factor-related gene expression in the rat and mouse uterus before and after implantation. *Journal of Reproduction and Fertility*, **88**, 721–31.

Potts, M. (1966). The attachment phase of ovoimplantation. *American Journal of Obstetrics and Gynecology*, **96**, 1122–8.

Probstmeier, R., Martin, R., and Schachner, M. (1990). Expression of J1/tenascin in the crypt–villus unit of the adult mouse small intestine: implications for its role in epithelial shedding. *Development*, **109**, 313–21.

Psychoyos, A. (1986). Uterine receptivity for nidation. *Annals of the New York Academy of Sciences*, **476**, 36–42.

Raboudi, N., Julian, J., Rohde, L.H., and Carson, D.D. (1992). Identification of cell-surface heparin/heparan sulfate-binding proteins of a human uterine epithelial cell line (RL95). *Journal of Biological Chemistry*, **267**, 11930–9.

Raines, E.W., Lane, T.F., Iruela-Arispe, M.L., Boss, R., and Sage, E.H. (1992). The extracellular glycoprotein SPARC interacts with the platelet derived growth factor (PDGF)-AB and -BB and inhibits the binding of PDGF to its receptors. *Proceedings of the National Academy of Sciences, USA*, **89**, 1281–5.

Rider, V., Carlone, D.L., Witrock, D., Cai, C., and Oliver, N. (1992). Uterine fibronectin mRNA content and localization are modulated during implantation. *Developmental Dynamics*, **195**, 1–14.

Roberts, D.D. and Ginsburg, V. (1988). Sulfated glycolipids and cell adhesion. *Archives of Biochemistry and Biophysics*, **267**, 405–15.

Robertson, W.B., Brosens, L.A., and Dixon, H.G. (1981). Maternal blood supply and fetal growth retardation. In *Fetal growth retardation*. (ed. F.A. Assche, W.B. Robertson, and M. Ranaer), pp. 126–38. Churchill Livingstone, London.

Robertson, S.A., Mayrhofer, G., and Seamark, R.F. (1992). Uterine epithelial cells synthesize granulocyte-macrophage colony-stimulating factor and interleukin-6 in pregnant and non-pregnant mice. *Biology of Reproduction*, **46**, 1069–79.

Rohde, L.H. and Carson, D.D. (1993). Heparin-like glycosaminoglycans participate in binding of a human trophoblastic cell line (JAR) to a human uterine epithelial cell line (RL95). *Journal of Cell Physiology*, **155**, 185–96.

Romagnano, L. and Babiarz, B. (1990). The role of murine cell surface galactosyltransferase in trophoblast laminin interactions *in vitro*. *Developmental Biology*, **141**, 254–61.

Saga, Y., Yagi, T., Ikawa, Y., Sakakura, T., and Aizawa, S. (1992). Mice develop normally without tenascin. *Genes and Development*, **6**, 1821–31.

Sage, E.H. and Bornstein, P. (1991). Extracellular proteins that modulate cell–matrix interactions: SPARC, tenascin and thrombospondin. *Journal of Biological Chemistry*, **266**, 14831–4.

Sage, H., Vernon, R.B., Funk, S.E., Everitt, E.A., and Angello, J. (1989a). SPARC, a secreted protein associated with cellular proliferation inhibits cell spreading *in vitro* and exhibits Ca^{+2}-dependent binding to the extracellular matrix. *Journal of Cell Biology*, **109**, 341–56.

Sage, H., Vernon, R.B., Decker, J., Funk, S., and Iruela-Arispe, M.L. (1989*b*). Distribution of the calcium-binding protein SPARC in tissues of embryonic and adult mice. *Journal of Histochemistry and Cytochemistry*, **37**, 819–29.

Salonen, E.M., Zitting, A., and Vaheri, A. (1984). Laminin interacts with plasminogen and its tissue type activator. *FEBS Letters*, **172**, 29–32.

Sanes, J.R., Engvall, E., Butowski, R., and Hunter, D.D. (1990). Molecular heterogeneity of basal laminae: isoforms of laminin and collagen IV at the neuromuscular junction and elsewhere. *Journal of Cell Biology*, **111**, 1685–99.

Santala, P. and Heino, J. (1991). Regulation of integrin-type cell adhesion receptors by cytokines. *Journal of Biological Chemistry*, **266**, 23505–9.

Sappino, A-P, Huarte, J., Belin, D., and Vassalli, J-D. (1989). Plasminogen activators in tissue remodelling and invasion: mRNA localization in mouse ovaries and implanting embryos. *Journal of Cell Biology*, **109**, 2471–9.

Sasaki, M. and Yamada, Y. (1987). The laminin B2 chain has a multichain structure homologous to the B1 chain. *Journal of Biological Chemistry*, **262**, 17111–17.

Sasaki, M., Kato, S., Kohno, K., Martin, G.R., and Yamada, Y. (1987). Sequence of the cDNA encoding the laminin B1 chain reveals a multi-domain protein containing cysteine-rich repeats. *Proceedings National Academy of Sciences, USA*, **84**, 935–9.

Sasaki, M., Kleinman, H.K., Huhes, H., Deutzman, R., and Y. Yamada. (1988). Laminin, a multi-domain protein. The A chain has a unique globular domain and homology with the basement membrane proteoglycan and laminin B chain. *Journal of Biological Chemistry*, **263**, 16536–44.

Sato, M., Muramatsu, T., and Berger, E.G. (1984). Immunological detection of cell surface galactosyltransferase in preimplantation mouse embryos. *Developmental Biology*, **102**, 514–18.

Schlafke, S. and Enders, A.C. (1975). Cellular basis of interaction between trophoblast and uterus at implantation. *Biology of Reproduction*, **12**, 41–65.

Schlafke, S., Welsh, A.O., and Enders, A.C. (1985). Penetration of the basal lamina of the uterine epithelium during implantation in the rat. *Anatomical Record*, **212**, 47–56.

Schlesinger, M. (1962). Uterus of rodents as sites for manifestation of transplantation immunity against transplantable tumours. *Journal of National Cancer Institute*, **56**, 221–34.

Senior, P.V., Critchley, D.R., Beck, F., Walker, R.A., and Varley, J.M. (1988). The localization of laminin mRNA and protein in the postimplantation embryo and placenta of the mouse: an *in situ* hybridization and immunocytochemical study. *Development*, **104**, 431–46.

Sephel, G.C., Tashiro, K-I., Sasaki, M., Kandel, S., Yamada, Y., and Kleinman, H.K. (1989). A laminin-pepsin fragment with cell attachment and neurite outgrowth activity at distinct sites. *Developmental Biology*, **135**, 172–81.

Sherman, M.I. and Wudl, L.R. (1976). The implanting mouse blastocyst. In *The cell surface in animal development* (ed. G., Poste and G.L., Nicolson), pp. 81–125. North Holland Publishers, Amsterdam.

Shur, B.D. (1984). The receptor function of galactosyltransferase during cellular interactions. *Molecular and Cellular Biochemistry*, **61**, 143–58.

Smith, A.F. and Wilson, I.B. (1974). Cell interaction at the maternal–embryonic interface during implantation in the mouse. *Cell Tissue Research*, **152**, 525–42.

Speeg, K.V., Jr., Azizkan, J.C., and Stromberg, K. (1976). The stimulation by methotrexate of human chorionic gonadotropin and placental alkaline phosphatase in cultured choriocarcinoma cells. *Cancer Research*, **36**, 4570–6.

Stamenkovic, I., Amiot, M., Pesandro, J.M., and Seed, B. (1989). A lymphocyte molecule implicated in lymph node homing is a member of the cartilage link protein family. *Cell*, **56**, 1057–62.

Stephens, R.W., Aumailley, M., Timpl, R., Reisberg, T., Tapiovaara, H., Myohanen, H., *et al.* (1992). Urokinase binding to laminin-nidogen: structural requirements and interactions with heparin. *European Journal of Biochemistry*, **207**, 937–42.

Stewart, I.J. and Mukhtar, D.D.Y. (1988). The killing of mouse trophoblast cells by granulated metrial gland cells *in vitro*. *Placenta*, **9**, 417–25.

Stoolman, L.M. (1989). Adhesion molecules controlling lymphocyte migration. *Cell*, **56**, 907–10.

Sun, X., Mosher, D.F., and Rapraeger, A. (1989). Heparan sulfate mediated binding of epithelial cell surface proteoglycan to thrombospondin. *Journal of Biological Chemistry*, **264**, 2885–9.

Sutherland, A.E., Calarco, P.G., and Damsky, C.H. (1988). Expression and function of cell surface extracellular matrix receptors in mouse blastocyst attachment and outgrowth. *Journal of Cell Biology*, **106**, 1331–48.

Sutherland, A.E., Anderson, R.D., Mayes, M., Seiberg, M., Calarco, P.G., Bernfield, M., and Damsky, C.H. (1991). Expression of syndecan, a putative low affinity fibroblast growth factor receptor, in the early mouse embryo. *Development*, **113**, 339–51.

Svalander, P.C., Odin, P., Nilsson, B.O., and Obrink, B. (1987). Trophectoderm surface expression of the cell adhesion molecule cell-CAM 105 on rat blastocysts. *Development*, **100**, 653–60.

Svalander, P.C., Hjortberg, M., Gronvik, K-O., and Nilsson, B.O. (1989). Mouse blastocyst surface expression of galactose-containing epitopes coinciding with trophoblast differentiation. *Cellular Differentiation and Development*, **26**, 191–200.

Svalander, P.C., Odin, P., Nilsson, B.O., and Obrink, B. (1990). Expression of cell CAM-105 in the apical surface of rat uterine epithelium is controlled by ovarian steroid hormones. *Journal of Reproduction and Fertility*, **88**, 213–21.

Tabibzadeh, S.S. (1991). Human endometrium: an active site of cytokine production and action. *Endocrine Reviews*, **12**, 272–89.

Tabibzadeh, S.S., Santhanasu, U., Seligal, P.B., and May, L.T. (1989). Cytokine-induced production of IFN-β_2/IL-6 by freshly explanted human endometrial cells. *Journal of Immunology*, **142**, 3134–9.

Tachi, C., Tachi, S., Knysznski, A., and Lindner, H.R. (1981). Possible involvement of macrophages in embryo-maternal relationships during ovum implantation in the rat. *Journal of Experimental Zoology*, **217**, 81–92.

Takeichi, M. (1987). Cadherins: a molecular family essential for selective cell–cell adhesion and animal morphogenesis. *Trends in Genetics*, **3**, 213–17.

Takeichi, M. (1991). Cadherin cell adhesion receptors as a morphogenetic regulator. *Science*, **251**, 1451–5.

Tamada, H., McMaster, M.T., Flanders, K.C., Andrews, G.K., and Day, S.K. (1990). Cell type-specific expression of transforming growth factor-β1 in the mouse uterus during the peri-implantation period. *Molecular Endocrinology*, **4**, 965–72.

Tang, J-P. and Carson, D.D. (1989). Estrogen induces N-linked glycoprotein expression by immature mouse uterine epithelial cells. *Biochemistry*, **28**, 8116–23.

Tang, J-P., Julian, J., Glasser, S.R., and Carson, D.D. (1987). Heparan sulfate proteoglycan synthesis and metabolism by mouse uterine epithelial cells *in vitro*. *Journal of Biological Chemistry*, **262**, 12832–42.

Taraboletti, G., Roberts, D., Liotta, L.A., and Giavazzi, R. (1990). Platelet thrombospondin modulates endothelial cell adhesion, motility and growth: a potential angiogenesis regulatory factor. *Journal of Cell Biology*, **111**, 765–72.

Tashiro, K-I, Sephel, G.C., Weeks, B., Sasaki, M., Martin, G.R., Kleinman, H.K., and Yamada, Y. (1989). A synthetic peptide containing the IKVAV sequence from the A chain of laminin mediates cell attachment, migration, and neurite outgrowth. *Journal of Biological Chemistry*, **264**, 16174–82.

Tienari, J., Lehtonen, E., Vartio, T., and Virtanes, I. (1989). Embryonal carcinoma cells adhere preferentially to fibronectin and laminin but their endodermal differentiation leads to a reduced adherence to laminin. *Experimental Cell Research*, **182**, 26–32.

Timpl, R. (1989). Review: Structure and biological activity of basement membrane proteins. *European Journal of Biochemistry*, **1804**, 87–502.

Tung, H-N., Parr, E.L., and Parr, M.B. (1988). Endocytosis in the uterine luminal and glandular epithelial cells of mice during early pregnancy. *American Journal of Anatomy*, **182**, 120–9.

Turpeenniemi-Hujanen, T., Thorgeirsson, U.P., Rao, C.N., and Liotta, L.A. (1986). Laminin increases the release of type IV collagenase from malignant cells. *Journal of Biological Chemistry*, **261**, 1883–9.

Turpeenniemi-Hujanen, T., Ronnberg, L., Kauppila, A., and Puistola, U. (1992). Laminin in the human embryo implantation: analogy to invasion by malignant cells. *Fertility and Sterility*, **58**, 105–13.

Tuszynski, G.P., Rothman, V., Murphy, A., Siegler, K., Smith, L., Karczewski, J., and Knudsen, K.A. (1987). Thrombospondin promotes cell-substratum adhesion. *Science*, **236**, 1570–3.

Valdizan, M.C., Julian, J., and Carson, D.D. (1992). WGA-binding, mucin glycoproteins protect the apical cell surface of mouse uterine epithelial cells. *Journal of Cell Physiology*, **151**, 451–65.

Van Blerkom, J. and Chavez, D.J. (1981). Morphodynamics of outgrowths of mouse trophoblast in the presence and absence of a monolayer of uterine epithelium. *American Journal of Anatomy*, **162**, 143–55.

Vestweber, D., Gossler, A., Boller, K., and Kemler, R. (1987). Expression and distribution of cell adhesion molecule uvomorulin in mouse preimplantation embryos. *Developmental Biology*, **124**, 451–6.

Vlodavsky, I., Korner, G., Ishai-Michaeli, R., Bashkin, P., Bar-Shavit, R., and

Fuks, Z. (1990). Extracellular matrix-resident growth factors and enzymes: possible involvement in tumour metastasis and angiogenesis. *Cancer Metastasis Reviews*, **9**, 203–26.

Vollmer, G., Siegel, G.P., Chiquet-Ehrismann, R., Lightner, V.A., Arnholdt, H., and Knuppen, R. (1990). Tenascin expression in the human endometrium and endometrial adenocarcinomas. *Laboratory Investigations*, **62**, 725–30.

Vukicevic, S., Luyten, F.P., Kleinman, H.K., and Reddi, A.H. (1990). Differentiation of canalicular cell processes in bone cells by basement membrane matrix components: regulation by discrete domains of laminin. *Cell*, **63**, 437–45.

Vukicevic, S., Kleinman, H.K., Luyten, F.P., Roberts, A.B., Roche, N.S., and Reddi, A.H. (1992). Identification of multiple active growth factors in basement membrane matrigel suggests caution in interpretation of cellular activity related to extracellular matrix components. *Experimental Cell Research*, **202**, 1–8.

Vuolteenaho, R., Chow, L.T., and Tryggvason, K. (1990). Structure of the human laminin B1 chain gene. *Journal of Biological Chemistry*, **265**, 15611–16.

Wassarman, P.M. (1987). Early events in mammalian fertilization. *Annual Reviews of Cell Biology*, **3**, 109–42.

Wegner, C.C. and Carson, D.D. (1992). Mouse uterine stromal cells secrete a 30-kilodalton protein in response to coculture with uterine epithelial cells. *Endocrinology*, **131**, 2565–72.

Welsh, A.O. and Enders, A.C. (1991). Chorioallantoic placenta formation in the rat II: Angiogenesis and maternal blood circulation in the mesometrial region of the implantation chamber prior to placenta formation. *American Journal of Anatomy*, **192**, 347–65.

Wewer, U.M., Faber, M., Liotta, L.A., and Albrechtsen, R. (1985). Immunochemical and ultrastructural assessment of the nature of the pericellular basement membrane of human decidual cells. *Laboratory Investments*, **53**, 624–33.

Wewer, U.M., Damjanov, A., Weiss, J., Liotta, L.A., and Damjanov, I. (1986). Mouse endometrial stromal cells produce basement membrane components. *Differentiation*, **32**, 49–58.

Wewer, U.M., Albrechtsen, R., Fisher, L.W., Young, M.F., and Termine, J.D. (1988). Osteonectin/SPARC/BM-40 in human decidua and carcinoma, tissues characterized by *de novo* formation of basement membrane. *American Journal of Pathology*, **132**, 345–55.

Wewer, U.M., Engvall, E., Paulsson, M., Yamada, Y., and Albrechtsen, R. (1992). Laminin A, B1, B2, S and M subunits in the postnatal rat liver development and after partial hepatectomy. *Laboratory Investigation*, **66**, 378–89.

Wilson, O., Jacobs, A.L., Stewart, S., and Carson, D.D. (1990). Expression of externally-disposed heparin/heparan sulfate binding sites by uterine epithelial cells. *Journal of Cell Physiology*, **143**, 60–7.

Wilson, I.B. and Potts, D.M. (1970). Melanoma invasion in the mouse uterus. *Journal of Reproduction and Fertility*, **22**, 429–34.

Wrana, J.L., Overall, C.M., and Sodek, J. (1991). Regulation of the expression of a secreted acidic protein rich in cysteine (SPARC) in human fibroblasts by transforming growth factor β: comparison of transcriptional and posttranscriptional control with fibronectin and type I collagen. *European Journal of Biochemistry*, **197,** 519–28.

Wu, T-C., Wan, Y-J., Chung, A.E., and Damjanov, I. (1983). Immunohisto-chemical localization of entactin and laminin in mouse embryos and fetuses. *Developmental Biology*, **100,** 496–505.

Yagel, S., Parhar, R.S., Jeffrey, J.J., and Lala, P.K. (1988). Normal nonmetastatic human trophoblast cells share *in vitro* invasive properties of malignant cells. *Journal of Cellular Physiology*, **135,** 455–62.

Yamada, K.M. (1985). Fibronectin and other structural proteins. In *Cell biology of the extracellular matrix* (ed. Hay, E.D. Hay, pp. 95–114. Plenum Press, New York.

Yamagata, M., Suyuki, S., Akiyama, S.K., Yamada, K.M., and Kimata, K. (1989). Regulation of cell–substrate adhesion by proteoglycans immobilized on extracellular substrates. *Journal of Biological Chemistry*, **264,** 8012–18.

Yurchenco, P.D. and Schittny, J.C. (1990). Molecular architecture of basement membranes. *FASEB Journal*, **4,** 1577–90.

Zabrenetzky, V.S., Kohn, E.C., and Roberts, D.D. (1990). Suramin inhibits laminin- and thrombospondin- mediated melanoma cell adhesion, migration and binding of these adhesive proteins to sulfatide. *Cancer Research*, **50,** 5937–42.

Zhu, B.C-R., Fisher, S-F., Pande, H., Calaycay, J., Shively, J.E., and Laine, R.A. (1984). Human placental (fetal) fibronectin: increased glycosylation and higher protease resistance than plasma fibronectin. *Journal of Biological Chemistry*, **259,** 3962–70.

4 Anti-Müllerian hormone: a masculinizing relative of TGF-β

NATHALIE JOSSO

I   INTRODUCTION

Anti-Müllerian hormone (AMH) produces regression of Müllerian ducts, the primordia for the uterus and Fallopian tubes, in the first step leading to male sex differentiation of the reproductive tract. At the beginning of the century, embryologists believed that testosterone, produced by fetal Leydig cells (Bouin and Ancel 1903), was responsible both for the development of male-specific attributes and for the disappearance of female-specific ones. By using fetal surgery, however, Alfred Jost (1953) demonstrated that crystallized testosterone, grafted near the fetal ovary, does not prevent Müllerian duct development, whereas a fragment of fetal testicular tissue leads to Müllerian regression in the vicinity of the graft. Thus the 'positive' aspects of male sex differentiation–development of vas deferens, seminal vesicles, prostate and penis — are androgen-dependent, while the 'negative' ones — repression of female reproductive organs — are under the control of another testicular product. The latter, AMH, also called Müllerian-inhibiting substance (MIS) or factor (MIF), was later identified as a glycoprotein (Picard *et al.* 1978) synthesized by fetal Sertoli cells (Blanchard and Josso 1974; Fig. 4.1).

II   AMH PURIFICATION AND CLONING

1   Protein purification

Partial purification of AMH was achieved by Budzik *et al.* (1983) using dye affinity chromatography. Bovine AMH was purified to homogeneity by immunoaffinity chromatography on a monoclonal antibody (Picard and Josso 1984), using incubation medium from fetal calf testicular tissue as starting material. Purified bovine AMH contains 13.5% carbohydrate, with both O-and N-oligosaccharide linkages, and a high proportion of hydrophobic amino acids (Picard *et al.* 1986*b*).

Recombinant DNA techniques are now preferentially used for AMH production. Chinese hamster ovary (CHO) cells are transfected with a pSV2 vector carrying the human AMH gene under the control of a viral promoter. Mature AMH is purified from the culture medium by a combination of ion-exchange, lentil-lectin and immunoaffinity chromatography on a monoclonal antibody (Wallen *et al.* 1989). Its electrophoretic pattern on polyacrylamide gels in the presence of sodium dodecyl sulphate (SDS-PAGE) is shown on Fig. 4.2. Under non-reducing conditions, in addition to a 140 kD moiety, other higher molecular weight polymers are present. These larger forms were also obtained after pulse-labelling with radioactive cysteine followed by direct analysis by SDS-PAGE, indicating

MALE DIFFERENTIATION OF GENITAL TRACT

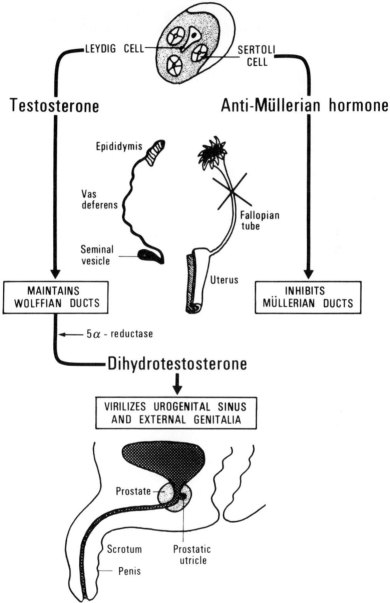

Fig. 4.1   Diagram of normal male sex differentiation. Testosterone, produced by Leydig cells, maintains the Wolffian ducts and, after reduction by 5α-reductase, virilizes the urogenital sinus and external genitalia. AMH, produced by Sertoli cells, inhibits the development of Müllerian derivatives, which otherwise would develop into uterus and tubes. From Josso, (1981) with permission.

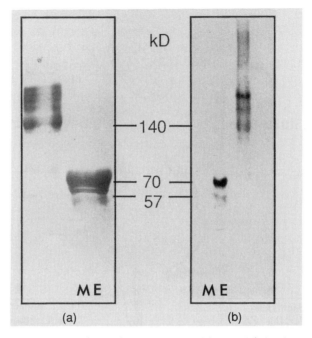

Fig. 4.2 PAGE of AMH (4–30% gradient gel, Pharmacia) in the presence of
SDS. (a) AMH purified from bovine fetal testes, 3 μg/ml. (b) Recombinant
human AMH in the presence (ME) or the absence of mercaptoethanol. In
non-reducing conditions (no mercaptoethanol), AMH migrates as a 140 kD dimer,
and higher molecular weight polymers. After reduction of disulphide bonds, a 70
kD monomer is seen, along with a fainter 57 kD band, representing the N-terminus
(see Fig. 4.6).

that they were not generated by the purification procedure (Pepinsky *et al.*
1988). Following cleavage of disulfide bonds by a reducing agent, the 140
kD recombinant dimer dissociates into 70 kD subunits. A minor species of
57 kD, also observed in native bovine AMH (Picard and Josso 1984), is
generated by post-translational processing (Pepinsky *et al.* 1988).

2 Cloning and sequencing of the AMH gene

Bovine AMH complementary DNA (cDNA) was cloned independently
by two laboratories, using different techniques. Picard *et al.* (1986*a*) took
advantage of the fact that, having purified bovine AMH to homogeneity
(Picard and Josso 1984), a specific polyclonal antibody against the native
protein was available to them. They used this polyclonal antibody to screen
a λgt 11 expression library constructed from fetal calf testicular cDNA,

Cate *et al.* (1986) electroeluted and sequenced AMH from partially purified bovine AMH run on a polyacrylamide gel, and cloned the bovine cDNA by screening a calf testicular library with degenerate probes. Reduction of probe degeneracy was obtained by Northern blot analysis; subpools recognizing only RNA from immature testicular tissue were split further and subjected to the same analysis. Probing of human genomic libraries with bovine cDNA probes easily yielded human clones due to the high homology between the two species. A bovine cDNA probe was also successful in isolating the rat gene (Haqq *et al.* 1992). The mouse gene was cloned by Münsterberg and Lovell-Badge (1991).

The AMH gene is relatively short, with a total of 2.75 kb in five exons. Overall, the gene is characterized by a surprisingly high guanine–cytosine (GC) content: 72% in the exons and 68% in the introns and flanking regions. The fifth exon, the most interesting because it exhibits the greatest interspecies homology and is the only part of the AMH gene to show homology to the transforming growth factor (TGF)-β family (Cate *et al.* 1986), contains up to 80% GC bases.

In contrast to the promoters of the AMH gene in animal species, which contain canonical or at least almost perfect TATA boxes, the human promoter has only a degenerate (TTAA) site directing the initiation of transcription, and exhibits several transcriptional initiation sites (Guerrier *et al.* 1990). Genes lacking a TATA box, as the so-called 'housekeeping' genes do, often have multiple transcription sites, but are also ubiquitously expressed. In contrast, the expression of the AMH gene is restricted to somatic gonadal cells, although the *cis*–regulating sequences have not been identified at the present time. Target sequences for the oestrogen receptor (Guerrier *et al.* 1990) and for the HMG motif contained in the SRY gene product (Nasrin *et al.* 1991; Harley *et al.* 1992) are present in the upstream flanking sequences.

Although involved in male sex differentiation, the AMH gene is not located on a sex chromosome, in keeping with the tenet that only the genetic trigger for gonadal sex determination needs to be Y-specific. The human AMH gene has been mapped to the tip of the short arm of chromosome 19 (Cohen-Haguenauer *et al.* 1987), and the mouse gene to the distal region of chromosome 10, between the phenylhydroxylase and the zinc finger autosomal genes (King *et al.* 1991).

III   AMH EXPRESSION AND ONTOGENY

AMH is produced exclusively by gonadal somatic cells in both sexes. However, there are major differences: immature Sertoli cells synthesize high amounts which accumulate in the rough endoplasmic reticulum (Tran and Josso 1982; Hayashi *et al.* 1984) and are slowly released from the cell

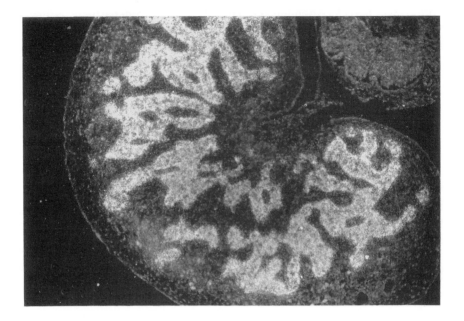

Fig. 4.3   *In situ* hybridization with an antisense riboprobe from the fifth exon of the AMH gene in testicular tissue from an 11-week-old human fetus. The seminiferous tubules contain abundant AMH transcripts. Dark field illumination, ×95.

(Vigier *et al.* 1985), while the granulosa cells synthesize the hormone only after birth and in low quantities (Vigier *et al.* 1984*a*).

1   Testicular production

In the testis, AMH is produced only by Sertoli cells, in contrast to inhibin and activin which are also synthesized by interstitial cells (Lee *et al.* 1989; Roberts *et al.* 1989; Shaha *et al.* 1989). In the fetus, AMH is the earliest cognate protein known to be expressed by developing Sertoli cells. Testicular anti-Müllerian activity is expressed at the time of formation of seminiferous tubules (Picon 1970; Tran *et al.* 1977; Vigier *et al.* 1983; Fig. 4.3). No AMH expression was found in the germ cells, outside the testis cords, or in any other fetal tissue. In mice expressing the *W* locus, in which the germ cells fail to reach and colonize the gonadal ridge, AMH expression is normal, indicating that it is not dependent upon interaction with germ cells (Münsterberg and Lovell-Badge 1991). AMH levels in fetal serum have been measured in the bovine by radioimmunoassay (Vigier *et al.* 1983) and more recently in humans by enzyme-linked immunoassay (Josso *et al.* 1993*c*). High levels were found in the serum of male fetuses at all ages. No AMH was detected in female sera, nor in unconcentrated amniotic fluid of either sex.

After birth, Sertoli cells continue to express high levels of AMH, then expression progressively decreases and drops sharply at pubertal maturation (Baker *et al.* 1990; Hudson *et al.* 1990; Josso *et al.* 1990; Rey *et al.* 1993). Nevertheless, low AMH concentrations are found in human adult serum (Hudson *et al.* 1990) and in bovine rete testis fluid (Vigier *et al.* 1983), indicating that the gene is not totally repressed in postpubertal Sertoli cells. In the rat testis Lee *et al.* (1992) have detected two discrete AMH transcripts, differing by the extent of polyadenylation, and report that the smaller one is more abundant postnatally. They suggest that differential polyadenylation may be a means of post-transcriptional regulation of AMH protein expression.

2   Ovarian production

Granulosa cells of the ovary and testicular Sertoli cells share many structural and functional characteristics. The first indication of ovarian AMH production was the demonstration by Hutson *et al.* (1981) of anti-Müllerian activity in ovarian tissue of hens. Follicular fluid extracted from small cow follicles contains significant concentrations of AMH, comparable with the levels found in the circulation of fetal males (Vigier *et al.* 1984*a*; Necklaws *et al.* 1986). Production of AMH in the mammalian ovary was localized to granulosa cells by Vigier *et al.* (1984*a*), but is restricted to postnatal developing follicles (Takahashi *et al.* 1986*a*; Bézard *et al.* 1987; Ueno *et al.*, 1989*a*). No AMH can be detected in fetal ovaries in the rat (Ueno *et al.* 1989*b*), where follicles form only after birth, or in the ewe, where growing follicles with two to six cell deep granulosa cell layers are present already at 120 days gestation (Bézard *et al.* 1987). In the mouse ovary, AMH transcripts, studied by *in situ* hybridization, are first detectable 6 days after birth (Münsterberg and Lovell-Badge 1991).

Ovarian AMH expression depends on the degree of follicular maturation rather than the age of the animal. As the follicle grows, the AMH immunoreactivity of granulosa cells close to the basal lamina progressively fades, while that of those lining the antrum and occupying the cumulus oophorus increases (Bézard *et al.* 1987; Ueno *et al.*, 1989*a*). AMH production appears confined to the stem cell population of granulosa cells, which is incapable of undergoing terminal differentiation under follicle–stimulating hormone (FSH) stimulation (Erickson 1983). Just before ovulation, when the oocyte completes the first meiotic division, AMH immunoreactivity (Ueno *et al.* 1989*a*) and transcripts (Münsterberg and Lovell-Badge 1991) become undetectable in cumulus cells.

Whatever the degree of follicular maturation or the age of the animal, the production of AMH by granulosa cells is low compared with that of immature Sertoli cells, as shown by Northern blots of messenger RNA extracted from gonads of both sexes (Picard *et al.* 1986*a*), and by differences in anti-Müllerian activity (Ueno *et al.* 1989*b*). In contrast, when tested at similar concentrations, AMH purified from either fetal testis or adult ovarian follicles has similar anti-Müllerian activity (Vigier *et al.* 1984*a*). Transcription initiation sites are identical in both tissues (Josso *et al.* 1993*a*), suggesting that the same promoter drives the gene in the testis and in the ovary.

IV   AMH REGULATION

1   AMH expression *in vitro*

Because AMH expression is so tightly regulated, both spatially and chrono-logically, the scarcity of information on the regulatory mechanisms involved is particularly galling. The cloning of the AMH promoter (Guerrier *et al.* 1990) has not significantly improved the situation because of the lack of a suitable cell model to test the influence of flanking sequences on gene expression. Bovine Sertoli cells rapidly cease to produce AMH when placed in culture (Vigier *et al.* 1985), even when they are obtained from immature testicular tissue, which would normally continue to produce AMH for several months *in vivo*. Extinction of AMH expression is not due to cell loss or damage, since after 3 days in culture, the amount of DNA remains stable and the Sertoli cells increase their cyclic adenosine monophosphate (AMP) production 10-fold in response to FSH stimulation. AMH expression can be reactivated by pharmacological doses of cAMP (Voutilainen and Miller 1987). Addition of FSH, testosterone (Vigier *et al.* 1985), oestrogen, progesterone, gonadotrophin-releasing hormone luteinizing hormone (LH) (LaQuaglia *et al.*, 1986) or human chorionic gonadotrophin (Voutilainen and Miller 1987) fails to restore testicular AMH expression. Cell interactions may perhaps be involved, since human fetal testicular fragments in organ culture retain anti-Müllerian activity up to 1 week after explantation (Josso 1974).

2   Hormonal regulation

Because AMH production and pubertal maturation are inversely corre-lated, the possibility that hypothalamic factors might play a role in AMH regulation has been entertained. FSH seemed a promising can-didate, in view of its pleiotropic actions upon Sertoli cells. Bercu *et al.* (1978) reported that intraperitoneal injection of anti-gonadotophin-releasing hormone serum into pregnant rats resulted in increased testicular anti-Müllerian activity in the progeny, but also observed a marked decrease in anti-Müllerian activity of the pups of females injected with normal serum, compared with controls born of non-injected mothers (Donahoe *et al.* 1976), making interpretation difficult. According to Kuroda *et al.* (1990), FSH, but not LH, treatment of newborn rats decreases the level of testicular AMH transcripts on Northern blots. They suggest that the drop in AMH production observed in the rat after birth is due to a rise in FSH. In the human, both FSH and LH serum levels are elevated during the first 3 months of life (Brook and Grumbach 1989), yet Sertoli cells produce significant amounts of AMH, though perhaps somewhat less than in older infants (Baker *et al.* 1990; Hudson *et al.* 1990). The difference is not easily documented, because venipuncture is seldom warranted in normal neonates.

An inverse correlation between serum levels of AMH and testosterone exists during normal and precocious pubertal development (Rey *et al.* 1993): AMH levels were uniformly low in subjects with testosterone levels over 2 ng^{-1}, except in androgen-insensitive ones, but including those with gonadotrophin-independent sexual precocity. Appropriate treatment restored AMH to prepubertal levels after 3–6 months, suggesting that repression of AMH production is the consequence of androgen-induced, reversible, Sertoli cell maturation.

V   BIOLOGICAL EFFECTS

1   Effects upon female genital primordia

i   *Anti-Müllerian activity*

AMH is known primarily for its eponymous function, namely the inhibition of Müllerian duct derivatives in the male fetus (Jost 1953). Male and female Müllerian ducts are equally sensitive to AMH action. However, in both cases the duct must be exposed to the hormone before the end of the 'critical period' of responsiveness, i.e. 15 days post-coitum in the rat (Picon 1969) and 8 weeks after the last menstrual period for human fetuses (Josso *et al.* 1977). After this window, the ducts no longer regress after AMH exposure. Histologically, Müllerian duct regression is characterized by the development of complex epithelial-mesenchymal interaction (Dyche 1979; Trelstad *et al.* 1982) perhaps triggered by the dissolution of the basement membrane. Epithelial cells become reoriented and extend into the mesenchyme, which condenses into a characteristic periductal ring. To determine whether AMH acts directly upon the Müllerian epithelium or via the adjacent mesenchyme, these tissues were isolated from $15^{1}/_2$-day-old fetal rat reproductive tracts and their ^3H thymidine labelling index was measured after 3 days in culture in the presence of AMH (Tsuji *et al.* 1992). The percentage of labelled cells was significantly decreased in the mesenchymal, but not the epithelial, cells, suggesting that the mesenchyme is the primary target for AMH. The effect of AMH upon the cell components of Müllerian ducts is rather similar to that of TGF-β, which is known to stimulate the growth of fibroblastic cells and promote the production of fibronectin, while acting as a potent growth inhibitor of epithelial cells (see Wakefield *et al.* 1990, for review). It is unlikely, however, that TGF-β could mediate the effect of AMH upon the Müllerian duct, since it is inactive in this system (Tran, unpublished).

The anti-Müllerian effect of AMH has been used to develop a bioassay, based upon the histological aspect of serially sectioned $14^{1/2}$-day-old rat fetal Müllerian ducts, exposed to AMH during 3 days in organ culture (Picon 1969). The test is quite reliable, and is not limited by species-specificity, but it is time-consuming, relatively insensitive and only semiquantitative.

ii Morphological masculinization of the fetal ovary

Jost and his associates were the first to suspect that AMH could be responsible for the ovarian stunting and masculinization observed in freemartins, bovine heterosexual twins united by placental anastomoses. Ovarian growth is arrested at the time the Müllerian duct regress in the freemartin and her male twin (Jost *et al.* 1972). AMH levels are correlated in both twins (Vigier *et al.* 1984*b*). To test the effect of AMH upon the fetal ovary, Vigier *et al.* (1987) cultured $14^{1/2}$-day-old rat fetal ovaries in the presence of purified bovine AMH. Treated ovaries were stunted, depleted of germ cells and contained cord-like structures resembling seminiferous tubules (Fig. 4.4). Similar lesions were observed in the ovaries of transgenic mice expressing the human AMH gene under the control of the metallothionein promoter (Behringer *et al.* 1990).

iii Sex-reversal of the ovarian steroidogenic pathway: the anti-aromatase test

The effects of AMH upon the fetal ovary are not limited to morphological masculinization of the gonadal blastema, they also extend to its endocrine function. Because of its high aromatase activity, the ovary synthesizes oestrogens from C_{19} precursors, which, in the testis, are metabolized to testosterone. AMH inhibits the synthesis of the enzyme $P450_{arom}$,' thereby reversing the steroidogenic pathway of fetal ovaries and causing them to secrete testosterone instead of oestradiol, as shown in sheep fetal gonads by Vigier *et al.* (1989). In the rat ovary, spontaneous aromatase activity is low up to birth, but can be stimulated by cAMP (Picon *et al.* 1985). This has allowed us to develop a quantitative AMH bioassay using fetal rat ovaries as target organs, which is known as the anti-aromatase test (di Clemente *et al.* 1992).

Briefly, 16-day-old rat fetal ovaries are explanted 3 days in organ culture in medium containing 1 mM dibutyryl cAMP (Bt_2cAMP). Aromatase

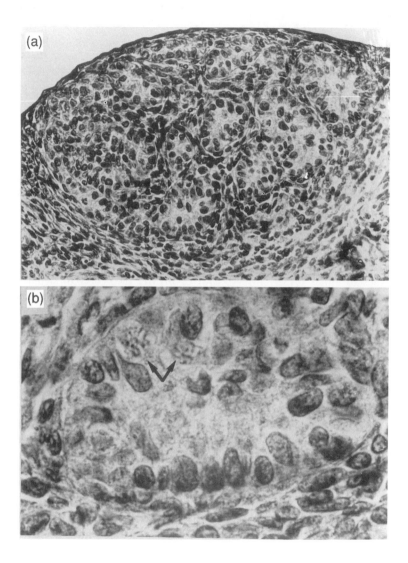

Fig. 4.4   Formation of seminiferous tubule-like structures in the ovary of a 14.5 day-old fetal rat, exposed to 5 μg/ml of bovine AMH. (a) Seminiferous tubules have developed in the ovarian parenchyma (×260). (b) At a higher magnification (×900), two surviving germ cells are seen, entering meîotic prophase. From Josso *et al.* (1993b), with permission.

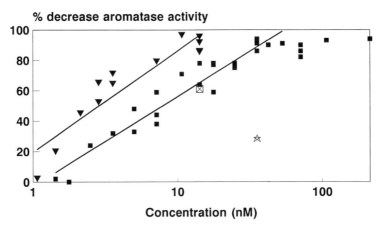

Fig. 4.5   The anti-aromatase test, a quantitative bioassay for AMH, based upon the inhibition of induction of aromatase activity of rat fetal ovaries. Inhibition of induction of aromatase activity of ovaries treated with cAMP and either bAMH or hAMH was compared with the aromatase activity of controlateral ovaries cultured only with cAMP. Each point represents the mean of triplicate experiments. Inhibition of induction of aromatase activity was proportional to the logarithm of the concentration of AMH in the culture medium. Monoclonal antibodies to either bAMH (MoAb 278) or hAMH (MoAb 10.6) significantly decreased hormone bioactivity. bAMH, ■; hAMH, ▼; bAMH + MoAb 278, ☆; hAMH + MoAb 10.6, ×. From di Clemente *et al.* (1992), with permission.

activity is measured at the end of the culture period by the tritiated water technique (Ackerman *et al.* 1981). Results of triplicate experiments are expressed as the percentage of decrease of aromatase activity in explants exposed to both Bt_2cAMP and AMH, compared with those receiving only Bt_2cAMP. Since a decrease in aromatase activity can also result from non-specific tissue injury, it is essential that suitable controls be used. A linear log/dose–response to AMH treatment was demonstrated between 1.5 and 30 nM of bovine AMH (bAMH; $r = 0.964$, $p < 0.01$) and between 1 and 10 nM for human AMH (hAMH; $r = 0.918$, $P < 0.01$). Intra-assay and interassay variations were, respectively, 12.3% ($n=3$) and 14.6% ($n=3$) for bAMH and 12.2% ($n=3$) and 19.2% ($n=4$) for hAMH. Monoclonal antibodies to bAMH and hAMH decreased the bioactivity of their respective antigens (Fig. 4.5).

2   Mapping the AMH bioactive site

The availability of a quantitative bioassay for AMH has allowed mapping of the bioactive site. A member of the TGF-β family, AMH is cleaved

by plasmin at a monobasic site (arginine427/serine428), generating a 25
kD C-terminal fragment with homology to the TGF-β active molecule
(Pepinsky *et al.* 1988). To map the AMH bioactive site, the C-and N-termini
of recombinant hAMH were purified by gel filtration in the presence of 1
M acetic acid (Fig. 4.6) and tested separately in two independent bioassays
(Wilson *et al.* 1993). The effect of AMH proteolytic fragments in the fetal
ovary aromatase assay is shown in Fig. 4.7: the C-terminus decreased

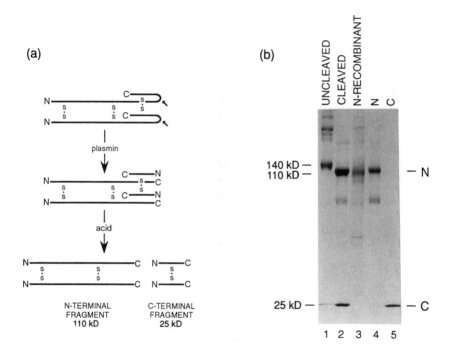

Fig. 4.6 Isolation of C- and N-terminal fragments of AMH. The diagram
shows full-length AMH (the small arrows indicate the plasmin cleavage site
between arginine 427 and serine 428), the non-covalently associated fragments of
plasmin-cleaved AMH, and the acid-dissociated fragments. The S–S designation
indicates that the N- and C-terminal dimers are disulphide-linked. (b) PAGE in the
presence of SDS on a 4–20% gradient gel of purified human AMH fragments.
Uncleaved: recombinant full-length AMH. Cleaved: plasmin-cleaved AMH, the
covalently-linked fragment has dissociated during electrophoresis. N-recombinant:
N-recombinant fragment, produced by a mutant cDNA with a stop codon in place
of the normal codon at serine 428. N and C fragments were purified from
plasmin-cleaved AMH by gel filtration in 1 M acetic acid and resuspended in
1 mM HCl to prevent aggregation. From Wilson *et al.* (1993), with permission.

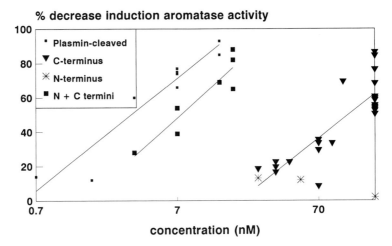

Fig. 4.7   Effect of AMH fragments in the anti-aromatase test. Fetal ovaries, treated with cAMP to induce aromatase activity were incubated in the presence or absence of AMH proteolytic fragments. See legend to Fig. 4.5 for details. The bioactivity of the C-terminus is greatly enhanced by addition of the N-terminus. From Wilson *et al.* (1993), with permission.

the aromatase activity of rat fetal ovaries in a dose-dependent fashion, but exhibited only 3% of the activity of the full-length, plasmin-cleaved molecule, unless the N-terminus was added back. Alone, the N-terminus had no significant activity. Similar results were obtained using the Müllerian duct assay (Fig. 4.8).

These data demonstrate that, as shown for other members of the TGF-β family, only the AMH C-terminus is endowed with bioactivity. The N-terminus, however, enhances the bioactivity of the AMH C-terminus. This is most unusual: the TGF-β proregion reimposes latency on the C-terminus (Gentry and Nash 1990) and dissociation is required for restoration of bioactivity. Our results differ from those of MacLaughlin *et al.* (1992), who observed no potentiating influence of the N-terminus, but the electrophoretic pattern of the proteolytic fractions obtained by these investigators clearly shows contamination of the C-terminus by the N-terminus. Wilson *et al.* (1993) have shown that C-terminal potentiation is accompanied by non-covalent reassociation. The C-terminal domain is highly hydrophobic and undergoes aggregation at neutral pH. Association with the N-terminus probably affects the conformation of the C-terminus, causing it to fold properly and helping to maintain it in solution, as has been demonstrated for activin (Gray and Mason 1990).

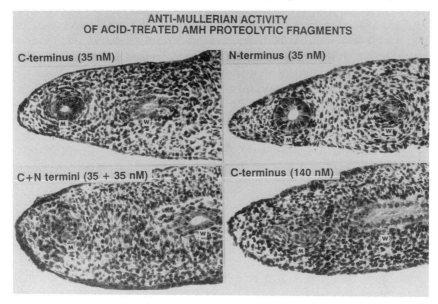

Fig. 4.8   Effect of AMH fragments in the Müllerian duct assay (Picon 1969). Anterior portion of 14^{1}/$_{2}$-day-old rat fetal reproductive tracts, cultured 3 days in the absence of fetal gonads and in the presence of AMH fragments at the indicated concentrations. Addition of N-terminus strongly enhanced the bioactivity of the C-terminus, up to the level of the plasmin-cleaved molecule. From Wilson *et al.* (1993), with permission.

3   Other proposed effects

i   Effect on testicular differentiation

Given the dramatic ovarian virilization produced by AMH, several investigators (Vigier *et al.* 1989; Behringer *et al.* 1990) have proposed that AMH might play a role in testicular differentiation. Repression of constitutive oestrogen production is believed to be crucial for testicular differentiation to occur (Dorizzi *et al.* 1991). Testicular tissue is normal in patients with the persistent Müllerian duct syndrome due to AMH gene mutations, indicating that AMH is not required for testicular development in humans; however, this does not formally rule out an auxiliary role for AMH in this process, since functionally redundant pathways may coexist in critical developmental processes (Shull *et al.* 1992).

ii   Effect on testicular descent

Hutson and Donahoe (1986) have produced circumstantial evidence for a role of AMH in testicular descent. Their contention is based upon the following arguments: testicular factors other than testosterone are involved in the first phase of testicular descent, patients with retained

Müllerian derivatives often suffer from cryptorchidism, and cryptorchid testes have lower anti-Müllerian activity than scrotal ones (Donahoe *et al*. 1977). None of these arguments are final. The mechanical restraint can explain failure of testicular descent in males with retained Müllerian ducts (Guerrier *et al*. 1989), cryptorchid testes are more likely to exhibit impaired Sertoli cell function than scrotal ones, and finally, a testicular factor thought to play a role in the first phase of testicular descent has been identified, but it is different from AMH (Fentener van Vlissingen *et al*. 1988). Testes of dogs affected with the persistent Müllerian duct syndrome are often located in the scrotum (Meyers-Wallen *et al*. 1989).

iii Effect on the adult ovary
Since AMH is produced by adult granulosa cells and is present at a relatively high concentration in follicular fluid, it was logical to look for a possible physiological effect of the hormone in the adult ovary. Because meiotic maturation of female germ cells does not progress beyond the diplotene phase until the germ cell is released from the follicle (Edwards 1965), a group of investigators has proposed that AMH is involved in the regulation of oocyte maturation. Rat ovaries exposed to an incompletely purified preparation of AMH exhibited a decreased rate of germinal vesicle breakdown, which reflects a partial inhibition of oocyte maturation (Takahashi *et al*. 1986*b*); however, this effect was not obtained with purified human recombinant AMH unless detergent was added, and detergent, on its own, is an inhibitor of oocyte maturation (Ueno *et al*. 1988).

iv Miscellaneous
Other biological activities of the molecule have been suggested, but not yet convincingly established. Antiproliferative effects on malignant cells (Fuller *et al*. 1982; Chin *et al*. 1991; Parry *et al*. 1992) and inhibition of protein phosphorylation (Hutson *et al*. 1984) have been obtained with some preparations but not with purified native (Rosenwaks *et al*. 1984) or recombinant (Wallen *et al*. 1989) AMH from other sources. Catlin *et al*. (1988) have reported that female lung fragments accumulated less phospholipid when cultured together with testis or partially purified bovine, or human recombinant AMH than do controls exposed to vehicle buffer. AMH has been reported to enhance the expression of major histocompatibility complex genes (Donahoe *et al*. 1989).

VI THE PERSISTENT MÜLLERIAN DUCT SYNDROME

Experiments of Nature often provide important physiological information, particularly in reproductive disorders which impair fertility but do not threaten the life of the sufferer. Thus, the freemartin model led to the discovery of the masculinizing effect of AMH upon the fetal ovary (see above, p.146). Conversely, investigation of the persistent Müllerian duct syndrome (PMDS) has yielded information relative to the role of AMH in normal sex differentiation. PMDS is a rare form of inherited male pseudo-hermaphroditism, characterized by the presence of uterus and tubes in otherwise normally virilized males. In the past, it was discovered mainly in adult or adolescent patients undergoing surgical repair of cryptorchidism or inguinal hernia. Because early treatment of undescended testes is now advocated, pseudo-hermaphroditism is now commonly diagnosed in childhood.

The testes are tightly linked to the Fallopian tubes (Fig. 4.9) and therefore the clinical picture depends upon the degree of mobility of the Müllerian derivatives. Usually, the Müllerian derivatives are not fixed in the pelvis and are dragged into the inguinal canal by the descending testis. The inguinal hernia, called '*hernia uteri inguinalis*' in the literature, contains the uterus and tubes. More rarely, the uterus and tubes are maintained in the pelvis by the round ligament, and the testes are in an 'ovarian' position: clinically, the patients are bilaterally cryptorchid, and no inguinal hernia is present. Irrespective of the clinical presentation, the testes are

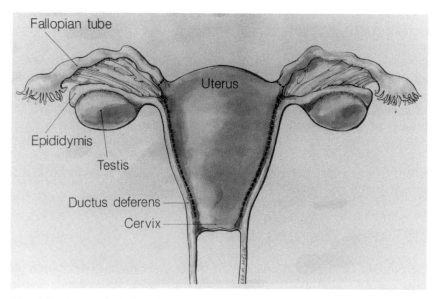

Fig. 4.9   Anatomical disposition of the internal reproductive organs in a PMDS affected patient. Note the close association of the testes with the Fallopian tubes, and of the vasa deferentia with the wall of the uterus. From Loeff *et al.* (1993), with permission.

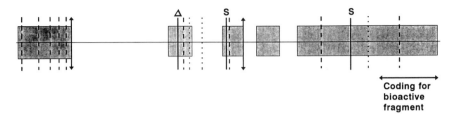

Coding for
bioactive
fragment

Fig. 4.10   Polymorphisms and/or mutations of the AMH gene observed in PMDS patients and their families. Misserse,– – – – –; Splicing, ←——→; stop, S; deletion, △; silent, ... From Josso *et al.* (1993*b*), with permission.

histologically normally differentiated, apart from secondary lesions due to longstanding cryptorchidism. A similar syndrome has been described in dogs (Meyers-Wallen *et al.* 1989), those with scrotal testes were fertile. Thus, in spite of its marked virilizing effect upon fetal ovaries, AMH seems to be required neither for testicular differentiation nor for germ cell maturation. The only proven role for AMH in male sex differentiation is Müllerian duct regression, which occurs very early in fetal life, 15 days post-coitum in the rat and 8 fetal weeks in the human, while AMH production by Sertoli cells continues up to puberty.

The molecular basis of PMDS is heterogeneous. Some patients fail to produce AMH: deletions and stop mutations of the AMH gene have been reported in such cases (Knebelmann *et al.* 1991; Carré-Eusèbe *et al.* 1992) as well as missense mutations (Imbeaud *et al.* 1994). However, the AMH gene is highly polymorphic (Fig. 4.10) making interpretation of the latter difficult in the absence of *in vitro* expression studies. Alternatively, AMH production may be normal for age, suggesting end-organ insensitivity. The latter is probably responsible for the PMDS syndrome observed in dogs (Meyers-Wallen *et al.* 1989), which is inherited as an autosomal recessive trait, as are most cases observed in humans. Expression is limited to homozygous males; heterozygous males and homozygous females are normal (Imbeaud *et al.* 1994).

VII   UNRESOLVED ISSUES

Purification, cloning, and elucidation of structure/function relationships have greatly added to our understanding of the AMH molecule itself. Gene regulation remains an obscure issue because of the lack of an appropriate cell model. Sertoli cells in primary culture are unable to continue expressing the AMH gene (Vigier *et al.* 1985), and established Sertoli cell lines are similarly inadequate (Peschon *et al.* 1992). Investigators are caught in a vicious circle: to understand AMH regulation, they need an appropriate cell model, and they are unable to devise one because they do not understand AMH regulation. The problem may eventually be solved by the use of transgenic mice.

However, the main unresolved issue at the present time is the nature of the AMH receptor. Most labelling methods destroy AMH bioactivity and the lack of a bioactive ligand has been a major obstacle. However, a glimmer of hope has appeared recently, with the cloning of the receptors for several of other members of the TGF-β superfamily (Massagué 1992). All are transmembrane serine–threonine kinases, and if the AMH receptor has a similar catalytic domain, it should be picked up by oligonucleotide hybridization using degenerate probes.

Indeed, using this approach, a transmembrane serine–threonine kinase expressed around the Müllerian duct, in granulosa cells and in immature Sertoli cells has been cloned from a fetal ovarian library (di Clemente et al., personal communication).

Cloning of the AMH receptor is a prerequisite for the understanding of the so-called 'AMH-positive' forms of PMDS, characterized by the production of normal amounts of bioactive AMH (Imbeaud et al. 1994). The coming year may witness significant developments in this field.

REFERENCES

Ackerman, G. E., Smith, M. E., Mendelson, C. R., MacDonald, P. C., and Simpson, E. R. (1981). Aromatization of androstenedione by human adipose tissue stromal cells in monolayer culture. *Journal of Clinical Endocrinology and Metabolism*, **53**, 412–7.

Baker, M. L., Metcalfe, S. A., and Hutson, J. M. (1990). Serum levels of Müllerian inhibiting substance in boys from birth to 18 years, as determined by enzyme immunoassay. *Journal of Clinical Endocrinology and Metabolism*, **70**, 11–5.

Behringer, R. R., Cate, R. L., Froelick, G. J., Palmiter, R. D., and Brinster, R. L. (1990). Abnormal sexual development in transgenic mice chronically expressing Müllerian inhibiting substance. *Nature*, **345**, 167–70.

Bercu, B. B., Morikawa, Y., Jackson, I. M. D., and Donahoe, P.K. (1978). Increased secretion of Müllerian inhibiting substance after immunological blockade of endogenous luteinizing hormone releasing hormone in the rat. *Pediatric Research*, **12**, 139–42.

Bézard, J., Vigier, B., Tran, D., Mauléon, P., and Josso, N. (1987). Immunocytochemical study of anti-Müllerian hormone in sheep ovarian follicles during fetal and postnatal development. *Journal of Reproduction and Fertility*, **80**, 509–16.

Blanchard, M. G. and Josso, N. (1974). Source of the anti-Müllerian hormone synthesized by the fetal testis: Müllerian-inhibiting activity of fetal bovine Sertoli cells in tissue culture. *Pediatric Research*, **8**, 968–71.

Bouin, P. and Ancel, P. (1903). Sur la signification de la glande interstitielle du testicule embryonnaire. *Comptes-Rendus de la Société de Biologie*, **55**, 1632–4.

Brook, G. D. and Grumbach, M. M., eds. (1989). *Clinical paediatric endocrinology*, 2nd edn. (Blackwell Scientific Publications, Oxford.

Budzik, G. P., Powell, S. M., Kamagata, S., and Donahoe, P. K. (1983). Müllerian inhibiting substance fractionation by dye affinity chromatography. *Cell*, **34**, 307–14.

Carré-Eusèbe, D., Imbeaud, S., Harbison, M., New, M. I., Josso, N., and Picard, J. Y. (1992). Variants of the anti-Müllerian hormone gene in a compound heterozygote with the persistent Müllerian duct syndrome and his family. *Human Genetics*, **90**, 389–94.

Cate, R. L., Mattaliano, R. J., Hession, C., Tizard, R., Farber, N. M., Cheung, A., *et al.* (1986). Isolation of the bovine and human genes for Müllerian inhibiting substance and expression of the human gene in animal cells. *Cell*, **45**, 685–98.

Catlin, E. A., Manganaro, T. F., and Donahoe, P. K. (1988). Müllerian inhibiting substance depresses accumulation *in vitro* of disaturated phosphatidylcholine in fetal rat lung. *American Journal of Obstetrics and Gynecology*, **159**, 1299–303.

Chin, T. W., Parry, R. L., and Donahoe, P. K. (1991). Human Müllerian inhibiting substance inhibits tumor growth *in vitro* and *in vivo*. *Cancer Research*, **51**, 2101–6.

di Clemente, N., Ghaffari, S., Pepinsky, R. B., Pieau, C., Josso, N., Cate, R. L., and Vigier, B. (1992). A quantitative and interspecific test for biological activity of anti-Müllerian hormone: the fetal ovary aromatase assay. *Development*, **114**, 721–7.

Cohen-Haguenauer, O., Picard, J. Y., Mattéi, M. G., Serero, S., Van Cong, N., de Tand, M. F., *et al.* (1987). The mapping of the gene for anti-Müllerian hormone to the short arm of chromosome 19. *Cytogenetics and Cell Genetics*, **44**, 2–6.

Donahoe, P. K., Ito, Y., Marfatia, S., and Hendren, W. H., III (1976). The production of Müllerian inhibiting substance by the fetal, neonatal and adult rat. *Biology of Reproduction*, **15**, 329–34.

Donahoe, P. K., Ito, Y., Price, J. M., and Hendren, W. H., III (1977). Müllerian inhibiting substance activity in bovine fetal, newborn and prepubertal testes. *Biology of Reproduction*, **16**, 238–43.

Donahoe, P. K., Catlin, E., Kuroda, T., Barksdale, E. M., Epstein, J., and MacLaughlin, D. T. (1989). Molecular modeling of Müllerian inhibiting substance and its actions in gonad, lung, and immune system. In *Development and function of the reproductive organs*, Vol. 3, (ed. N. Josso), pp. 99–110. Raven Press, New York.

Dorizzi, M., Mignot, T. M., Guichard, A., Desvages, G., and Pieau, C. (1991). Involvement of oestrogens in sexual differentiation of gonads as a function of temperature in turtles. *Differentiation*, **47**, 9–17.

Dyche, W. J. (1979). A comparative study of the differentiation and involution of the Müllerian duct and Wolffian duct in the male and female mouse. *Journal of Morphology*, **162**, 175–210.

Edwards, R. G. (1965). Maturation *in vitro* of mouse, sheep, cow, pig, rhesus monkey and human ovarian oocytes. *Nature*, **208**, 349–51.

Erickson, G. F. (1983). Primary cultures of ovarian cells in serum-free medium as models of hormone-dependent differentiation. *Molecular and Cellular Endocrinology*, **29**, 21–49.

Fentener van Vlissingen, F. M., Van Zoelen, E. J. J., Ursem, P. J. F., and Wensing, C. J. G. (1988). *In vitro* model of the first phase of testicular descent: identification of a low molecular weight factor from fetal testis involved in proliferation of gubernaculum testis cells and distinct from specified polypeptide growth factors and fetal gonadal homones. *Endocrinology*, **123**, 2868–77.

Fuller, A. F., Guy, S., Budzik, G. P., and Donahoe, P. K. (1982). Müllerian inhibiting substance inhibits colony growth of a human ovarian carcinoma cell line. *Journal of Clinical Endocrinology and Metabolism*, **54**, 1051–5.

Gray, A.M. and Mason, A.J. (1990). Requirement for activin A and transforming growth factor-β1 proregions in homodimer assembly. *Science*, **247**, 1328–30.

Gentry, L. E. and Nash, B. W. (1990). The pro domain of pre-pro- transforming growth factor-β1 when independently expressed is a functional binding protein for the mature growth factor. *Biochemistry*, **29**, 6851–7.

Guerrier, D., Tran, D., VanderWinden, J. M., Hideux, S., Van Outryve, L., Legeai, L., *et al.* (1989). The persistent Müllerian duct syndrome: a molecular approach. *Journal of Clinical Endocrinology and Metabolism*, **68**, 46–52.

Guerrier, D., Boussin, L., Mader, S., Josso, N., Kahn, A., and Picard, J. Y. (1990). Expression of the gene for anti-Müllerian hormone. *Journal of Reproduction and Fertility*, **88**, 695–706.

Haqq, C., Lee, M. M., Tizard, R., Wysk, M., DeMarinis, J., Donahoe, P. K., and Cate, R. L. (1992). Isolation of the rat gene for Müllerian inhibiting substance. *Genomics*, **12**, 665–9.

Harley, V. R., Jackson, D. I., Hextall, P. J., Hawkins, J. R., Berkovitz, G. D., Sockanathan, S., *et al.* (1992). DNA binding activity of recombinant SRY from normal males and XY females. *Science*, **255**, 453–6.

Hayashi, H., Shima, H., Hayashi, K., Trelstad, R. L., and Donahoe, P. K. (1984). Immunocytochemical localization of Müllerian inhibiting substance in the rough endoplasmic reticulum and Golgi apparatus in Sertoli cells of the neonatal calf testis using a monoclonal antibody. *Journal of Histochemistry and Cytochemistry*, **32**, 649–54.

Hudson, P. L., Dougas, I., Donahoe, P. K., Cate, R. L., Epstein, J., Pepinsky, R. B., and MacLaughlin, D. T. (1990). An immunoassay to detect human Müllerian inhibiting substance in males and females during normal development. *Journal of Clinical Endocrinology and Metabolism*, **70**, 16–22.

Hutson, J. M. and Donahoe, P. K. (1986). The hormonal control of testicular descent. *Endocrine Reviews*, **7**, 270–83.

Hutson, J. M., Ikawa, H., and Donahoe, P. K. (1981). The ontogeny of Müllerian inhibiting substance in the gonads of the chicken. *Journal of Pediatric Surgery*, **16**, 822–7.

Hutson, J. M., Fallat, M. E., Kamagata, S., Donahoe, P. K., and Budzik, G. P. (1984). Phosphorylation events during Müllerian duct regression. *Science*, **233**, 586–9.

Imbeaud, S., Carré-Eusèbe, D., Rey, R., Belville, C., Josso, N., and Picard, J.Y. (1994). Molecular genetics of the persistent Müllerian duct synsdrome: a study of 19 families. *Human Molecular Genetics*, (in press).

Josso, N. (1974). Müllerian inhibiting activity of human fetal testicular cells deprived of germ cells by *in vitro* irradiation. *Pediatric Research*, **8**, 755–8.

Josso, N. (1981). Physiology of sex differentiation: a guide to the understanding and management of the intersex child. In *The intersex child* (ed. N. Josso), pp. 1–13. (Karger, Basel.)

Josso, N., Picard, J. Y., and Tran, D. (1977). The anti-Müllerian hormone. *Recent Progress in Hormone Research*, **33**, 117–60.

Josso, N., Legeai, L., Forest, M. G., Chaussain, J. L., and Brauner, R. (1990). An enzyme-linked immunoassay for anti-Müllerian hormone: a new tool for the evaluation of testicular function in infants and children. *Journal of Clinical Endocrinology and Metabolism*, **70**, 23–7.

Josso, N., Cate, R. L., Picard, J. Y., Vigier, B., di Clemente, N., Wilson, C., Imbeaud, S., Pepinsry, R. B., Guerrier, D., Boussin, L., Legeai, L., and Carré-Eusèbe, D. (1993*a*). Anti-Müllerian hormone, the Jost factor. *Recent Progress in Hormone Research*, **48**, 1–59, San Diego Academic Press.

Josso, N., Imbeaud, S., Picard, J.Y., and Cate, R. L. (1993*b*). The gene for anti-Müllerian hormone. In (ed. S. Wachtel), pp. 439–55. *Molecular genetics of sex determination* Academic Press, Orlando, FL.

Josso, N., Lamarre, I., Picard, J. Y., Berta, P., Davies, N., Morichon, N., *et al.* (1993*c*). Anti-Müllerian hormone in early human development. *Early Human Development*, **33**, 91–9.

Jost, A. (1953). Problems of fetal endocrinology: the gonadal and hypophyseal hormones. *Recent Progress in Hormone Research*, **8**, 379–418.

Jost, A., Vigier, B., and Prépin, J. (1972). Freemartins in cattle: the first steps of sexual organogenesis. *Journal of Reproduction and Fertility*, **29**, 349–79.

King, T. R., Lee, B. K., Behringer, R. R., and Eicher, E. M. (1991). Mapping anti-Müllerian hormone (Amh) and related sequences in the mouse: identification of a new region of homology between MMU10 and HSA19p. *Genomics*, **11**, 273–83.

Knebelmann, B., Boussin, L., Guerrier, D., Legeai, L., Kahn, A., Josso, N., and Picard, J. Y. (1991). Anti-Müllerian hormone Bruxelles: a nonsense mutation associated with the persistent Müllerian duct syndrome. *Proceedings of the National Academy of Sciences, USA*, **88**, 3767–71.

Kuroda, T., Lee, M. M., Haqq, C. M., Powell, D. M., Manganaro, T. F., and Donahoe, P. K. (1990). Müllerian inhibiting substance ontogeny and its modulation by follicle-stimulating hormone in the rat testes. *Endocrinology*, **127**, 1825–32.

LaQuaglia, M., Shima, H., Hudson, P., Takahashi, M., and Donahoe, P. K. (1986). Sertoli cell production of Müllerian inhibiting substance *in vitro*. *Journal of Urology*, **136**, 219–24.

Lee, M. M., Cate, R. L., Donahoe, P. K., and Waneck, G. L. (1992). Developmentally regulated polyadenylation of two discrete messenger ribonucleic acids for Müllerian inhibiting substance. *Endocrinology*, **130**, 847–53.

Lee, W., Mason, A. J., Schwall, R., Szonyi, E., and Mather, J. P. (1989). Secretion of activin by interstitial cells in the testis. *Science*, **243**, 396–7.

Loeff, D. S., Imbeaud, S., Reyes, H. M., Meller, J. L., and Rosenthal, I. M. (1993). Surgical and genetic aspects of persistent Müllerian duct syndrome. *Journal of Pediatric Surgery*, (In press).

MacLaughlin, D. T., Hudson, P., Graciano, A. L., Kenneally, M. K., Ragin, R. C., Manganaro, T. F., and Donahoe, P. K. (1992). Müllerian duct regression and antiproliferative activities of Müllerian inhibiting substance reside in its carboxy-terminal domain. *Endocrinology*, **313**, 291–6.

Massagué, J. (1992). Receptors for the TGF-beta family. *Cell*, **69**, 1067–70.

Meyers-Wallen, V. N., Donahoe, P. K., Ueno, S., Manganaro, T. F., and Patterson, D. F. (1989). Müllerian inhibiting substance is present in testes of dogs with persistent Müllerian duct syndrome. *Biology of Reproduction*, **41**, 881–8.

Münsterberg, A. and Lovell-Badge, R. (1991). Expression of the mouse anti-Müllerian hormone gene suggests a role in both male and female sex differentiation. *Development*, **113**, 613–24.

Nasrin, N., Buggs, C., Kong, X. F., Carnazza, J., Goebl, M., and Alexander-Bridges, M. (1991). DNA-binding properties of the product of the testis-determining gene and a related protein. *Nature*, **354**, 317–20.

Necklaws, E. C., LaQuaglia, M. P., MacLaughlin, D., Hudson, P., Mudgett-Hunter, M., and Donahoe, P. K. (1986). Detection of Müllerian inhibiting substance in biological samples by a solid phase sandwich radioimmunoassay. *Endocrinology*, **118**, 791–6.

Parry, R. L., Chin, T., Epstein, J., Hudson, P. L., Powell, D. M., and Donahoe, P. K. (1992). Recombinant human Müllerian inhibiting substance inhibits human ocular melanoma cell lines *in vitro* and *in vivo*. *Cancer Research*, **52**, 1182–6.

Pepinsky, R. B., Sinclair, L. K., Chow, E. P., Mattaliano, R. J., Manganaro, T. F., Donahoe, P. K., and Cate, R. L. (1988). Proteolytic processing of Müllerian inhibiting substance produces a transforming growth factor-β- like fragment. *Journal of Biological Chemistry*, **263**, 18961–5.

Peschon, J. J., Behringer, R. R., Cate, R. L., Harwood, K. A., Idzerda, R. L., Brinster, R. L., and Palmiter, R. D. (1992). Directed expression of an oncogene to Sertoli cells in transgenic mice using Müllerian inhibiting substance regulatory sequences. *Molecular Endocrinology*, **6**, 1403–11.

Picard, J. Y. and Josso, N. (1984). Purification of testicular anti-Müllerian hormone allowing direct visualization of the pure glycoprotein and determination of yield and purification factor. *Molecular and Cellular Endocrinology*, **34**, 23–9.

Picard, J. Y., Tran, D., and Josso, N. (1978). Biosynthesis of labelled anti-Müllerian hormone by fetal testes: evidence for the glycoprotein nature of the hormone and for its disulfide-bonded structure. *Molecular and Cellular Endocrinology*, **12**, 17–30.

Picard, J. Y., Benarous, R., Guerrier, D., Josso, N., and Kahn, A. (1986*a*). Cloning and expression of cDNA for anti-Müllerian hormone. *Proceedings of the National Academy of Sciences, USA*, **83**, 5464–8.

Picard, J. Y., Goulut, C., Bourrillon, R., and Josso, N. (1986*b*). Biochemical analysis of bovine testicular anti-Müllerian hormone. *FEBS Letters*, **195**, 73–6.

Picon, R. (1969). Action du testicule foetal sur le développement *in vitro* des canaux de Müller chez le rat. *Archives d'Anatomie Microscopique et de Morphologie Expérimentale*, **58**, 1–19.

Picon, R. (1970). Modifications, chez le rat, au cours du développement du testicule, de son action inhibitrice sur les canaux de Müller *in vitro*. Comptes-Rendus de l'Académie des Sciences, Série D, Paris, **271**, 2370–2.

Picon, R., Pelloux, M. C., Benhaim, A., and Gloaguen, F. (1985). Conversion of androgen to estrogen by the rat fetal and neonatal female gonad: effects of dcAMP and FSH. *Journal of Steroid Biochemistry*, **23**, 995–1000.

Rey, R., Lordereau-Richard, I., Carel, J.C., Barbet, P., Cate, R.L., Roger, M., Chaussain, J.L., and Josso, N. (1993). Anti-Müllerian hormone and testosterone serum levels are inversely related during normal and precocious pubertal development. *Journal of Clinical Endocrinology and Metabolism*, **77**, 1220–6.

Roberts, V., Meunier, H., Sawchenko, P. E., and Vale, W. (1989). Differential production and regulation of inhibin subunits in rat testicular cell types. *Endocrinology*, **125**, 2350–9.

Rosenwaks, Z., Liu, H. C., Picard, J. Y., and Josso, N. (1984). Anti-Müllerian hormone is not cytotoxic to human endometrial cancer in tissue culture. *Journal of Clinical Endocrinology and Metabolism*, **59**, 166–9.

Shaha, C., Morris, P. L., Chen, C. L. C., Vale, W., and Bardin, C. W. (1989). Immunostainable inhibin subunits are in multiple types of testicular cells. *Endocrinology*, **125**, 1941–50.

Shull, M. M., Ormsby, I., Kier, A. B., Pawlowski, S., Diebold, R. J., Yin, M. Y., et al. (1992). Targeted disruption of the mouse transforming growth factor-β gene results in multifocal inflammatory, disease. *Nature*, **359**, 693–9.

Takahashi, M., Hayashi, M., Manganaro, T. F., and Donahoe, P. K. (1986a). The ontogeny of Müllerian inhibiting substance in granulosa cells of the bovine ovarian follicle. *Biology of Reproduction*, **35**, 447–54.

Takahashi, M., Koide, S. S., and Donahoe, P. K. (1986b). Müllerian inhibiting substance as oocyte meiosis inhibitor. *Molecular and Cellular Endocrinology*, **47**, 225–34.

Tran, D. and Josso, N. (1982). Localization of anti-Müllerian hormone in the rough endoplasmic reticulum of the developing bovine Sertoli cell using immunocytochemistry with a monoclonal antibody. *Endocrinology*, **111**, 1562–7.

Tran, D., Meusy-Dessole, N., and Josso, N. (1977). Anti-Müllerian hormone is a functional marker of foetal Sertoli cells. *Nature*, **269**, 411–2.

Trelstad, R. L., Hayashi, A., Hayashi, K., and Donahoe, P. K. (1982). The epithelial mesenchymal interface of the male rat Müllerian duct: loss of basement membrane integrity and ductal regression. *Developmental Biology*, **92**, 27–40.

Tsuji, M., Shima, H., Yonemura, C. Y., Brody, J., Donahoe, P. K., and Cunha, G. R. (1992). Effect of human recombinant Müllerian inhibiting substance on isolated epithelial and mesenchymal cells during mullerian duct regression in the rat. *Endocrinology*, **131**, 1481–8.

Ueno, S., Manganaro, T. F., and Donahoe, P. K. (1988). Human recombinant Müllerian inhibiting substance inhibition of rat oocyte meiosis is reversed by epidermal growth factor *in vitro*. *Endocrinology*, **123**, 1652–9.

Ueno, S., Kuroda, T., MacLaughlin, D. T., Ragin, R. C., Manganaro, T. F., and Donahoe, P. K. (1989a). Müllerian inhibiting substance in the adult rat ovary during various stages of the estrous cycle. *Endocrinology*, **125**, 1060–6.

Ueno, S., Takahashi, M., Manganaro, T. F., Ragin, R. C., and Donahoe, P. K. (1989b). Cellular localization of Müllerian inhibiting substance in the developing rat ovary. *Endocrinology*, **124**, 1000–6.

Vigier, B., Tran, D., Du Mesnil du Buisson, F., Heyman, Y., and Josso, N. (1983). Use of monoclonal antibody techniques to study the ontogeny of bovine anti-Müllerian hormone. *Journal of Reproduction and Fertility*, **69**, 207–14.

Vigier, B., Picard, J. Y., Tran, D., Legeai, L., and Josso, N. (1984a). Production of anti-Müllerian hormone: another homology between Sertoli and granulosa cells. *Endocrinology*, **114**, 1315–20.

Vigier, B., Tran, D., Legeai, L., Bézard, J., and Josso, N. (1984b). Origin of anti-Müllerian hormone in bovine freemartin fetuses. *Journal of Reproduction and Fertility*, **70**, 473–9.

Vigier, B., Picard, J. Y., Campargue, J., Forest, M. G., Heyman, Y., and Josso, N. (1985). Secretion of anti-Müllerian hormone by immature bovine Sertoli cells in primary culture, studied by a competition type radio-immunoassay: lack of modulation by either FSH or testosterone. *Molecular and Cellular Endocrinology*, **43**, 141–50.

Vigier, B., Watrin, F., Magre, S., Tran, D., and Josso, N. (1987). Purified bovine AMH induces a characteristic freemartin effect in fetal rat prospective ovaries exposed to it *in vitro*. *Development*, **100**, 43–55.

Vigier, B., Forest, M. G., Eychenne, B., Bézard, J., Garrigou, O., Robel, P., and Josso, N. (1989). Anti-Müllerian hormone produces endocrine sex-reversal of fetal ovaries. *Proceedings of the National Academy of Sciences, USA*, **86**, 3684–8.

Voutilainen, R. and Miller, W. L. (1987). Human Müllerian inhibitory factor messenger ribonucleic acid is hormonally regulated in the fetal testis and in adult granulosa cells. *Molecular Endocrinology*, **1**, 604–8.

Wakefield, L. (1990). Growth factors: an overview. In *Hormonal communicating events in the testis*, Serono Symposia Publications, 70 (ed. A. Isidori, A. Fabbri, and M. L. Dufau), pp. 181–90. Raven Press, New York.

Wallen, J., Cate, R. L., Kiefer, D. M., Riemen, M. W., Martinez, D., Hoffman, R. M., *et al.* (1989). Minimal anti-proliferative effect of recombinant Müllerian inhibiting substance on gynecological tumor cell lines and tumour explants. *Cancer Research*, **49**, 2005–11.

Wilson, C., di Clemente, N., Ehrenfels, C., Pepinsky, R. B., Josso, N., Vigier, B., and Cate, R. L. (1993). Müllerian inhibiting substance requires its N-terminal domain for maintenance of biological activity, a novel finding within the TGF-β superfamily. *Molecular Endocrinology*, **7**, 247–57.

5   The female thymus and reproduction in mammals

MARION D. KENDALL and ANN G. CLARKE

I   Introduction

II   The thymus gland
1   Embryological origins
2   Thymocyte development
3   Cells of the microenvironment
4   Thymic hormones and other thymic factors
5   Communication between the thymus and the periphery
6   Activity of the thymus during life

III   Neuroendocrine–immune axis
1   The pineal
2   The hypothalamus
3   The pituitary
4   The adrenal glands
5   The gonads
6   The placenta

IV   Maternal immunity during pregnancy
1   The steroid hormones
2   Pregnancy-specific proteins
3   Antigens of paternal and fetal origin

V   Pregnancy and the maternal thymus
1   Thymic involution
2   Immunology
3   Changes in subsets of thymocytes
4   Changes in non-thymocytes
5   The effects of pseudopregnancy
6   The significance of thymic changes for pregnancy

VI   The recovery of the thymus after birth

VII   Rheumatoid arthritis (RA) and pregnancy

I   INTRODUCTION

The thymus gland is first found in lower vertebrates, and its microscopic structure is remarkably similar in all vertebrates despite different embryological origins and gross morphology in various classes. Its function in the production of the thymocytes is fundamental to the immune response. The thymus of laboratory rodents and humans is the most studied, and forms the basis of this chapter. In lower vertebrates, seasonal changes of the thymus relate to the process of reproduction, but it is in mammals, where the embryo presents a potential immunological challenge to the mother, that most studies on reproductive immunology have been focused.

One of the most important advances in immunology was the discovery that the mammalian thymus in the adult continues to play a central role in the differentiation and processing of lymphocytes (Zinkernagel and Doherty 1975). Its presence is essential for the development and maintenance of cell-mediated immunity. To become T cells (expressing a T-cell receptor, TCR), prothymocytes interact with other cells in the thymus, with soluble factors and with thymic hormones to acquire a series of surface antigens that allow them to become immunocompetent, and determine their functions in the periphery.

The thymus both influences, and is influenced by, the process of reproduction. The structure of the gonads is affected during development by the activity of the neonatal thymus. In adult life sex steroids act on the thymus, as indeed do many neuroendocrine factors. Dramatic changes occur in the thymus during pregnancy, but surprisingly, apart from anatomical and histological descriptions, these changes have, until recently, received little attention from reproductive physiologists or immunologists.

Today, it is the interplay of neuroendocrine and immunological factors that is attracting attention as we struggle to understand the complexity of immunity. For reproductive immunologists, how the fetus is preserved from immunological attack, why some diseases such as autoimmune conditions may either start or remit during pregnancy, and how immunity and reproduction is influenced by the use and abuse of drugs, are all important questions. In this chapter, we will consider these questions in conjunction with the function of the thymus.

II   THE THYMUS GLAND

1   Embryological origins

The thymus forms bilaterally from the interaction of branchial ectoderm, pharyngeal endoderm, and mesoderm in the embryonic cervical region.

A neuroectodermal component was clearly demonstrated in birds (Le Douarin and Jotereau 1975). Failure of the three elements to develop together properly results in an athymic condition where the organ is represented only by non-lymphoid epithelial and mesenchymal components (Cordier and Haumont 1980). Thymic tissue originates from different branchial pouches in the various vertebrate classes but generally contains very similar cell types arranged in a homologous manner.

Each anlage is at first devoid of lymphoid cells but they soon enter, attracted by soluble products of the epithelial cells (e.g. thymotaxin or β_2-microglobulin in mice; Dargemont et al. 1989) before the thymic blood vessels invade the anlage. In the embryo, the thymus is formed before any other lymphoid organ, and the lymphoid progenitor cells come initially from extra-embryonic (yolk sac) or para-aortic sites, and then from fetal liver. These cells may be multipotential. In adult life, in animals with bone marrow, thymocyte progenitor cells derive from that organ. They are probably T-lineage restricted. The thymus also contains uncommitted pluripotent stem cells (de Vries et al. 1992), which may account for intrathymic erythropoiesis, granulopoiesis, or mastopoiesis.

As the bilateral thymus rudiments grow and develop, they gain blood vessels and nerves, a cortex well populated with thymocytes and, lastly, a central medulla containing antigen-presenting cells of the macrophage lineage. During this time, the two primitive thymuses migrate caudally to lie over the heart in the superior anterior mediastinum in humans and rodents. The two organs are united in the mid-line only by connective tissue, without shared lymphatics, nerves or blood vessels. This bilobed structure, with cervical extensions sometimes remaining close to, or even embedded in the parathyroid and thyroid glands, is called 'the thymus', in the singular. In some lower vertebrates such as birds, the thymus is multilobed and extends along either side of the neck.

2   Thymocyte development

The sequence of thymocyte maturation (Fig. 5.1) leading to fully functional post-thymic T cells has been extensively studied, as have the functional capabilities of cells of the thymic microenvironment in influencing these events. Current views are summarized in a special issue of *Immunology Today* (1991, **12**), and in a recent book by Ritter and Crispe (1992).

Progenitor cells entering the thymus have not yet developed the TCR, and are CD4$^-$ CD8$^-$ (double negative or DN cells). In mouse they are heat stable antigen (HSA) positive, and in humans CD7$^+$. There is a transient expression of various epitopes before the majority of cortical thymocytes become TCR$^+$ and both CD4$^+$ and CD8$^+$ (double positive or DP cells). The prothymocytes express receptors for lectins (e.g. peanut, wheat germ, and soya bean agglutinins–PNA, WGA, and SBA, respectively) and they

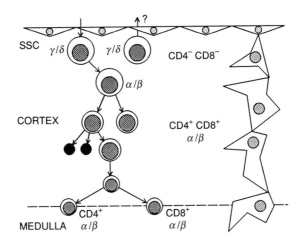

Fig. 5.1   A simplified diagram to show the major steps in thymocyte differentiation. SSC = subcapsular epithelium. Each region contains epithelial cells (spiky
outlines). Cells dying by apoptosis are indicated by a solid black circle. (Reproduced
with permission from Kendall 1990.)

decrease as the cells mature through the DP stages. In mice, as lectin
binding abilities fall, so the expression of the surface epitope Thy1 rises.
During development of the TCR, the CD3 epitope is first associated with
γ/δ receptors, and then the α/β component. While the cells are DP they
differentiate towards self-tolerant single positive (SP) thymocytes with a
TCR capable of recognizing antigens associated with molecules of the
major histocompatibility complex (MHC). SP CD4 cells (CD4+ CD8−)
recognize MHC class II and SP CD8 (CD4− CD8+) cells recognize MHC
class I. In development, the genetic loci of the TCR are rearranged to give
genes with functional diversity, but only some combinations are retained.
Only cells potentially tolerant to self-MHC or to ubiquitous non-MHC
self-molecules are required and the others are deleted (negative selection).
The use of antibodies to the Vβ regions of the TCR has shown that the
cells appear to be deleted from the medulla, probably in conjunction with
interdigitating cells (IDCs). Positive selection, or the retention of potentially
useful thymocytes, is thought to be enacted through signals from the cortical
microenvironment that 'rescue' thymocytes from cell death. Such signals
can be given before cells become SP, or can alter the choice of the CD4
or CD8 SP cell pathway. Very few thymocytes are selected in this manner
and the majority die by apoptosis in the cortex.
    As a result of intrathymic differentiation, SP cells located in the medulla,
are exported from the thymus to become mature immunocompetent T
cells. There is a strong homeostatic control of peripheral T-cell (and

B-cell) numbers and the thymic contribution may vary at different times of life. Lymphocyte lifespans, homeostasis, selection, and competition have recently been reviewed by Freitas and Rocha (1993).

The sequential expression of γ/δ and then α/β components of the TCR form important markers for thymocytes, enabling studies *in vitro* on the colonization of thymic epithelial cultures (Jenkinson *et al.* 1987), and the sequential appearance of different subsets of thymocytes during embryonic development. Of relevance to reproduction is the expression of the γ/δ TCR component. The expression of these epitopes occurs early in thymocyte development, and is transient on cells destined to become DP thymocytes. However, there are numerous γ/δ lymphocytes in skin, and in the endodermally derived epithelia of the alimentary, respiratory, and reproductive systems. Whether these cells have passed through the thymus is not yet established. Their numbers suggest that, even if they have done so, there must be clonal expansion in the periphery, but it is not known where this occurs.

3   Cells of the microenvironment

The influence of the thymic microenvironment on thymocyte development cannot be underestimated, and many *in vitro* studies are now emerging showing the release of cytokines and other signals after thymocytes bind to epithelial cells. The major supporting cells of the thymus have been well described. These are a series of interconnected epithelial cells between which lodge haemopoietic cells (mainly thymocytes, but also B cells, plasma cells, eosinophils, mast cells, macrophage lineage cells, and fibroblasts). The epithelial cells in humans belong to six morphologically distinct cell types (Wijngaert *et al.* 1984), which are seen in all species. Antibodies (Abs) developed to thymic epithelial cell components, and grouped into clusters of epithelial cell staining (CTES) patterns, are valuable for characterizing thymic cells (Kampinga *et al.* 1989; Ladyman *et al.* 1989) and revealing molecular heterogeneity in non-lymphoid cells.

Monocytes enter the thymus and form cortical macrophages (often containing apoptotic thymocytes and debris), another population at the corticomedullary junction (CMJ), and IDCs in the medulla. All these cells express class I MHC and about half of them express MHC class II. γ-Interferon (IFN-γ) induces MHC class II on epithelial cells and it could have this effect on macrophages too.

B cells are found in the thymus (some 0.2–0.3% in mice; Michie and Rouse 1989) mainly near the CMJ and in the medulla. Increased numbers of B cells and plasma cells have been found in hypothyroid-induced thymic atrophy (Abou-Rabia and Kendall, in press), and plasma cells were more frequent in mouse pregnancy (Clarke and Kendall 1989). In these instances, the medulla may be regarded as a secondary lymphoid organ (i.e. antigen driven) which could have an important influence on the selection of newly formed mature cells in the medulla.

4   Thymic hormones and other thymic factors

The thymic hormones [principally thymulin, thymosins, thymopentin, and thymic humoral factor (THF)] have paracrine, and/or exocrine functions. Only some of the first described thymic extracts have been purified and characterized. Thymulin, originally called 'facteur thymique sérique' (FTS) relies on zinc for its biological activity (Dardenne and Bach 1988). The thymosins are a large family isolated from thymosin fraction 5 (Goldstein *et al*. 1977). The precursor of thymosin-α_1, prothymosin, is found in highest quantities in the thymus, but is also secreted elsewhere (Haritos *et al*. 1984) as is parathymosin. Thymosin-β_1 is ubiquitin (Schlesinger *et al*. 1975), and thymosins-β_4 and -β_{10} have recently been shown to be sequestering components of connective tissue (Yu *et al*. 1993). Thymopoietin, although originally studied for its neuromuscular effects, has its biological and immunological activity in residues 32–36 (thymopentin or TP5; Goldstein *et al*. 1979). THF has also been sequenced (Burstein *et al*. 1988) and no homology has been found between any of the thymic hormones.

The type 6 medullary epithelial cell (as well as type 1 subcapsular/ perivascular cells) both *in vivo* and *in vitro* has been shown by immuno-cytochemistry to contain thymulin, thymopoietin, and thymosin-α_1 (Savino and Dardenne 1984) as well as progestagen and oestrogen receptors (Kawashima *et al*. 1992).

The peptide hormones of the thymus have a range of immunomodulatory effects on lymphocyte maturation within the thymus and in the periphery (reviewed in Trainin 1974; Dardenne and Bach 1988; Safieh-Garabedian *et al*. 1992). Most will induce markers of early differentiation on lymphoid cells lacking such markers, and enhance various T-cell functions. Thymulin's action does not necessarily involve proliferation (Chang and Marsh 1993). The injection of most thymic hormones restores immunological competence to neonatally thymectomized mice, modulates surface epitopes in patients with immune deficiencies, and improves immunocompetence in humans and animals. Thymulin is a potent inhibitor of suppressor T cells (Bach *et al*. 1981). The peripheral blood CD8+ cell numbers are altered in patients with both high and low levels of plasma thymulin (Kendall *et al*. 1991). A single injection of physiological levels of thymulin into mice significantly alters thymocyte phenotype within 2 h (M. D. Kendall, personal observations) and induces thymocyte surface markers within 1 day (Dardenne *et al*. 1978). Thymosins stimulate cellular activity in T cells and macrophages, and enhance the production of interferon. Commercially prepared thymopentin (TP5) has been used *in vivo* for the rejection of carcinomas, the prevention of autoimmunity, and the generation of T cytotoxic cells. THF protects mice from graft-versus-host disease, and allows them to reject tumours and skin grafts.

The action of cytokines within the thymus is very important but complex, and not yet fully understood. Ritter and Crispe (1992) give a synopsis of the current knowledge. Interleukin (IL)-1, IL-2, IL-4, and IL-6 are secreted by thymocytes (as well as other cell types), and IL-1, IL-3, IL-4, IL-6, and IL-7 by thymic epithelium. Cells bearing receptors for all of these cytokines, as well as for tumor necrosis factor-α (TNFα), colony-stimulating factors for granulocytes and macrophages (GM-CFS, M-CSF or CSF-1) and γ-IFN, occur in the thymus. Almost all studies have been performed *in vitro* and potent synergistic actions affecting the differentiation of thymocytes have been demonstrated. However, how all this relates to the myriad of microenvironments in the thymus is still speculative (Hadden 1992).

5   Communication between the thymus and the periphery

Stem cells enter the thymus through the capsule in the early stages of development until blood vessels develop, when transendothelial passage becomes important. The actual site of entry of these and other cells (e.g. monocytes, T and B cells), may be determined by the course and structure of thymic blood vessels. Major blood vessels (with substantial muscle coats) travel in large septa up to the CMJ before serving either the cortex or medulla. Around major septa and cortical capillaries there are anatomically well defined perivascular sheaths formed by type 1 epithelial cells. In the medulla the epithelial cell sheath is incomplete (Kendall 1989), and high walled postcapillary endothelial vessels may also exist. Thus blood-borne cells or factors most probably cross endothelia and enter the thymus at the CMJ or medulla, rather than the cortex. Cellular egress seems to be mainly from the medulla. Cells leaving the gland could pass directly into blood vessels or into draining lymphatics, or remain in the connective tissue of the perivascular spaces (PVSs), especially around major septa (macrophages and mature plasma cells are frequently seen here).

The passage of cells in blood vessels may be affected by sympathetic noradrenergic innervation. Sympathetic stimulation certainly increases reticulocyte egress from bone marrow (Webber *et al.* 1970), and catecholamines release activated lymphocytes from spleen (Ernström and Söder 1975) and lymph nodes (Moore 1984). However, smooth muscle contractions are reduced after β-adrenergic stimulation and lymphatic pumping is reduced by isoprenaline (Allen *et al.* 1986).

Blood-borne factors may also enter and leave the thymus through fenestrated capillaries. These have been observed in the capsule, in the subcapsular cortex, and at the CMJ.

The concept of a blood–thymic barrier was developed because antigen has to be directly injected into the thymic cortex for an antigenic response to be seen there (Marshall and White 1961). Morphologically, the barrier is a complete type 1 epithelial sheath around cortical blood vessels. This concept has to be seriously questioned since Nieuwenhuis (1990) demonstrated that self-antigen injected into rats could diffuse across the thymic cortex from the capsule to the medulla. The effect of non-self-antigens at the site where the primary events leading to the recognition of self occur, could influence the reactivity of newly generated T cells.

6   Activity of the thymus during life.

The functional capabilities of the thymus at birth are pertinent to reproduction because of the development of immunity to foreign antigens and its interactions with the developing gonads. In the rat, although thymocyte selection starts in the embryo, neither thymocytes nor epithelial cells are phenotypically identical to those of adults until the postnatal period (von Boehmer 1986; Adkins *et al.* 1987; Kampinga and Aspinall 1990). This is also probably true for humans.

The neonatal thymus, in addition to thymocyte development, has other important non-immunological functions (Rebar *et al.* 1983). In particular, the absence of an active thymus at a limited time in neonatal life reduces gonadotrophin levels and the growth and development of the ovary. Neonatal thymectomy of mice between 2 and 4 days postnatally results in a high frequency of ovarian dysgenesis with lymphocytic infiltration, a rapid loss of oocytes, a subsequent decrease in the numbers of follicles and corpora lutea (Nishizuka and Sakakura 1969), and delayed vaginal opening (Allen *et al.* 1984). Fetal thymectomy of rhesus monkeys inhibits oogenesis and induces abnormal ovarian differentiation (Healy *et al.* 1985). Girls with ataxia telangiectasia (including thymic aplasia) have dysgenic ovaries (Miller and Chatten 1967). Neonatally thymectomized mice have circulating anti-ovarian antibodies and low levels of gonadotrophins (Michael 1983). Gonadotrophin exogenously administered to 1-month-old genetically athymic mice can restore the pool of actively growing follicles to the levels found in heterozygotes (Lintern-Moore and Pantelouris 1975), and if thymuses are transplanted into such mice at birth, then luteinizing hormone (LH) and follicle–stimulating hormone (FSH) are raised to normal levels. Nude or athymic mice have similar deficiencies (Rebar *et al.* 1981), as well as reduced fertility, and delayed vaginal opening, all of which can be corrected by thymic transplants at birth.

The isolation of a 28-kD thymic fraction, and the demonstration that this factor *in vitro* causes a dose-dependent decrease in human chorionic gonadotrophin (hCG)-stimulated production of progesterone, oestradiol, and testosterone in 1-day-old rats, has led to the suggestion that there is a growth factor-like activity in the thymus in addition to the action of the characterized thymic hormones (Aguilera and Romano 1989).

The major role of the fully formed and functional thymus is the generation of T cells for immune reactivity in the periphery. The T-cell repertoire is formed before puberty when exposure to new antigens is maximal. Opinions vary as to the importance of the thymus thereafter. Generally, the thymus is largest, relative to body weight, shortly after birth (reviewed in Anderson 1932; Kendall 1991). Some authors report that the organ begins to involute after puberty, but in humans it starts at birth (Kendall, unpublished observations; Steinmann 1986). Fatty infiltration commences under the capsule and spreads towards the medulla, thereby reducing the cortex and numbers of thymocytes. In humans, thymic size and weight may not change greatly, but rodents may lose 90% of thymic weight with age. The medulla is spared from fatty infiltration and even in old people it contains many cells immunoreactive to antithymulin antibodies (Kendall *et al.* 1991). Thus the thymus still produces thymic factors that may interact with the neuroendocrine axis late in life.

An accelerated loss of cortical cellularity occurs with stress and disease, and can be induced by a wide variety of drugs, hormones, and conditions such as starvation. Many of these situations are times of reduced immunocompetence, which may or may not originate in the thymus. Thymic involvement is discussed below (Section V 6). Acute changes in thymocyte numbers can probably be reversed with time, and age-related changes are restored to pubertal levels in castrated ageing rodents.

III   NEUROENDOCRINE–IMMUNE AXIS

The thymus in reproduction (Marchetti 1989) must be considered in the context of interactions with non-thymic factors such as those of the neuroendocrine axis, especially as the actions of many of them are fundamental to the course of reproduction. This section of the chapter will therefore draw attention to several potential interactions with the thymus in reproduction.

1   The pineal

The pineal gland is of major regulatory importance, usually acting through inhibitory mechanisms to modify the activity of the pituitary, and most other endocrine organs. There is a circadian release of the pineal hormones, melatonin and serotonin, which is greatest during darkness. Both are known to interact with the immune system, and since melatonin affects the release of prolactin (PRL), pineal activity could be important in pregnancy. Pineal hormones influence the onset of puberty by an inhibitory effect on LH and FSH.

Melatonin (reviewed by Cassone 1990), acts through a pineal-immuno-opioid network and can antagonize the immunosuppressive effect of corti-costeroids. Its release enhances primary antibody responses and counteracts the effects of acute stress on antibody production, thymus cellularity, and antiviral resistance in mice. It also modulates the generation of natural killer (NK) cells, expression of IL-2 receptors, and IL-2 production (Maestroni and Conti 1992). Serotonin is an important monoamine neurotransmitter in the brain and a major component of mast cells. It suppresses lymphocyte proliferation, their DNA synthesis, and NK cell activity in human peripheral blood leucocytes, albeit only at quite high concentrations *in vitro*.

Very little is known about the effects of pineal hormones on the thymus apart from studies correlating melatonin, cortisol, and thymulin levels in melanoma patients and healthy adults (Grinevitch and Labunetz 1986). The effects of raised PRL levels on the thymus will be considered below.

2   The hypothalamus

It is generally accepted that the hypothalamic gonadotrophin-releasing hormone (GnRH otherwise known as LH-releasing hormone, LHRH) is the principal factor controlling and directing reproductive function. Its primary action is on cells of the pituitary (Jennes and Conn 1988) that release LH and FSH. In the female, FSH stimulates ovarian follicular development and oestrogen secretion and LH triggers ovulation, corpus luteum formation, and progesterone synthesis. In the male FSH stimulates spermatogenesis, and LH stimulates androgen synthesis by Leydig cells of the testis. Administration of oestrogens or androgens leads to a fall in plasma gonadotrophin levels through negative feedback to the hypothalamus or pituitary.

Within the thymus, low affinity LHRH receptors have been identified although they may differ from those of the ovary or pituitary (Marchetti *et al.* 1989). In rats, a direct effect of LHRH on the thymus was shown histologically, and by thymocyte mitogenic responses to concanavalin A (ConA) in control and hypophysectomized animals, and after LHRH treatment of both groups. Hypophysectomy lowered thymic weight, reduced cellularity, and depressed the mitogenic response. Treatment with LHRH improved all three responses, but did not restore them to control values.

Neither pituitary LH or LHRH were found in the thymus by radio-immunoassay (Zaidi *et al.* 1988), but this does not preclude the presence of other LH- or LHRH-like peptides. Certainly, LHRH-like molecules have been detected in other tissues. Antibodies against FSH, β-FSH, LH, β-LH and other anterior pituitary hormones, gave positive reactions in thymic epithelial cells, especially those of the medulla. FSH reactivity, was however, predominantly in the cortex (Batanero *et al.* 1992). The medullary epithelial cells giving positive reactions are probably those that secrete thymic hormones. This secretion may be functionally important since Rebar *et al.* (1981) demonstrated *in vitro* that thymosin fraction 5 and thymosin-β_4 (but not thymosin-α_1) stimulate the secretion of LHRH from the hypothalamus, and LH from pituitaries when they were superfused in sequence. Treatment of rats with LHRH agonists caused thymic enlargement, but only increased the levels of thymosin- α_1, not thymosin-β_4 (Ataya *et al.* 1989).

Other hypothalamic factors to consider in neuroendocrine–immune interactions are thyrotrophin-releasing hormone (TRH), corticotrophin-releasing hormone (CRH), and growth hormone-releasing factor (GHRF), which act on the pituitary. TRH increases the release of thyrotrophin (TSH) and PRL, and CRH increases adrenocorticotrophic hormone (ACTH) release. GHRF increases growth hormone (GH) levels while somatotrophin release-inhibiting factor (somatostatin) inhibits. Inhibition of TSH and CRH release has been shown at the hypothalamus *in vitro* with thymosin-α_1 (Milenković *et al.* 1992).

3   The pituitary

The anterior pituitary hormones are TSH, somatotrophic hormone (STH), GH, FSH, LH, PRL and peptides derived from pro-opiomelanocortin (POMC), which include ACTH, endorphins, preproenkephalin and enkephalins. The posterior pituitary secretes oxytocin and vasopressin, and the intermediate lobe is generally associated with the release of melatonin-stimulating hormone (MSH). All of these hormones are active in neuroimmune interactions and many influence thymus activity, either directly or through other endocrine organs, but not all have been studied in pregnancy.

Both hyperthyroidism and raised levels of thyroid hormone in the plasma are associated with an increase in thymic weight. Hypothyroidism results in thymic cortical atrophy (Abou-Rabia and Kendall, in press). Thymic epithelial cells bear nuclear receptors for triiodothyronine (Villa Verde *et al.* 1992). There may be interactions during reproduction since the thyroid appears to influence oxytocin levels in the thymus. Epithelial cells in the subcapsule and medulla synthesize oxytocin and vasopressin, and contain neurophysin (Geenen *et al.* 1991), but immunostaining for oxytocin gives different reactions in thymus and brain (Geenen *et al.* 1991). Oxytocin levels in the thymus are very high (Argiolas *et al.* 1990 *a,b*; Jeremović *et al.* 1990) and oxytocin receptors have been detected in rat thymic membrane preparations and thymocytes (Elands *et al.* 1990). The induction of hypothyroidism perinatally in rats causes a rise in thymic oxytocin content, but not in hypothyroid adults with atrophied thymuses. Atrophy due to dexamethasone increases the oxytocin content. While a major function of oxytocin during reproduction is to cause uterine contractions and milk ejection, many other neuroendocrine roles have been assigned to it, especially during stress since it is a cofactor for the release of ACTH (see the review by Jenkins and Nussey 1991). ACTH, GH, and PRL positive cells have been found in the thymus (Batanero *et al.* 1992).

The importance of PRL in immunity has recently been reviewed by Chikanza and Panayi (1991) but apart from its control of thymulin release, little is known of its effect on thymic function, although it has been suggested as a pituitary thymotrophic factor (Hadden, 1992). Raised PRL, GH, and thyroid hormone levels cause the release of thymulin *in vivo* and *in vitro* and PRL promotes an increase in the numbers of KL1[+] medullary epithelial cells (Dardenne and Savino 1990). PRL receptors are present on mature T and B cells (Russell *et al.* 1985) and GH receptors are present on thymocytes (Arrenbrecht 1974). GH has a direct action on mitogen induced proliferation of thymocytes (Grossman and Roselle 1983). The effect of GH is probably mediated through insulin-like growth factor one (IGF-1) and the effect of thyroid hormones primarily through T_3 (Mocchegiani and Fabris 1992), and high levels of either will increase thymulin production. The increased release of thymulin is pleiotrophic and quite slow (over days *in vitro*). It is also effected by Met-enkephalin, β-endorphin as well as adrenal and sex steroids (reviewed in Dardenne and Savino 1990). A rapid release of thymulin, however, is achieved by elevated physiological levels of ACTH *in vitro* and *in vivo* and its action is potentiated by glucocorticoids (Buckingham *et al.* 1992).

Thymosin fraction 5 has been shown *in vitro* to release TSH, PRL, and GH from pituitaries. Its PRL and GH releasing activity does not seem to reside in thymosin-α_1 but in a sequenced protein (MB-35; Badamchian *et al.* 1991). Thymosin action on ACTH release is more confusing: some authors find no release with thymosin-α_1 (McGillis *et al.* 1988) whereas others do (Milenković *et al.* 1992).

Thymulin, but not thymosin-α_1 or thymopoietin, can also act *in vitro* on the pituitary by releasing LH (Zaidi *et al.* 1988). Mendoza and Romano (1989) found a 28-kD factor in thymus extracts that had a dose-related LHRH potentiating action on the release of FSH and LH from cultured pituitary cells, without influencing basal gonadotrophin secretion levels. Similarly thymopoietin and thymopentin stimulate secretion of ACTH, β-endorphin, and β-lipotrophin from cultured pituitary cells, without affecting basal gonadotrophin release (Malaise *et al.* 1987).

4   The adrenal glands

The main hormones of the adrenal are cortical glucocorticoids, medullary mineralocorticoids, with some production of sex steroids. The glucocorticoids, mainly cortisol in humans and corticosterone in rodents, are important in reproductive immunity and are released under the action of ACTH from the anterior pituitary on a circadian basis, and after stress. They are immunoregulatory, anti-inflammatory, have endocrine actions notably on gonadotrophin secretion, and direct actions on the gonads depressing testosterone and oestrogen secretion and elevating progesterone levels (Moberg 1987). There are major differences in sensitivity to corticosteroids between species, rodents being 'sensitive', and humans 'resistant'. In sensitive species, low levels of the hormone cause proliferation and activation of lymphocytes, and high levels not only induce lymphocytopenia, primarily due to the redistribution of lymphocytes from the blood into lymphoid organs, but also inhibit immune function in T cells and macrophages. Within the thymus corticosteroids preferentially affect cortical thymocytes, causing death by apoptosis (reviewed in Kendall 1990). There are, however, more corticosteroid receptors on epithelial cells than on thymocytes (Dardenne et al. 1986). This may account for other effects of corticosteroids, e.g. the potentiating thymopoietic effect of low levels in the embryo (Ritter 1977) and the enhanced migration of pre-thymocytes into the thymus (Bomberger and Haar 1992).

5 The gonads

The gonads, especially in the female during reproduction, are the major source of the sex steroids, principally oestrogen, progesterone, and testosterone. The regulation of the immune system by sex steroids has been widely documented (reviewed by Dougherty 1952; Grossman 1984). Grossman based his views on the importance of sex steroids to the immune reaction by studying sexual dimorphism in the immune response, gonadectomy, and sex hormone replacement, changes during pregnancy, and the presence of receptors in the thymus and bursa of Fabricius.

It has been known for many years that high levels of sex steroids cause the thymus to atrophy, even after adrenalectomy or gonadectomy (Persike 1940; Selye 1946; Dougherty 1952; Greenstein et al. 1986). Progesterone has little, if any, effect. The picture is complicated, however, as thymocytes contain enzymes for the catabolism of progesterone and testosterone. Thymic sensitivity to sex steroids only develops postnatally (Barr et al. 1984), which is interesting as this is the period of time during which thymic activity influences gonadal development. Recent evidence has shown that oestrogen and progesterone control the production of IFN-γ by T lymphocytes within the ovary (Grasso and Muscettola 1992).

Receptors for all of the sex steroid hormones are found on thymic epithelial cells (Grossman et al. 1979; Pearce et al. 1983) and the particular type of oestrogen receptor may be unique to the thymus. In rats oestradiol injections stimulate the cortical epithelium (Glucksman and Cherry 1968). Oestrogen administration results in a loss of MHC class II antigen expression and thymosin-β_4 positive cells, and a gain in vascular permeability and macrophage numbers (Moreno and Zapata 1991). Recent work using more sensitive methods has shown oestrogen receptors in thymocytes, but at a much lower level than on epithelial cells (Kawashima *et al*. 1992). Thus a direct action of sex steroids on thymocytes cannot be ruled out. The finding of oestrogen receptors in blast cells of the thymus is consistent with the fact that high levels of oestrogens increase the numbers of thymocytes in early and late stages of differentiation (Screpanti *et al*. 1989).

Grossman (1984) drew attention to the interactions between sex steroids and thymic hormones. He reported that oestrogen (or testosterone metabolized to oestrogen) inhibits the release of thymic serum factors. The treatment of mice with oestradiol reduced the levels of thymosin-α_1 (Allen *et al*. 1984). These findings, however, are in contrast to those of Stimson and Hunter (1980) where oestradiol treatment caused the appearance of thymus-dependent factors in the serum. Progesterone and oestradiol applied over several days induced thymulin release from cultured thymic epithelial cells (Savino *et al*. 1988). Earlier work, however, had shown a very complex picture of thymulin release during adrenalectomy and gonadectomy. Conflicting results could be due, in part, to the concentrations of steroids used (see Section IV 1).

6   The placenta

While the placenta is the main source of steroids in pregnancy, it also contains many factors that are identical with, or similar to, those of the hypothalamo-pituitary-gonadal axis (Krieger 1982; Smith and Thomson 1991; Zhang *et al*. 1991), numerous cytokines (Buyon *et al*. 1993), and prostaglandins (Embrey 1993). Their actions on the maternal thymus would depend on the ability to cross the placenta, and at least CRH has been shown to do so (Goland *et al*. 1992).

Other placental factors, such as hCG and a transcortin-like protein, can also alter immunity. However, the lower percentage of circulating SP CD4+ cells that coincides with raised hCG levels, as early as the fourth week of gestation after artificial insemination in humans (Degenne *et al*. 1988) could result from hCG action in the thymus or on mature circulating T cells. There is evidence that hCG is produced by medullary epithelial cells (Fukayama *et al*. 1990), but its actions there are as yet unexplored.

IV   MATERNAL IMMUNITY DURING PREGNANCY

There are several influences on maternal immunity that are particular to pregnancy. They are the hormones (gonadotrophins, steroid hormones, and placental lactogens), the pregnancy-specific proteins (α_2-glycoproteins, placental proteins, α-fetoprotein, and progesterone-induced blocking factor), and the antigens of paternal and fetal origin (see reviews by Carter

1984 and Pope 1990). The cytokines are considered to have important effects in pregnancy (Hill 1992) but it is not known whether the altered levels in pregnancy influence thymus function.

1   The steroid hormones

Serum levels of steroids rise dramatically during pregnancy and return to normal at birth or just before it. These hormones may be immunosuppressive but their effects depend on their concentration. It has been generally assumed that the higher levels of steroidal hormones are responsible for the changes in maternal immunity during reproduction. However, there are problems with this interpretation. First, all the hormones are metabolized within the tissues that they influence, and it is not yet clear which of them have the strongest effects on the immune response. Second, they are difficult to measure, so that their levels are uncertain. Third, most experimental studies have used high (pharmacological) concentrations of hormones, rather than normal physiological levels. Although the hormones can profoundly affect immune reactivity, their available concentrations in the circulation are probably too low to have significant direct effects on lymphocytes (Schiff *et al.* 1975; Pavia *et al.* 1979). Thus it is possible that they are not relevant to the immunology of normal pregnancy except where levels rise locally within tissues, as they do at the placenta (Kasakura 1971). None the less, some comments on these hormones are appropriate.

During pregnancy oestrogens reach very high levels in the circulation of certain species (e.g. woman) and in the placenta, which is the main site of their synthesis. The fetus also synthesizes them. Of possible relevance to reproduction is the fact that they block cellular responses to antigens, inhibit the proliferation of lymphocytes and inhibit the activity of cytotoxic T cells. However, they are reported also to stimulate T-cell responses (Lahita 1992) and to promote phagocytic activity in the uterus, especially at implantation and just before birth.

Progestagens are synthesized early in pregnancy by the corpus luteum. Later the placenta may become the major source with contributions from the preimplantation embryo (Dickman *et al.* 1976). In mice and humans the concentrations of progesterones rise early in pregnancy, and in mice they reach a maximum at term (Barkley *et al.* 1979). The binding of progesterone and oestradiol by rat thymocytes (in the cytosol incubation assay) reaches a peak early in pregnancy. Maximal binding of oestradiol occurs before that of progesterone (Symonds 1981; Fig. 5.2.). After day, 10, the levels of both steroids fall to non-pregnant values. Serum progesterone levels correlate with the degree of thymic involution (Chambers and Clarke 1979). In the rat, the infiltration of lymphocytes into the placenta after ovariectomy can be prevented by high doses of progesterone (Clemens *et al.* 1977) which is consistent with its known pharmacological action in

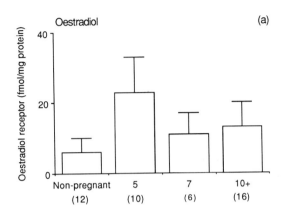

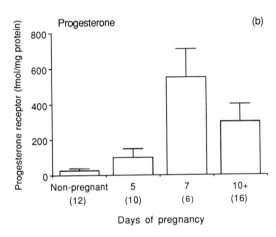

Fig. 5.2   (a) Oestradiol and (b) progesterone binding in rat thymocytes at different stages of pregnancy. Mean ± 1 SEM (number of animals). (Reproduced with permission from Symonds 1981.)

reducing inflammatory responses. Just before parturition uterine mono-cytes increase as progesterone levels fall, but normal cellularity does not return if progesterone is given (Padykula and Tansey 1979).

2   Pregnancy-specific proteins

Although about 30 pregnancy-associated proteins have been described, little is known of their effects on the immune response (see review by Stimson 1983). These proteins include the early pregnancy factor (found in the sera of mice, humans, and sheep), which is synthesized by the ovary. It appears a few hours after fertilization but its function is obscure and the field is controversial. Among the pregnancy-associated plasma proteins (PAPP) is PAPP-A, an incompletely characterized α_2-macroglobulin synthesized by the placental trophoblast. Its concentration rises throughout pregnancy. These proteins which include PAPP-A, pregnancy-associated α_2-glycoprotein, (α_2-PAG), α-fetoprotein (AFP), hCG, and human placental lactogen (hPL) were once thought to have immunosuppressive effects, but their systematic removal from pregnancy serum does not alter its inhibitory properties (Stimson 1983). AFP, a component of amniotic fluid and serum, is thought to inhibit CD4+ T-cell helper activity (Murgita et al. 1977). Determining precisely the activity and specificity of many of these proteins awaits their purification.

3   Antigens of paternal and fetal origin

Many substances of paternal and embryonic origin are potentially antigenic to the mother, and her immune response to them could damage the embryo. They include the antigens on the sperm and other cells in the ejaculate, the proteins and other molecules of the semen, the paternal MHC and non-MHC antigens on the fetus, the fetal-specific antigens, and the antigens of fetal origin on trophoblast and placenta. Although trophoblastic tissue fails to express the polymorphic class I MHC antigens, both human and rat trophoblasts have been shown to possess other non-classical MHC class I molecules, such as HLA-G (Kouvats et al. 1990) and Pa (Kanour-Shakir et al. 1990). In the mouse, MHC class I expression in the trophoblast is still being defined (Hedley et al. 1989) but the expression of class Ib molecules (Qa and TL) has been found in extra-embryonic tissues (Transey et al. 1987; Philpott et al. 1988).

V   PREGNANCY AND THE MATERNAL THYMUS

1   Thymic involution

A dramatic involution of the thymus has long been known to be a feature of pregnancy in mammals, including humans (Hammar 1926; Persike 1940; Pepper 1961). At the end of gestation, the thymus becomes repopulated with lymphocytes and regains its original size. In the mouse 75% of the weight and cells of the thymus are lost by the time of birth. Involution during pregnancy is not restricted to mammals. It also occurs in the viviparous lizard Chaladia ocellatus (Saad 1989).

Involution in mammals has been attributed both to cell death (Dougherty and White 1945) and to an exodus of small lymphocytes from the thymic cortex (Ernström 1970). Evidence from three sources has led to the belief that involution is caused solely by elevated levels of sex steroids. First, the injection of steroidal hormones causes a transient involution similar to that found in pregnancy. Second, in pseudopregnant mice, without embryos but with the hormonal conditions of pregnancy, involution occurs to the same extent as it does in normal pregnancy (Chambers and Clarke 1979). Third, thymic involution is similar in the presence and absence of genetic differences between the mother and her fetus (Maroni and de Sousa 1973; Chatterjee-Hasrouni et al. 1980). However, although the ultimate causes are clearly hormonal, recent work suggests that involution is not solely a response to hormones, but involves reactions to foreign paternal and fetal antigens. Neurological factors may also be important.

Involution during pregnancy in wild mammals is complicated by seasonal events, and the difficulties of obtaining large enough samples of animals. The difficulties are particularly acute when studying small mammals. Life expectation and period of gestation are short, and there are often many pregnancies in one year. It is among such small mammals that reproductive failure, population crashes and a greater susceptibility to disease are well documented. A 'stress hypothesis' was developed to account for these events (formulated by Christian 1950, and reviewed by Lee and McDonald 1985). This clearly implicates the neuroendocrine axis, and alterations in the thymus become very relevant.

Thymic weight changes were studied over 4 years in a small rodent, the bank vole, a species known for population crashes (Kendall and Twigg 1981). Maximum weight was achieved by maturity with a subsequent age-related decline to a low each winter. In females, thymus weight rose slightly in anoestrus, fell in pregnancy, and rose again between pregnancies. In the red fox, Twigg and Harris (1982) found the greatest thymus weights before puberty. A subsequent loss of weight resulted in low values before breeding began. Recovery of the thymus structure and weight was delayed in females until the end of lactation. Similar changes but with a lower amplitude were seen in subsequent years. Chapman and Twigg (1990), studying the thymus in many different species of deer (but only a few were pregnant or lactating females), found among the seasonal breeding fallow deer that thymus weights begin to rise after conception and may be low during the later stages of pregnancy. Among the aseasonally breeding Muntjac deer, there was no clear-cut relationship between thymus involution and sexual state as was seen in foxes. In the Muntjac with a post-partum oestrus, where pregnancy and lactation may be concurrent, the stress of breeding might be expected to be associated with very low thymus weights. While there was a tendency for thymus weight to fall in pregnancy, it did not always do so. Despite the difficulties of obtaining data from the field, such studies are valuable as the differing patterns of reproduction can sometimes enable basic concepts to be identified.

2 Immunology

Several experiments with rodents have suggested that immunological factors affect the thymus during pregnancy. C57BL female mice, immunized to CBA antigens before mating to CBA males, do not show the normal involution of the thymus. Immunization to a non-related T-dependent antigen, or immunization to the paternal strain in the absence of pregnancy, has no effect (Clarke 1979). Other experiments show that earlier pregnancies by CBA males also prevent involution (Clarke 1984, 1988). However, in the reciprocal mating (CBA females mated to C57BL males) involution is normal regardless of immunization or of earlier pregnancy, so the effect appears to be strain-dependent. In any event these experiments show that the physical response of the thymus can be altered by maternal immunity to paternal antigens, whether it is gained naturally by mating or experimentally by injection. However, there is a paradox. Involution can also be prevented by exposure to non-paternal antigens but only in allogeneic, not in syngeneic matings. C57BL females were given skin grafts from C57BL females, C57BL males, CBA females, or BALB/c males, or were left ungrafted, before mating to CBA males. Grafting from self (C57BL female) or from a third party (BALB/c) significantly reduced involution. Immunization to C57BL male or CBA female tissues, however, did not significantly alter thymus weight. In matings within the C57BL strain, females grafted with self or with male tissues showed a normal involution of the thymus (Clarke 1984). At this stage we cannot explain these results. If the prevention of thymic involution is a non-specific effect of skin grafting, it is only associated with an allogeneic pregnancy.

Studies comparing the degree of thymic involution in allogeneic and syngeneic pregnancies have given conflicting results. Although there is agreement that thymic involution occurs during pregnancy, its time of onset has been disputed. In the mouse some experiments show that it starts at the beginning of pregnancy (Pepper 1961; Miller et al. 1973; Maroni and DeSousa 1973; Phuc et al. 1981a), while others, some of which did not examine early pregnancy, report that it only occurs during the second half (Persike 1940; Ito and Hoshino 1962; Chambers and Clarke 1979; Chatterjee-Hasrouni et al. 1980; Gambel et al. 1980). In rats, most workers have not seen involution during the first half of pregnancy.

In mice there is a transient enlargement of the thymus before implantation. On day 4 the thymus is significantly heavier, and cell numbers are greater, in allogeneically mated than in syngeneically mated females. The effect is greater in multiparity than primiparity (Fig. 5.3). The enlargement is apparently not strain-specific; as it occurs in C57BL, CBA, and (on day

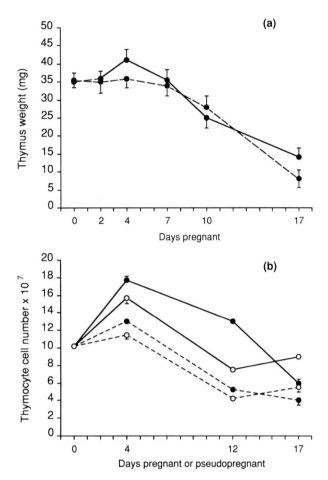

Fig. 5.3   (a) Thymus weight (± SEM) during pregnancy in allogeneically (●——●) and syngeneically mated (●– – –●) CBA mice. (b) Total number of thymocytes (± SEM) during normal multiparity and after several sterile matings to an allogeneic or syngeneic strain male Allogeneic mating, (○——○); syngeneic mating, (○– – –○; allogeneic pregnancy, (●——●); syngeneic pregnancy (●– – –●. (Reproduced with permission from Clarke 1984.)

3) in outbred BKTO mice (Clarke 1984, 1988). The earlier difference in outbred mice may be related to their faster embryonic development. The enlargement is independent of at least one of the hormones that normally cause involution. Levels of circulating progesterone are similar in both syngeneic and allogeneic pregnancies on all days measured. This suggests that early in allogeneic pregnancies some other mechanism is overriding

the effects of elevated progesterone levels. The early enlargement of the thymus is transient, and by day 7 the amount of involution is the same in both kinds of pregnancy [as found by Maroni and de Sousa (1973) who did not look earlier than day 7].

Studies of the rat also show a greater thymic weight and cellularity in early pregnancy in some strain combinations (Leeming *et al.* 1984). It was not associated with the presence of an allogeneic pregnancy, although more thymocytes were present (Leeming *et al.* 1985). However, later work did not find changes in thymic weight (Habbal and McLean 1992*a*).

In late pregnancy, thymus weight and cellularity are much lower in allogeneically than in syngeneically mated C57BL female mice, but this is not true of CBA females (Clarke 1984). Differences in the amount of involution between the two kinds of pregnancy were not found by Maroni and de Sousa (1973), but were found by others (Miller *et al.* 1973; Baines *et al.* 1977; Chambers and Clarke 1979; Chatterjee-Hasrouni *et al.* 1980; Chaouat *et al.* 1982). In rats the degree of involution seems to depend on the combination of mating strains and not on whether the mating is allogeneic or syngeneic (Leeming *et al.* 1984). So the mouse and rat may behave differently in both early and late pregnancy. Chaouat *et al.* (1982) suggested that an immunological response to alloantigens is responsible for the late difference; however, there may be another explanation. In both CBA and C57BL mice there is a negative correlation on day 17 between the number of embryos and thymic weight (Clarke 1984). Perhaps a larger number of embryos (and placentas) raises the level of circulating steroids, and thereby causes more involution. Circulating progesterone levels are negatively correlated with thymic weight (Chambers and Clarke 1979).

3   Changes in subsets of thymocytes

Until recently it has been thought that the only thymocytes affected by pregnancy are the small DP cortisone-sensitive cells of the cortex. However, virtually all the cell populations are affected. In mice the loss of thymocytes begins on day 4 after syngeneic matings, with focal depletions in the cortex. This does not happen, or at any rate happens to a much slighter extent, in allogeneic pregnancy when there is either a greater volume of cortex with more thymocytes (CBA mice), or a thymus resembling that in non-pregnant controls (C57BL mice) with no focal depletions (Clarke 1984). In syngeneically mated rats, there was a significantly greater proliferation of lymphocytes, measured by the incorporation of thymidine, on days 3 and 4 (Habbal and McLean 1992*a*).

Studies of mice using monoclonal antibodies to thymocyte antigens, analysed by flow cytometry, show more DP cells early in pregnancy. This is consistent with the presence of an enlarged cortex. At the same time the numbers of large precursor DN cells and mature SP CD4+ and SP CD8+

cells are higher than those found in virgins (Clarke 1989; Clarke and Miller 1991). Later in pregnancy, involution is accompanied by the loss of many DP cortical cells, and this is consistent with the smaller cortex. SP cells and large DN cells are retained throughout pregnancy.

These findings are supported and extended by recent studies using multiple colour analyses, where the numbers of CD3[+]SP CD4+, CD3[+] SP CD8+, and DN cells rose early in pregnancy and those of DP cells fell later (Brunelli *et al.* 1992). It is not clear to which subset the DN cells belong. When they are large, they are thought to be immature TCR[-] precursor cells just arrived from the bone marrow. When they are small, they are thought to be mature cells expressing the a/β or γ/δ TCR epitope (von Bohmer 1988). In the Brunelli *et al.* (1992) study the DN cells of pregnancy were small and expressed low levels of Thy[-1], suggesting maturity, whereas the larger cells, found by Clarke and Miller (1991) in mice and by Habbal and McLean (1992*a*) in rats, suggested immaturity. Perhaps both types of DN cell increased. In mice, the mortality of thymocytes fall early in pregnancy, followed by a sharp rise. The numbers of large blast cells show an early rise before implantation and then a later and larger rise (Clarke 1988). The numbers of dead cells rise steeply. More Mel-14[+] cells are present from the fourth day until the end of pregnancy. These cells are believed to emigrate to the periphery (Clarke 1989). Large cells accumulate in the subcapsular region of the cortex early in pregnancy. They are negative for CD4, CD8, and Mel-14 markers. Plasma cells and mitotic figures are retained throughout pregnancy (Clarke and Kendall 1989). Proliferating cells increase in number (Brunelli *et al.* 1992). In the medulla of the mouse, mature thymocytes seem to increase after mid-pregnancy.

The thymus of pregnant mice shows a higher percentage of PNA[-] cells (which belong to a mature phenotype; Chaouat *et al.* 1982). The percentage is greater in allogeneic than in syngeneic pregnancies, and greater in multiparity than in primiparity. Interestingly, the proliferation of PNA cells could be mimicked by injections of placental extracts. Extracts from allogeneic placentae caused the largest increase. This study found no changes in the proportions of cells reacting with anti-Lyt1 or anti-CD4 antibodies. However, the analysis with single markers does not allow the four major functional thymocyte subsets to be separated. In the context of a rise in PNA[-] cells in pregnancy, the finding that the interaction of PNA[-] and PNA[+] cells is involved in the generation of suppressor T cells (at least *in vitro*; Eisenthal *et al.* 1982) is intriguing, particularly in combination with the earlier results of Phuc *et al.* (1981 *a,b*) who found fewer PNA[+] cells late in syngeneic pregnancies. They also showed that after hydrocortisone treatment of pregnant mice, the remaining cells of the medulla had a lower response to the lectins phytohaemagglutinin (PHA) and ConA than did similarly treated cells from virgin mice, where responses to lectins are strongly stimulated. However, co-culturing medullary thymocytes from virgin and pregnant mice gave no evidence of T-suppressor cells, or of any alteration in thymic function. Phuc *et al.* (1981*a*) did not report changes in other thymocyte subsets, but again double labelling was not used. By contrast, polyclonal stimulation by mitogens is greater in thymocytes from syngeneically pregnant mice than in those from virgin controls. The activity is correlated with the amount of involution, and is located in mature SP CD4+ (Lyt1) cells, presumably of medullary origin (Hooper *et al.* 1987).

4   Changes in non-thymocytes

The non-thymocyte cells of the thymus are much affected by pregnancy (Clarke 1988; Clarke and Kendall 1989; Clarke *et al.* in press). As the numbers of thymocytes in the cortex fall so there appear to be more epithelial cells, macrophages, and mast cells. Macrophages and some

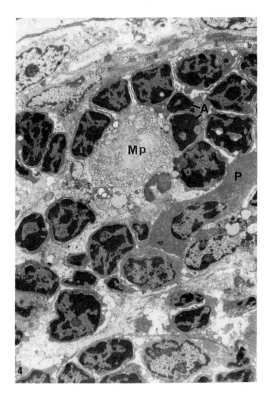

Fig. 5.4   An electron micrograph of the subcapsular cortex of the mouse thymus in late pregnancy to show a macrophage rosette (Mp) with apoptotic thymocytes (A). Nearby is a mature plasma cell (P). (Reproduced with permission from Clarke and Kendall 1989.)

epithelial cells, mainly in the subcapsular region and the cortex, become extensively phagocytic for apoptotic thymocytes (Fig. 5.4). Sometimes these cells can be observed completely surrounded by rosettes of thymocytes. Late in pregnancy, after phagocytosis, the epithelial cells become electron dense, suggesting that they are dying. Macrophages with numerous phagocytic vacuoles are often seen in the perivascular spaces, as if they were leaving the gland. Other cells, notably B and plasma cells, invade the gland.

In virgins and early in pregnancy, channels filled with thymocytes and devoid of epithelial cells run through the cortex, sometimes as far as the medulla. The channels are greatly reduced by the middle of pregnancy, when there is a loss of cells. These structures have been reported before in other situations but their function is unclear (Rozing et al. 1989).

The changes in the murine thymic medulla are very striking (Fig. 5.5). Staining with antibodies shows that, by mid-pregnancy, type 6 'medullary epithelial cells form large structures surrounding spherical masses of mononuclear cells, and sometimes encircling blood vessels. We have called these structures, which apparently have not been described before, medullary epithelial rings' (MERs) (Clarke et al. in press). As pregnancy proceeds, the central mononuclear cells fall in number and the MERs collapse. None the less, the numerous fibroblasts, epithelial cells, and extracellular components of connective tissue are retained until late in pregnancy. Occasional small MERs are seen in virgin mice, and in those at early stages of pregnancy.

The MERs are not visible when the thymus is stained with haematoxylin and eosin. This may explain why they had been overlooked, and why the full extent of medullary changes had not been appreciated (Clarke 1988; Clarke and Kendall 1989). However, a larger medulla had already been noted (Miller et al. 1973; Levett 1981), as had a rise in the number of mature medullary (PNA-) cells late in pregnancy (Chaouat et al. 1982).

The origins of the cells within the MERs can only be guessed. They could have developed in the cortex and then entered the medulla. They could be recent immigrants from the periphery, or clonal expansions of medullary subsets. Since many cortical thymocytes are dying at this stage, survivors may be a special subset. Dolgova (1985) has suggested that in rat pregnancy the medulla harbours thymocytes accumulated from the cortex.

Trafficking to the thymus occurs in pregnancy, as at other times (see Section II 5). Enlarged PVS containing cells have been seen early in pregnancy, but they shrink later when the MERs are large (Clarke 1988; Clarke and MacLennan, unpublished observations). Shrinkage of PVS may be peculiar to pregnancy, since enlarged PVS are associated with thymic involution in old age and under stress (see Clarke and McLennan 1989).

From mid-pregnancy the epithelial cells forming the MERs, and those remaining after the MERs collapse, secrete thymulin. This is unexpected, since Phuc et al. (1981b) found that concentrations of active plasma thymulin

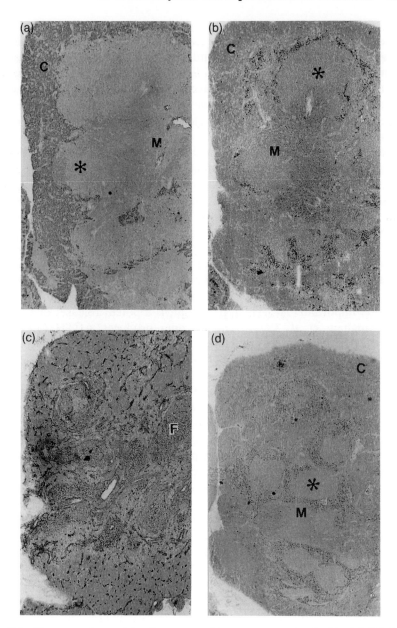

Fig. 5.5   Frozen sections of the thymus at mid-pregnancy stained with: (a) 4F1 for cortical epithelial cells; (b) ERTR5 for medullary epithelial cells; (c) ERTR7 for fibroblasts and connective tissue components; and (d) A2B5 for neuroendocrine cells. C = cortex, M = medulla, F = fibroblasts and *= MERs. Reproduced with permission from Clarke et al. (in press).

in pooled sera do not change from days zero to 16 of mouse gestation. They concluded that epithelial functions are unaltered during thymic involution. Levels of thymosin-α_1 are significantly higher after the second trimester in human pregnancy (McClure *et al*. 1982), and higher circulating levels of cyclic adenosine monophosphate have been found early in mouse pregnancy (Clarke and Jequier, unpublished observations). Raised concentrations of thymosin have been held responsible for the inhibition of the sheep red blood cell rosette and cell-transformation assays by sera from pregnant animals (Kasakura 1971, Blackstock and Stimson 1976). Osoba (1965) suggested that these hormones can come from the fetus into the maternal circulation, and thereby contribute to the high levels found late in pregnancy. Synthetic serum thymic factor is a potent stimulator of suppressor T cells (Bach *et al*. 1981).

What causes the hormone-secreting epithelial cells to proliferate? There are many candidates, especially among the sex steroids (see Section III 5). A high local concentration of thymic hormones could affect the development of thymocytes in the MERs. In stimulating the release of thymic hormones, both intrathymic factors and placental hormones may be important. ACTH, in particular, is a potent releaser of thymulin (Buckingham *et al*. 1992). The placenta produces ACTH (and other POMC-derived peptides) in the second and third trimesters of pregnancy (Smith and Thomson 1991) when the numbers of medullary epithelial cells increase.

As well as the development of MERs, there are other cellular changes in the medulla. Flow cytometric and histological studies show β_7 integrins, detected by the monoclonal antibodies M290 and M293 (Kilshaw and Murant 1991) and thought to have a role in adhesion processes in the mucosa, are up-regulated on thymic medullary lymphocytes early in pregnancy (Clarke and Kilshaw, unpublished observations). The greater expression of 4F1 antigen on epithelial cells at the same time (Clarke *et al*., in press) supports this finding, as 4F1 is also believed to be an adhesion molecule involved in T-lymphocyte differentiation (Imami *et al*. 1992). As Moreno and Zepata (1991) have suggested, epithelial cells probably contribute to the immunological changes within the thymus. Their hyperplasia in pregnancy suggests a role in the immune interactions during gestation. The functional significance of the changes remains to be explored.

Fibroblasts may also be important to cell interactions in pregnancy. Recently, Anderson *et al*. (1993) have showed that they are needed for some steps in the maturation of T cells. Whether the higher numbers of medullary fibroblasts in late pregnancy are active in this way needs to be investigated.

It is clear the architecture of the medulla greatly changes during pregnancy and we suggest the following sequence of events (Fig. 5.6). After an initial increase in cortical tissue and cells they are then lost. At

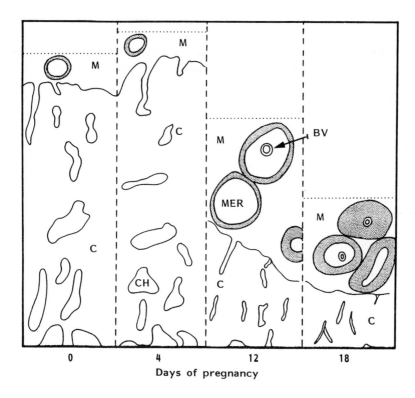

Fig. 5.6   Schematic diagram of thymic changes during mouse pregnancy. C = cortex, M = medulla, CH = epithelial-free channels of the cortex, BV = blood vessel and MER = medullary epithelial ring. The dotted line at the top indicates the centre of the medulla. Reproduced with permission from Clarke *et al.* (in press).

mid-pregnancy the medulla enlarges and MERs develop. In late pregnancy the medulla reduces in size leaving collapsed MERs and large numbers of epithelial cells and fibroblasts.

5   The effects of pseudopregnancy

During a normal pregnancy the levels of hormones and the immunological stimulation by paternal and embryonic cells are interrelated, and the influence of each alone cannot be assessed. Attempts to separate them have used pseudopregnant rodents. Pseudopregnancy lasts for 10–12 days, and is dependent on a prolactin surge with LH, FSH and oestrogens. The way it is induced determines the degree of maternal exposure to paternal and fetal antigens. Ligation of the oviducts and mating to normal males gives an exposure to the whole ejaculate. Mating a normal female to a vasectomized male excludes sperm but exposes the female to all the other components of the ejaculate. Mechanical stimulation during oestrus produces a pseudopregnancy without an ejaculate. All methods show levels of circulating hormones similar to those found in pregnancy.

The use of sterile females shows that the early response of the thymus is independent of the embryo and the sperm, but could be caused by some other component of the ejaculate (Clarke 1984). Mated mice, whether they are sterilized or not, have significantly larger thymuses than virgins soon after mating. The enlargement also occurs after several matings to a vasectomized male, and is greater with allogeneic males than with syngeneic males. It is always less than in normal pregnancies, but is greater after many sterile matings than after a single one (Fig. 5.3). In mice, the effect of pseudopregnancy by mechanical stimulation has not been studied. In rats, only mechanical stimulation has been used. With some strains, the numbers of thymocytes, and mitotic activity are all greater than in non-pregnant animals (Habbal and McLean 1992*b*). Since in mice the proliferation of thymocytes and the secretion of oestrogen both reach a peak at about day 3 of pseudopregnancy, it has been suggested that the early cellular response depends on the oestrogen surge after mating (Shelesnyak *et al.* 1963). The numbers of macrophages and polymorphonuclear cells in the developing deciduomata are equal to those found in normal pregnancy (Choudhuri and Wood 1992). Taken together, all these results suggest that events in pseudopregnancy resemble those in pregnancy but are less marked. The neuroendocrinological factors and responses to the non-sperm components of the ejaculate probably act together to induce the early rise in thymus weight and in the number of thymocytes.

The thymocyte subsets in pseudopregnant mice (mated to vasectomized males) show differences from, and similarities to, those in normal pregnancy (Clarke 1989). The rise in SP CD4[+] and SP CD8[+] cells soon after mating resembles that in normal pregnancy. These cells are retained after pseudopregnancy (17 days after mating) whereas after normal pregnancy they are lost. The numbers of Mel-14[+] cells are similar in pseudopregnancy and pregnancy until day 12, when they fall in pseudopregnancy but continue to rise in pregnancy.

Oestradiol probably plays an important role in these changes. The rise in the numbers of cycling cells, and DN, SP CD4[+], and SP CD8[+] cells, is as great in virgin mice injected with low doses of oestradiol as it is in pregnant mice (Brunelli *et al.* 1992). The higher numbers of mature cells in the treated virgins compared with untreated ones suggest that the changes are not merely due to the lysis of immature cells, but to an accelerated maturation of thymocytes. Although oestrogen causes similar phenotypic changes in lymphocytes as are found in pregnancy, its effects on their function are different (Shinomiya 1991).

An alternative stimulus to the thymus could be through the afferent nerves from the genitalia during mating. This stimulus is essential for the initiation of pseudopregnancy (and pregnancy) in both the rat and mouse. Neither state can be elicited after bilateral pelvic neurectomy unless exogenous progesterone is given (Kollar 1953). Stimulation of the central nervous system has multiple effects on the release of hormones, neuropeptides and other factors, all of which affect the immune system (see Section III).

6   The significance of thymic changes for pregnancy

The significance of all these changes in the cells of the thymus during pregnancy and pseudopregnancy remains to be discovered. It seems inevitable, however, that such radical rearrangements will have large effects on the maternal immune system.

There have been many suggestions to explain how the embryo manages to escape the harmful consequences of maternal immunity but the precise mechanisms are still unknown. One of the more promising approaches is to enquire if the maternal immune system actively suppresses harmful cellular immune responses. This is the only mechanism that we will consider here. The possible roles of B cell immunity, of enhancing antibodies, and of the diminution of trophoblastic and other antigens have been discussed elsewhere (Loke 1989; Herrera–Gonzalez and Dresser 1993). It has been suggested that several kinds of suppressor cells act at the maternal–fetal interface and some of these cells could originate in the thymus. As first pointed out by Clarke (1984), if the protection of the embryo is aided by a specific inhibition of maternal immune responses to paternal and fetal antigens, then it should start as soon as possible after mating. The mother would then have ample warning of antigens carried on the embryo. We believe that the changes in the thymus during pregnancy may reflect corresponding changes in the populations of cells that promote clonal deletion or anergy within the thymus or the periphery, processes that could help suppress maternal responses to embryonic or paternally derived epitopes, and help the embryo to survive despite its antigenic differences from the mother.

In considering such a role for the thymus, we must take account of evidence suggesting that it is not necessary for a successful first pregnancy. Nude mice, which lack a functional thymus and do not generate T cells, can become pregnant if they are kept under germ-free conditions. Placental weight and litter size are unaffected by the absence of a thymus, but fetal weight is smaller, probably because they are less healthy (Hetherington 1978). In rats, the average body weight of the young born to thymectomized mothers is also significantly less than normal (Csaba *et al.* 1975). Most nude female mice are fertile, but only 20% of them rear their litters to weaning age (Hetherington and Hegan 1975). Thymectomized mice can carry a first allogeneic pregnancy to term, even after prior immunization to paternal alloantigens, but the number of reabsorbed embryos is higher, and there is a concordant reduction in litter size (Clarke, unpublished observations). SCID mice, which lack T cells (as well as B cells, IgGs, and NK cells) are able to breed (Hetherington and Dresser 1990). Croy and Chapeau (1990) have examined the establishment of, and maintenance of, pregnancy in *scid/scid bg/bg* mice. They have normal pregnancies and litters and contain the usual populations of polymorph leucocytes and metrial glands in the uterus. This has been taken as evidence against an immune involvement in mammalian pregnancy.

Although pregnancy is possible when the mother lacks a thymus, it is apparently compromised to some extent. An important point is that immunologically deprived mice readily accept allografts. It has been argued that in the absence of an effective immune system there is no need to develop the means to inactivate it (Clarke 1984). If both the functioning of immunity and the development of specific unresponsiveness (i.e. positive and negative selection) reside in the thymus, its removal will take away both the problem and its solution. Furthermore, pregnancy can restore competence to immunologically deprived mice. The restoration may come about by thymic factors from the embryos passing through the placenta into the maternal circulation (Osoba 1965).

One of the most fascinating areas in immunology involves the types of cells that apparently occur only in pregnancy. Many such cells have been isolated from tissues such as the decidua, but rarely have they been precisely characterized, nor do we know their role in non-pregnant animals. We are concerned here with the possibility that some of these cells originate in the thymus, even though information on the subject is sparse. Some of the cells are certainly thymus-derived CD3+ conventional T cells. Others that are CD3− could either be independent of the thymus, or derived from a minor subpopulation of thymocytes. We will therefore give brief descriptions of the most important cell groups so far described, and refer the reader to reviews or recent papers.

By far the best-documented are the granular endometrial gland (GMG) cells, which have been reviewed by Bulmer (1989). They are also called uterine NK cells, and are thought to be a subset of the NK cells found circulating in non-pregnant animals (Kiso *et al.* 1992). Their phenotype (CD3− CD4− NK1.1+ Asialo-GM1+ Thy1+ FcR+ Fcγ− LCA+ IgM− IgG+ LGL-1+ CD45+) does not immediately suggest a thymic origin, and they are often supposed to have come from the bone marrow. However, the phenotype is not characteristic of B-derived cells either, and they might well be atypical T cells.

Null cells (Thy1− SIgM−) are not regarded as T cells, B cells, or macrophages. They are not unique to pregnancy, but occur in neonates, in old mice, and in normal mice after parasitic infections (Hoskin *et al.*

1989). They may be identical with some of the natural suppressor cells (NS) described by Clark and McDermott (1981). NS cells can be classified as phase A or phase B. Phase A cells (phenotype not known) occur early in pregnancy. They are large, and lack cytoplasmic granules. They are believed to act on the afferent arm of the cellular immune response without influencing the production of antibodies. Phase B cells occur in the second half of pregnancy. They are small and associate with other cells containing eosinophilic cytoplasmic granules. Their phenotype is $Thy1^-$ $Lyt1^-$ $Lyt2^-$ $Mac1^-$ NK^-. They produce soluble suppressor factors that block the proliferation of T cells in response to IL-2. They also block the production of antibodies, and may be involved in the rejection of allografts. Their absence is associated with spontaneous abortion. They are probably independent of the thymus, since they are present in pregnant nude mice (see review by Clark et al. 1987).

Also with suppressor activity are the six or more types of T cells that are γ/δ^+, and many of which are $IL-2R^+$ activated cells. They have been described in the uterus of the mouse (Heybourne et al. 1992) and the human (Mincheva-Nilsson et al. 1992). They occur early and late in pregnancy, and they suppress the immune response by inhibiting a/β-T-cytotoxic cells. At the maternal-fetal interface in the uterus their numbers are increased 100-fold. They recognize the MHC class Ib (Qa and TL) epitopes uniquely found on mouse embryonic tissues (Ito et al. 1990).

A better characterized group comprises the pregnancy-associated natural suppressor cells (SPAN cells) found in the spleen of pregnant mice (Brooks-Kaiser et al. 1992). They are thought to be immature T cells, and have the phenotype $CD3^+$ $CD4^-$ $CD8^-$ WGA^+ SBA^+ $CD45R^+$ $J11d.2^+$ $NK2.1^-$ $Asialo-GM1^-$. They produce a soluble suppressor in pregnancy, and are also found after irradiation and treatment with cyclophosphamide. They are present in the spleens of mice immunized to Mls[a] (Bruley-Rosset et al. 1990), suggesting that they play a part in the induction of tolerance. They also occur in human cord blood (Kawano et al. 1990).

Antigen-specific T cells (SP $CD4^+$ or SP $CD8^+$) with suppressor activity have been reported in the uterus and spleen of pregnant mice (Thomas and Erickson 1986).

It is not clear whether the cells found within the medullary MERs during mid-pregnancy (described above) contribute to any of these populations of peripheral cells. If they do, there are two possibilities–either they differentiate within the thymus perhaps in response to paternal/fetal antigens that enter the organ soon after mating, or they are precursor cells that may develop further after being exported to the periphery.

The ability of antigen to enter the thymus, and of activated peripheral T cells to migrate back into it, has been discussed earlier in Section II 5. As suggested by Clarke (1984), antigens of paternal origin (in some non-sperm component of the ejaculate), and/or fetal antigens, can enter the thymus, and may induce a shift in the thymocyte populations that is reflected in the peripheral lymphoid tissues and uterus following their emigration. The similarity in timing of the ability of effector cells to return to the thymus and the early response of the thymus to paternal antigens suggests a connection between the two processes. We have proposed that clones reacting to paternal and embryonic antigens in the thymus may be deleted by negative selection (Clarke 1984, 1988; Clarke and Kendall 1989). Involution and cell death in the cortex could reflect a temporary 'shut-down' that inhibits the production of clones active to these antigens (Clarke *et al.* in press). This idea extends a suggestion by Aronson (1991) when discussing the significance of thymic involution in old age and after stress. He proposed that thymic involution reduces the risk of autoimmunity when there is an increased exposure to altered self (see Section VII). In pregnancy, fetal tissue in contact with the mother is half self (maternal epitopes) and half foreign (paternal epitopes), or else is modified in its antigenic expression (e.g. in sperm and trophoblast). Both alternatives may represent a situation of altered self leading to the down-regulation, both non-specific and specific to the antigens of the paternal strain, consistently found in pregnancy (Munroe 1971; Smith and Powell 1977). On the other hand, thymocytes responding to paternal epitopes might be rendered anergic within the thymus. The recognition of self-antigens on epithelial cells by TCRs is believed to be responsible for clonal anergy within the thymus (Roberts *et al.* 1990). A similar mechanism might operate for altered self. Obviously, some other mechanism must be responsible for the inactivation or removal of paternally reactive clones already in the periphery. The atypical suppressor cells associated with pregnancy, and described above, may well be components of this mechanism.

VI   THE RECOVERY OF THE THYMUS AFTER BIRTH

The regeneration of the maternal thymus soon after birth was first noticed in 1913 and later well documented by others. The organ immediately starts to be repopulated with lymphocytes and, in rodents, 3 weeks later regains its original weight and numbers of cells. If birth is followed by lactation, however, involution is maintained throughout the suckling period in rodents, i.e. except the guinea-pig. Grégoire (1947a), using rats, neatly demonstrated that the suckling stimulus maintains involution. When litters are removed from mothers at birth, regeneration of the thymus rapidly follows. When litters are kept with mothers, involution continues during lactation beyond the level found at birth. The maintenance of thymic involution during suckling and its regeneration afterwards are independent of the ovarian hormones. They both occur in spayed rats. Furthermore, the rate of regeneration is negatively related to the number of young and of functional nipples. Neurological stimuli from the nipples are clearly involved.

There are several hormonal changes that might also affect involution post-partum. PRL reaches high levels while progesterone and oestrogen fall. However, the injection of PRL at physiological doses fails to maintain involution in rats whose young have been removed at birth (Grégoire 1947b). Possibly the lower levels of oestrogen and progesterone after birth are sufficient stimuli for regeneration. There have been few studies on thymic subsets of cells during regeneration post-partum. PNA$^+$ cells increase, from 2.8 to 67.8%, probably reflecting the recovery of immature cortical cells (Phuc et al. 1981a). The death of thymocytes falls from about 12% late in murine pregnancy to normal levels (5%) three weeks post-partum. Over the same period, the percentage of large blast cells also falls (42–29%; Clarke 1988). T cells from the spleen of mice on day 3 post-partum are Lyt1$^+$ Lyt2$^-$ (Yokoyama et al. 1988). Since these cells are absent from the spleens of thymectomized mice, they are thought to come from the thymus. It has been suggested that they induce T-suppressor cells to inhibit the production of IgG1, which is found at high concentrations in the serum early post-partum, but disappears later. The ability to transfer the inhibition adoptively by cells supports the idea. Consistent with suppression post-partum are experiments showing that the graft-versus-host reaction in mouse spleens is reduced at this time (Gambel et al. 1980). It has been suggested that mobilization of cells from the thymus depends on oestrogens. However, the numbers of thymocytes in the thymus, and oestrogen levels in the circulation, are very low post-partum, and there may not be enough of either for such a process. Furthermore, since the numbers of Mel-14$^+$ cells are low at this time (Clarke 1989), rates of emigration are unlikely to be high.

VI   RHEUMATOID ARTHRITIS (RA) AND PREGNANCY

The effects of pregnancy on patients with rheumatic diseases are currently attracting great interest, and have recently been the subject of an international symposium (reviewed by Spector and Da Silva 1992). RA improves in 75% of pregnant patients and exacerbates after delivery. The factors and cells potentially responsible for the changes have recently been reviewed by Pope (1990). The mechanisms are still obscure, but the disease is believed to be mediated by T cells, suggesting that the effects of pregnancy on T lymphocytes may be clinically important. Recent studies have consistently found a fall in circulating CD3$^+$T lymphocytes, particularly the SP CD4$^+$ subset, as early as the fourth week of gestation (Castilla et al. 1989).

Collagen-induced arthritis (CIA) is produced by injecting collagen into susceptible strains of mice, and closely mimics the symptoms and progress of RA in humans. It too is eliminated or reduced during pregnancy, and returns post-partum (Waites and Whyte 1987). Its onset is delayed in both syngeneic and allogeneic pregnancy, but allogeneic is reported to be more beneficial. There is a decline in collagen auto-antibodies that is reversed post-partum (Whyte et al. 1988).

Oestradiol is important in the remission of CIA and RA (Holmdahl et al. 1987) although its benefits as a treatment to women may be limited (Bijlsma et al. 1992). Exacerbation of CIA is prevented by oestradiol injections (Mattsson et al. 1991). Glucocorticoids are less likely to be involved. There is little correlation between disease and plasma corticosterone levels (Persillin 1981). Because they modulate autoimmunity, thymic hormones are believed to play a part in CIA and RA. Thymopentin delays the appearance of antibodies when autoreactive cells are adoptively transferred to naive recipients (Lau et al. 1980). An unconfirmed observation, based on a non-quantitative assay, suggests that the level of thymulin is raised in RA (Dardenne and Bach 1981), but treating RA patients with synthetic thymulin produces clinical improvement (Amor et al. 1987). PRL, which increases the secretion of thymulin in vivo and in vitro (Dardenne and Savino 1990) has been reported to promote CIA in rats (Berczi and Nagi 1982). Bromocriptine (which inhibits PRL release) delays the return of the disease after birth in mice, and induces resistance to the disease in rats (Whyte and Williams 1988). In CIA (Mattsson et al. 1991) although not in RA (Ostensen et al. 1983) there is evidence that lactation, which increases prolactin, none the less prevents the exacerbation post-partum. There is conflicting evidence for other roles of the thymus in RA and CIA. Diseased individuals show higher levels of autoreactive T cells. CIA can be transferred by SP CD4+ T helper cells (Holmdahl et al. 1985). Injections of monoclonal antibodies against SP CD4+ cells prevents the disease (Herzog et al. 1989). It is difficult to induce arthritis in athymic nude rats and chronic disease does not develop (Renz et al. 1989). Also pristane-induced arthritis does not develop in athymic nude mice (Wooley et al. 1989). These facts point to a role for the thymus in CIA, at least in generating the T-cell repertoire. However, in rats, the suppression of CIA by oestrogen is apparently independent of the thymus. Oestrogen treatment of thymectomized and castrated females was as effective as it was in intact females (Larssen et al. 1989). An opposite conclusion was reached by Luster et al. (1984), whose experiments with mice indicated that the suppression of cell mediated immune responses by oestrogen is reduced in thymectomized mice, again supporting a role for the thymus in oestrogen-mediated suppression.

The consequences for arthritis of losing cortical thymocytes during pregnancy is unknown. Although loss of the cortex is generally thought to reflect a fall in thymic function and lymphocyte output (Schuurman *et al.* 1991), a beneficial effect has been proposed by Aronson (1991), who argues that when the risk of autoimmunity is high (e.g. in old age, and under chronic stress) the death of cells in the cortex, where negative selection is believed to take place, could permanently or transiently render the animal less immunoreactive and less autoreactive. We have suggested that this may also happen in pregnancy (see Section V 6). A relatively inactive cortex could be of advantage if potentially autoimmune thymocytes are down-regulated and could help to explain the remission of many autoimmune diseases in pregnancy. Since the disease exacerbates post-partum it is unlikely that the autoreactive clones are deleted. They are more likely to be reversibly suppressed or inhibited for the duration of pregnancy, probably by oestrogen suppressive factors and/or one or more of the pregnancy-specific factors discussed above. However, other processes must be involved in those autoimmune diseases that arise or are exacerbated during pregnancy (e.g. gestational diabetes and myasthenia gravis).

ACKNOWLEDGEMENTS

We would like to thank the Volkswagen Stiftung for funding, and MK is also grateful to the Welton Foundation for support. We are indebted to Professor Bryan Clarke for helpful criticisms of the manuscript, and to Julie Brown for typing.

REFERENCES

Abou-Rabia, N. and Kendall, M.D. *Cell and Tissue Research*, in press. Involution of the rat thymus in experimentally induced hypothyroidism.

Adkins, B., Meuller, C., Okada, C.Y., Reichert, R.A., Weissman, I.L., and Spangrude, G.J. (1987). Early events in T-cell maturation. *Annual Review of Immunology*, **5**, 325–65.

Aguilera, G. and Romano, M.C. (1989). Influence of the thymus on steroidogenesis by rat ovarian cells *in vitro*. *Journal of Endocrinology*, **123**, 367–73.

Allen, L.S., McClure, J.E., Goldstein, A.L., Barkley, M.S., and Michael, S.D. (1984). Estrogen and thymic hormone interactions in the female mouse. *Journal of Reproductive Immunology*, **6**, 25–37.

Allen, J.M., Iggulden, H.L.A., and McHale, N.G. (1986). β-adrenergic inhibition of bovine mesenteric lymphatics. *Journal of Physiology*, **374**, 401–11.

Amor, B., Dougados, M., Mery, C., Dardenne, M., and Bach, J.F. (1987). Nonathymulin in rheumatoid arthritis: two double blind, placebo controlled trials. *Annals of the Rheumatic Diseases*, **46**, 549–54.

Anderson, D.H. (1932). The relationship between the thymus and reproduction. *Physiological Reviews*, **12**, 1–22.

Anderson, G., Jenkinson, E.J., Moore, N.C., and Owen, J.J.T. (1993). MHC class II-positive epithelium and mesenchyme cells are both required for T-cell development in the thymus. *Nature*, **362**, 70–3.

Argiolas, A., Melis, M.R., Stancampiano, R., Mauri, A., and Gessa, G.L. (1990*a*). Hypothalamic modulation of immunoreactive oxytocin in the rat thymus. *Peptides*, **1**, 539–43.

Argiolas, A., Gessa, G.L., Melis, M.R., Stancampiano, R., and Vaccari, A. (1990*b*). Effects of neonatal and adult thyroid dysfunction on thymic oxytocin. *Neuroendocrinology*, **52**, 556–9.

Aronson, M. (1991). Hypothesis: Involution of the thymus with aging – programmed and beneficial. *Thymus*, **18**, 7–13.

Arrenbrecht, S. (1974). Specific binding of growth hormone to thymocytes. *Nature*, **252**, 255–7.

Ataya, K.M., Sakr, W., Blacker, C.M., Mutchnick, M.G., and Latif, Z.A. (1989). Effect of GnRH agonists on the thymus in female rats. *Acta Endocrinologica*, **121**, 833–40.

Bach, J-F., Bach, M.A., Charreire, J., Dardenne, M., Erard, D., and Riveau-Kaiserlian, D. (1981). The effect of the serum thymic factor (FTS) on suppressor T cells. In *Advances in immunopharmacology*. (ed. J. Hodder, L. Chedid, P. Mullen, and F. Spreafico), pp. 77–81. Pergamon Press, Oxford.

Badamchian, M., Spangelo, B.L., Damavandy, T., MacLeod, R.M., and Goldstein, A.L. (1991). Complete amino acid sequence analysis of a peptide isolated from the thymus that enhances release of growth hormone and prolactin. *Endocrinology*, **128**, 1580–8.

Baines, M.G., Pross, H.F., and Millar, K.S. (1977). Effects of pregnancy on the maternal lymphoid system in mice. *Obstetrics and Gynaecology*, **50**, 457–61.

Barkley, M.S., Geschwind, I.F., and Braeford, G.E. (1979). The gestational pattern of oestradiol, testosterone and progesterone secretion in selected strains of mice. *Biology of Reproduction*, **20**, 733–8.

Barr, I.G., Pyke, K.W., Pearce, P., Toh, B.H., and Funder, J.W. (1984). Thymic sensitivity to sex hormones develops post-natally; an *in vivo* and *in vitro* study. *Journal of Immunology*, **132**, 1095–9.

Batanero, E., de Leeuw, F.E., Jansen, G.H., van Wichen, D.F., Huber, J., and Schuurman, H.J. (1992). The neural and neural–endocrine component of the human thymus. II. Hormone immunoreactivity. *Brain Behaviour and Immunity*, **6**, 249–64.

Berczi, I. and Nagi, E. (1982). A possible role of prolactin in adjuvant arthritis. *Arthritis and Rheumatology*, **25**, 591–6.

Bijlsma, J.W.J. and Van der Brink, H.R. (1992). Estrogens and rheumatoid arthritis. *American Journal of Reproduction and Immunology*, **28**, 231–4.

Blackstock, J.C. and Stimson, W.H. (1976). Evidence for immunoregulation by a new thymic hormone during pregnancy. *Proceedings of the Society of Endocrinology*, **69**, 41–2.

von Boehmer, H.R. (1986). Thymus development. *Current Topics in Microbiology and Immunology*, **126**, 19–25.

von Boehmer, H. (1988). The developmental biology of T lymphocytes. *Annual Reviews of Immunology*, **6**, 309–26.

Bomberger, C.E. and Haar, J.L. (1992). Dexamethasone and hydrocortisone enhance the *in vitro* migration of prethymic stem cells to thymus supernatant. *Thymus*, **20**, 89–99.

Brooks-Kaiser, J.C., Murgita, R.A., and Hoskin, D.W. (1992). Pregnancy–associated suppressor cells in mice: functional characteristics of CD3+4−8−45R+ T cells with natural suppressor activity. *Journal of Reproduction and Immunology*, **21**, 103–25.

Bruley-Rosset, M., Miconnet, L., Charon, L., and Halle-Pannenko, O. (1990). Mls generated suppressor cells. I. Suppression is mediated by double-negative (CD3+ CD5+ CD4-CD8−) α/β T cell receptor-bearing cells. *Journal of Immunology*, **145**, 4046–52.

Brunelli, R., Frasca, D., Baschieri, S., Spano, M., Fattorossi, A., Mosiello, F. *et al.* (1992). Changes in thymocyte subsets induced by estradiol administration or pregnancy. *Annals of the New York Academy of Sciences*, **650**, 109–14.

Buckingham, J.C., Safieh, S., Singh, S., Arduino, L.A., Cover, P.O., and Kendall, M.D. (1992). Interactions between the hypothalamo-pituitary axis and the thymus in the rat: a role for corticotropin in the control of thymulin release. *Journal of Neuroendocrinology*, **4**, 295–301.

Bulmer, J.N. (1989). Decidual cellular responses. *Current Opinion in Immunology*, **1**, 1141–7.

Burstein, Y., Buchner, V., Pecht, M., and Trainin, N. (1988). Thymic humoral factor gamma 2: purification and amino acid sequence of an immunoregulatory peptide from calf thymus. *Biochemistry*, **27**, 4066–71.

Buyon, J.P., Yaron, M. and Lockshin, M.D. (1993). First International conference on rheumatic diseases in pregnancy. *Arthritis and Rheumatism*, **36**, 59–64.

Carter, J. (1984). The maternal immunological response during pregnancy. In *Oxford Reviews in Reproductive Biology* (ed. Clarke, J.R.), pp. 47–128. Clarendon Press, Oxford.

Cassone, V.M. (1990). Melatonin! Time in a bottle. *Oxford Reviews of Reproductive Biology*, **12**, 319–67.

Castilla, J., Rueda, R., Vargas, M.L., González-Gómez, E., and García-Olivates, E. (1989). Decreased levels of circulating CD4+ T lymphocytes during normal human pregnancy. *Journal of Reproduction and Immunology*, **15**, 103–111.

Chambers, S.P. and Clarke, A.G. (1979). Measurements of thymus weight, lumbar node weight and progesterone levels in syngeneically pregnant and allogeneically pregnant and pseudopregnant mice. *Journal of Reproduction and Fertility*, **55**, 309–15.

Chang, W.-P. and Marsh, J.A. (1993). The effect of synthetic thymulin on cell surface marker expression by avian T-cell precursors. *Developmental and Comparative Immunology*, **17**, 85–96.

Chaouat, G., Fowlkes, B.J., Leiserson, W.L., and Asofsky, R. (1982). Modification of thymocyte subsets during pregnancy analysed by flow microfluorometry: role of the allogeneic status of the conceptus. *Thymus*, **4**, 299–308.

Chapman, N. and Twigg, G.I. (1990). Studies on the thymus gland of British Cervidae, particularly muntjac, *Muntiacus reevesi*, and fallow, *Dama dama*, deer. *Journal of Zoology*, **222**, 653–75.

Chatterjee-Hasrouni, S., Santer, V., and Lala, P.K. (1980). Characterization of the maternal small lymphocyte cell subsets during allogeneic pregnancy in the mouse. *Cellular Immunology*, **50**, 290–304.

Chikanza, I.C. and Panayi, G.S. (1991). Hypothalamic–pituitary mediated modulation of immune function: prolactin as a neuroimmune peptide. *British Journal of Rheumatology*, **30**, 203–7.

Choudhuri, R. and Wood, G.W. (1992). Leukocyte distribution in the pseudopregnant mouse uterus. *American Journal of Reproduction and Immunology*, **27**, 69–76.

Christian, J.J. (1950). The adreno-pituitary system and population cycles in mammals. *Journal of Mammology*, **31**, 247–59.

Clark, D.A. and McDermott, M.R. (1981). Active suppression of host versus graft reaction in pregnant mice. III. Developmental kinetics, properties and mechanism of induction of suppressor cells during first pregnancy. *Journal of Immunology*, **127**, 1267–73.

Clark, D.A., Chaput, A., Slapsys, R.M., Brierley, J., Daya, S., and Allardyce, R. (1987). In *Immunoregulation and fetal survival* (ed. T.J. Gill III, T.G. Wegmann, and E. Nisbet-Brown), pp. 63–77. Oxford University Press, New York.

Clarke, A.G. (1979). Pregnancy induced involution of the thymus can be prevented by immunizing with paternal skin grafts: a strain dependent effect. *Clinical Experimental Immunology*, **35**, 421–4.

Clarke, A.G. (1984). Immunological studies on pregnancy in the mouse. In *Immunological aspects of reproduction in mammals* (ed. D.B. Crighton), pp. 153–82, Butterworths, London.

Clarke, A.G. (1988). Structural and cellular changes in the mouse thymus during pregnancy. In *Histophysiology of the immune system. Advances in Experimental Biology*, **237**, (ed. S. Fossum and B. Rolstad), pp. 363–8. Plenum Press, New York.

Clarke, A.G. (1989). The thymus in mouse pregnancy. *Journal of Developmental and Comparative Immunology*, **13**, 359.

Clarke, A.G. and Kendall, M.D. (1989). Histological changes in the thymus during mouse pregnancy. *Thymus*, **14**, 65–78.

Clarke, A.G., Gill, A.L. and Kendall, M.D. (in press). The effect of pregnancy on the mouse thymic epithelium. *Cell and Tissue Research*.

Clarke, A.G. and MacLennan, K.A. (1989). The many facets of thymic involution. *Immunology Today*, **7**, 204–5.

Clarke, A.G. and Miller, N.G.A. (1991). Major thymocyte subpopulations of the thymus in pregnant mice using flow cytometry. In *Lymphatic tissues and in vivo immune responses* (ed. B.A. Imhof, S. Berrih-Aknin, and S. Ezine), pp. 189–98. Marcel Dekker, New York.

Clemens, L.E., Contopoulos, A.N., Ortiz, S., Stites, D.P., and Siiteri, P.K. (1977). Inhibition of progesterone (P) of leucocyte migration *in vivo* and *in vitro*. *59th Meeting of the Endocrine Society*, p.175. Abstr. No. 237. Endocrine Society, Bethesda, MD.

Cordier, A.C. and Haumont, S.J. (1980). Development of thymus, parathyroids and multibrachial bodies in NMRI and nude mice. *American Journal of Anatomy*, **157**, 227–63.

Croy, B.A. and Chapeau, G. (1990). Evaluation of the pregnancy immunotrophism hypothesis by assessment of the reproductive performance of young adult mice of genotype *skid/skid/bg/bg*. *Journal of Reproduction and Fertility*, **88**, 231–9.

Csaba, G., Hajpal, A., and Juvancz, L. (1975). Influence of maternal thymectomy on development of the offspring in the rat. *Acta Paediatrica Academiae Scientiarum Hungaricae*, **16**, 243–8.

Dardenne, M. and Bach, J-F. (1981). Thymic hormones. In *The thymus gland* (ed. M.D. Kendall), pp. 113–31. Academic Press, London.

Dardenne, M. and Bach, J-F. (1988). Functional biology of thymic hormones. *Thymus Update*, **1**, 101–16.

Dardenne, M. and Savino, W. (1990). Neuroendocrine control of the thymic epithelium: modulation of thymic endocrine function, cytokeratin expression and cell proliferation by hormones and neuropeptides. *Progress in NeuroEndocrine-Immunology*, **3**, 18–25.

Dardenne, M., Charreire, J., and Bach, J-F. (1978). Alterations in thymocyte surface markers after *in vivo* treatment by serum thymic factor. *Cellular Immunology*, **39**, 47–54.

Dardenne, M., Itoh, T. and Homo-Delarche, F. (1986). Presence of glucocorticoid receptors in cultured thymic epithelial cells. *Cellular Immunology*, **100**, 112–18.

Dargemont, C., Dunon, D., Deugnier, M., Denoyelle, M., Girault, J-M., Lederer, F. *et al.* (1989). Thymotaxin, a chemotactic protein is identical to $\beta 2$ microglobulin. *Science*, **246**, 803–6.

De Vries, P., Brasel, K.A., McKenna, H.J., Williams, D.E., and Watson, J.D. (1992). Thymus reconstitution by c-*kit* expressing haematopoietic stem cells purified from adult mouse bone marrow. *Journal of Experimental Medicine*, **176**, 1503–9.

Degenne, D., Canepa, S., Lecomte C., Renoux, M., and Bardos, P. (1988) Serial study of T-lymphocyte subsets in women during very early pregnancy. *Clinical Immunology and Immunopathology*, **48**, 187–91.

Dickman, Z., Dey, S.K., and Gupta, J.S. (1976). A new concept, control of early pregnancy by steroid hormones originating in the preimplantation embryo. *Vitamin and Hormones*, **34**, 215–42.

Dolgova, M.A., Kuľbakh, O.S., and Podosinnikov, I.S. (1985). Structure of the thymus, iliac and mesenteric lymph nodes in the rat at different stages of pregnancy. *Arkhiv. Anatomii, Gistologii Embriologii*, **89**, 71–6.

Dougherty, T.F. (1952). Effect of hormones on lymphatic tissue. *Physiological Reviews*, **32**, 379–401.

Dougherty, T.F. and White, A. (1945). Functional alterations in lymphoid tissue induced by adrenal cortical secretion. *American Journal of Anatomy*, **77**, 81–116.

Eisenthal, A., Nachtigal, D., and Feldman, M. (1982). The *in vitro* generation of suppressor lymphocytes involves interactions between PNA+ and PNA− thymocyte populations. *Immunology*, **46**, 697–705.

Elands, J., Resnik, A., and De Kloet, R.E. (1990). Neurophypophyseal hormone receptors in the rat thymus, spleen and lymphocytes. *Endocrinology*, **126**, 2703–10.

Embrey, M. (1983). Prostaglandins in human reproduction. *Oxford Reviews of Reproductive Biology*, **5**, 62–105.

Ernström, U. (1970). Hormonal influences on thymic release of lymphocytes into the blood. In *Hormones and the immune response*. Ciba Foundation Study

Group No. 36 (ed. G.E. Wolstenholme and I. Knight), pp. 53–60. Churchill Livingstone, London.

Ernström, U. and Söder, O. (1975). Influence of adrenaline on the dissemination of antibody-producing cells from the spleen. *Clinical and Experimental Immunology*, **21**, 131–40.

Freitas, A.A. and Rocha, B.B. (1993). Lymphocyte lifespans: homeostasis, selection and competition. *Immunology Today*, **14**, 25–9.

Fukayama, M., Hayashi, Y., Shiozawa, Y., Maeda, Y., and Koike, M. (1990). Human chorionic gonadotrophin in the thymus. An immunocytochemical study on discordant expression of subunits. *American Journal of Pathology*, **136**, 123–9.

Gambel, P.I., Cleland, A.W., and Ferguson, F.G. (1980). Alterations in thymus and spleen cell populations and immune reactivity during syngeneic pregnancy and lactation. *Journal of Clinical and Laboratory Immunology*, **3**, 115–19.

Geenen, V., Robert, F., Martens, H., Benhida, A., Giovanni, G. de, Defresne, M-P. *et al.* (1991). Biosynthesis and paracrine/cryptocrine actions of 'self' neurohypophyseal-related peptides in the thymus. *Molecular and Cellular Endocrinology*, **76**, C27–31.

Glucksman, A. and Cherry, C.P. (1968). The effect of castration, oestrogens, testosterone and the oestrous cycle on the cortical epithelium of the thymus in male and female rats. *Journal of Anatomy*, **103**, 113–33.

Goland, R.S., Conwell, I.M., Warren, W.B., and Wardlow, S.L. (1992). Placental corticotropin-releasing hormone and pituitary–adrenal function during pregnancy. *Neuroendocrinology*, **56**, 742–9.

Goldstein, A.L., Low, T.K.L., McAdoo, M., McClure, J., Thurman, G.B., Rossio, J.J. *et al.* (1977). Thymosin α1. Isolation and sequence analysis of an immunologically active thymic peptide. *Proceedings of the National Academy of Sciences, USA*, **74**, 725–9.

Goldstein, G., Scheid, M.P., Boyse, E.A., Schlesinger, D.H., and van Vauwe, J. (1979). A synthetic pentapeptide with biological activity characteristic of the thymic hormone thymopoietin. *Science*, **204**, 1309–10.

Greenstein, B.D., Fitzpatrick, F.T.A., Adcock, I.M., Kendall, M.D., and Wheeler, M.J. (1986). Reappearance of the thymus in old male rats after orchidectomy: inhibition of regeneration by testosterone. *Journal of Endocrinology*, **110**, 417–22.

Grégoire, Ch. (1947*a*) Failure of lactogenic hormone to maintain pregnancy involution of the thymus. *Journal of Endocrinology*, **5**, 115–20.

Grégoire, Ch. (1947*b*). Factors involved in maintaining involution of the thymus during suckling. *Journal of Endocrinology*, **5**, 68–87.

Grinevich, Y.A. and Labunetz, I.F. (1986). Melatonin, thymic serum factor, and cortisol levels in healthy subjects of differing age and patients with skin melanoma. *Journal of Pineal Research*, **3**, 263–75.

Grasso, G., and Muscettola, M. (1992) Possible role of interferon–gamma in ovarian function. *Annals of the New York Academy of Sciences*, **650**, 191–6.

Grossman, C.J. (1984). Regulation of the immune system by sex steroids. *Endocrine Reviews*, **5**, 435–55.

Grossman, C.J. and Roselle, G.A. (1983). The interrelationship of the HPG-thymic axis and immune system regulation. *Journal of Steroid Biochemistry*, **19**, 461–7.

Grossman, C.J., Sholiton, L.J., and Nathan, P. (1979). Rat thymic estrogen receptor. I. Preparation, location and physicochemical properties. *Journal of Steroid Biochemistry*, **11**, 1233–40.

Habbal, O. and McLean, J. (1992*a*). An increase in thymocyte proliferation during pseudopregnancy in two inbred strains of rat. *Acta Anatomica*, **144**, 296–303.

Habbal, O. and McLean, J. (1992*b*). Lymphoid tissue changes during early syngeneic pregnancy in the RTIu rat. *Acta Anatomica*, **144**, 363–70.

Hadden, J.W. (1992). Thymic endocrinology. *International Journal of Immuno-pharmacology* **14**, 345–52.

Hammar, J.A. (1926). Die Menschenthymus in Gesundheit und Krankheit. I. Das normale Organ. *Zeitschrift für Mikroskopisch-Anatomische Forschung*, **6**, 1–570.

Haritos, A.A., Tsolas, O., and Horecker, B.L. (1984). Distribution of prothymosin alpha in rat tissues. *Proceedings of the National Academy of Sciences, USA*, **81**, 1391–5.

Healy, D.L., Bacher, J., and Hodgen, G.D. (1985). Thymic regulation of primate fetal ovarian–adrenal differentiation. *Biology of Reproduction*, **32**, 1127–33.

Hedley, M.L., Drake, B.L., Head, J.R., Tucker, P.W., and Forman, J. (1989). Differential expression of the Class I MHC genes in the embryo and placenta during midgestational development in the mouse. *Journal of Immunology*, **142**, 4046–53.

Herrera-Gonzalez, N.E. and Dresser, D.W. (1993). Fetal-maternal immune interaction: blocking antibody and the survival of the fetus. *Development and Comparative Immunology*, **17**, 1–18.

Hetherington, C.M. (1978). Placental and foetal weight in the nude mouse. *Journal of Immunogenetics*, **5**, 193–5.

Hetherington, C.M. and Dresser, D.W. (1990). Establishment of functional T cells in SCID mice does not lead to termination of pregnancy. *Immunology*, **71**, 449–53.

Hetherington, C.M. and Hegan, M.A. (1975). Breeding nude (nu/nu) mice. *Laboratory Animals*, **9**, 19–20.

Herzog, C., Walker, C. Müller, W., Rieber, P., Reiter, C., Riethmüller, G. *et al.* (1989). Anti-CD4 antibody treatment of patients with rheumatoid arthritis. I. Effect on clinical course and circulating T cells. *Journal of Autoimmunity*, **2**, 627–42.

Heyborne, K.D., Cranfill, R.L., Carding, S.R., Born, W.K., and O'Brien, R.L. (1992). Characterization of γδ T lymphocytes at the maternal–fetal interface. *Journal of Immunology*, **149**, 2872–8.

Hill, J. (1992). Cytokines considered critical in pregnancy. *American Journal of Reproduction and Immunology*, **28**, 123–6.

Holmdahl, R., Klareskog, L., Rubin, K., Larsson, E., and Wigzell, H. (1985). T lymphocytes in collagen II-induced arthritis in mice. Characterization of arthritogenic collagen II-specific T cell lines and clones. *Scandinavian Journal of Immunology*, **22**, 295–306.

Holmdahl, R., Jansson, L., Meyerson, B., and Klareskog, L. (1987). Oestrogen induced suppression of collagen arthritis: 1. Long term oestradiol treatment of DBA/1 mice reduces the severity and incidence of arthritis and decreases the anti-type II collagen immune response. *Clinical Experimental Immunology*, **70**, 372–8.

Hooper, D.C., Chantry, D.H., and Billington, W.D. (1987). Murine pregnancy-associated modulation in lymphocyte reactivity to mitogens: identification of the cell populations affected. *Journal of Reproduction and Immunology*, **11**, 273–86.

Hoskin, D.W., Grönvik, K-O., Hooper, D.C., Reilly, B.D., and Murgita, R.A. (1989). Altered immune response patterns in murine syngeneic pregnancy. Presence of natural null suppressor cells in maternal spleen identifiable by monoclonal antibodies. *Cellular Immunology*, **120**, 42–60.

Imami, N., Ladyman, H.M., Spanopoulou, E., and Ritter, M.A. (1992). A novel adhesion molecule in the murine thymic environment: functional and biochemical analysis. *Developmental Immunology*, **2**, 161–73.

Ito, T. and Hoshino, T. (1962). Histological changes of the mouse thymus during involution and regeneration following administration of hydrocortisone. *Zeitschrift für Mikroskopisch-Anatomische Forschung*, **56**, 445–64.

Ito, K., Kaer, L.V., Bonneville, M., Hsu, S., Murphy, D.B., and Tonegawa, S. (1990). Recognition of the product of a novel MHC TL region gene (27b) by a mouse δ T cell receptor. *Cell*, **62**, 549–61.

Jenkins, J.S. and Nussey, S.S. (1991). The role of oxytocin: present concepts. *Clinical Endocrinology*, **34**, 515–25.

Jenkinson, E.J., Kingston, R., and Owen, J.T. (1987). Importance of IL-2 receptors in intrathymic generation of cells expressing T cell receptors. *Nature*, **329**, 160–2.

Jennes, L. and Conn, P.M. (1988). Mechanism of gonadotropin releasing hormone action. In *Hormones and their actions. Part II.* (ed. B.A. Cooke, R.J.B. King, and H.J. van der Molen), pp. 135–53. Elsevier Science Publishers, Cambridge.

Jeremović, M., Barbijeri, M., Kovacević, D., Arambasić, M., Kartaljević, G., Natalić, D.J., and Pazin, S. (1990). Identification of neuroendocrine oxytoxic activity of the human fetal thymus. *Thymus*, **15**, 181–5.

Kampinga, J. and Aspinall, R. (1990). Thymocyte differentiation and thymic microenvironment development in the fetal rat thymus: an immunohistological approach. *Thymus Update*, **3**, 149–86.

Kampinga, J., Berges, S., Boyd, R.L., Brekelmans, P., Colić, M., van Ewijk, W., et al. (1989). Thymic epithelial antibodies: Immunohistological analysis and introduction of nomenclature. *Thymus*, **13**, 165–73.

Kanour-Shakir, A., Zhang, X., Rouleau, A., Armstrong, D.T., Kunz, H.W., MacPherson, T.A., and Gill III, T.J. (1990). Gene imprinting and major histocompatibility complex Class I antigen expression in the rat placenta. *Proceedings of the National Academy of Sciences USA*, **87**, 444–8.

Kasakura, S. (1971). A factor in maternal plasma during pregnancy that suppresses the reactivity of mixed leucocyte cultures. *Journal of Immunology*, **107**, 1296–301.

Kawano, Y., Norma, T., and Yata, J. (1990). Identification of a cord blood T cell subset of CD3+ 4− 8− 45R+ suppressing interleukin 2 production in the autologous mixed lymphocyte reaction and the mode of action of exogenous IL 2 in the induction of IL 2 production. *Cellular Immunology*, **131**, 27–40.

Kawashima, I., Seiki, K., Sakabe, K., Ihara, S., Akatsuka, A., and Katsumata, Y. (1992). Localization of estrogen receptors and estrogen receptor-mRNA in female mouse thymus. *Thymus*, **20**, 115–21.

Kendall, M.D. (1980). Avian thymus glands: a review. *Developmental and Comparative Immunology*, **4**, 191–210.

Kendall, M.D. (1989). The morphology of perivascular spaces in the thymus. *Thymus*, **13**, 157–64.

Kendall, M.D. (1990). The cell biology of cell death in the thymus. *Thymus Update*, **3**, 53–76.

Kendall, M.D. (1991). Functional anatomy of the thymic microenvironment. *Journal of Anatomy*, **117**, 1–29.

Kendall, M.D. and Twigg, G.I. (1981). The weight of the thymus gland in a population of wild Bank voles, *Clethrionomys glareolus*, from Wicken Fen, Cambridgeshire. *Journal of Zoology*, **194**, 323–39.

Kendall, M.D., Safieh, B., Sareen, A., Venn. G., Matheson, L., and Ritter, M. (1991). Thymulin secreting cells in man: distribution, LM histochemistry and plasma thymulin levels. In *Lymphatic tissues and in vivo immune responses* (ed. B. Imhof, S. Berrih-Aknin, and S. Ezine), pp. 41–3. Marcel Dekker, New York.

Kilshaw, P.J. and Murant, S.J. (1991). Expression and regulation of B$_7$ (βp) integrins on mouse lymphocytes: relevance to the mucosal immune system. *Experimental Journal of Immunology*, **21**, 2591–7.

Kiso, Y., Pollard, J.W., and Croy, B.A. (1992). A study of granulated metrial gland cell differentiation in pregnant, macrophage-deficient osteopetrotic (opop) mice. *Experimentia*, **48**, 973–5.

Kollar, E.J. (1953). Reproduction in the female rat after pelvic nerve neurectomy. *Anatomical Record*, **115**, 641–58.

Kouvats, S., Main, E.K., Librach, C., Stubblebine, M., Fisher, S.J., and DeMars, R. (1990). A Class I antigen HLA-G, expressed in human trophoblasts. *Science*, **248**, 220–3.

Krieger, D.T. (1982). Placenta as a source of 'brain' and 'pituitary' hormones. *Biology of Reproduction*, **26**, 55–72.

Ladyman, H., Boyd, R., Brekelmans, P., Colić, M., van Ewijk, W., von Gaudecker B., *et al.* (1989). Monoclonal antiepithelial antibody workshop. Flow cytometric and biochemical analysis of thymic epithelial cell heterogeneity. In *Lymphatic tissues and in vivo immune responses* (ed. B.A. Imhof, S. Berrih-Aknin, and S. Ezine), pp. 15–20. Marcel Dekker, New York.

Lahita, R.G. (1992). The effects of sex hormones on the immune system in pregnancy. *American Journal of Reproduction and Immunology*, **28**, 136–7.

Larssen, P., Goldschmidt, T.J., Klareskog, L., and Holmdahl, R. (1989). Oestrogen-mediated suppression of collagen-induced arthritis in rats. Studies on the role of the thymus and of peripheral CD8+T lymphocytes. *Scandinavian Journal of Immunology*, **30**, 741–7.

Lau, C.Y., Freestone, J.A., and Goldstein, G. (1980). Effect of thymopoietin pentapeptide (TP5) on autoimmunity. I. TP5 suppression of induced erythrocyte autoantibodies in C3H mice. *Journal of Immunology*, **125**, 1634–8.

Le Douarin, N.M. and Jotereau, F.V. (1975). Tracing cells of the avian thymus through embryonic life in interspecific chimeras. *Journal of Experimental Medicine*, **142**, 17–40.

Lee, A.K. and McDonald, I.R. (1985). Stress and population regulation in small mammals. *Oxford Reviews of Reproductive Biology*, **7**, 261–304.

Leeming, G., McLean, J.M., and Gibbs, A.C. (1984). Thymic and body weight changes during first syngeneic and allogeneic pregnancy in the rat and the effect of strain differences. *Thymus*, **6**, 153–65.

Leeming, G., McLean, J.M., and Gibbs, A.C. (1985). The cell content and proliferative response of the rat thymus during first syngeneic and allogeneic pregnancy and the effect of strain difference. *Thymus*, **7**, 247–55.

Levett, S. (1981). A study of the histological changes taking place in the thymus during syngeneic and allogeneic pregnancy in mice. Thesis. BSc Honours Research Project, Genetics Department, University of Nottingham.

Lintern-Moore, S. and Pantelouris, E.M. (1975). Ovarian development in athymic nude mice. The size and composition of the follicle population. *Mechanisms of Ageing and Development*, **4**, 385–90.

Loke, Y.W. (1989). Trophoblast antigen expression. *Current Opinion in Immunology*, **1**, 1131–4.

Luster, M.I., Hays, H.T., Korach, K., Tucker, A.N., Dean, J.H., Greenlee, W.F., and Broomsman, G.A. (1984). Oestrogen immunosuppression is regulated through oestrogen responses in the thymus. *Journal of Immunology*, **133**, 110–16.

Maestroni, G.J.M. and Conti, A. (1992). The pineal–immuno-opioid network. *Annals of the New York Academy of Sciences*, **650**, 56–9.

Malaise, M.G., Hazee-Hagelstein, M.T., Renter, A.M., Vrinds-Gevaert, Y., Goldstein, G., and Franchimont, P. (1987). Thymopoietin and thymopentin enhance the levels of ACTH, β-endorphin and β-lipotropin from rat pituitary cells *in vitro*. *Acta Endocrinologica*, **115**, 455–60.

Marchetti, B. (1989). Involvement of the thymus in reproduction. *Progress in NeuroEndocrinImmunology*, **2**, 64–9.

Marchetti, B., Guarcello, V., Morale, M.C., Bartoloni, G., Farinella, Z., Cordaro, S., and Scapagnini, U. (1989). Luteinizing hormone-releasing hormone-binding sites in the rat thymus: characteristics and biological function. *Endocrinology*, **125**, 1025–36.

Maroni, E.S. and De Sousa, M. (1973). The lymphoid organs during pregnancy in the mouse. A comparison between syngeneic and an allogeneic mating. *Clinical Experimental Immunology*, **31**, 107–24.

Marshall, A.H.E. and White, R.G. (1961). The immunological reactivity of the thymus. *British Journal of Experimental Pathology*, **42**, 379–85.

Mattsson, R., Mattsson, A., Holmdahl, R., Whyte, A., and Rook, G.A.W. (1991). Maintained pregnancy levels of oestrogen afford complete protection from post-partum exacerbation of collagen-induced arthritis. *Clinical Experimental Immunology*, **85**, 41–7.

McClure, J.E., Lameris, N., Wara, D.W., and Goldstein, A.L. (1982). Immunochemical studies of thymosin: radioimmunoassay of thymosin α1. *Journal of Immunology*, **128**, 368–75.

McGillis, J.P., Hall, N.R., and Goldstein, A.L. (1988). Thymosin fraction 5 stimulates secretion of adrenocorticotropic hormone (ACTH) from cultured rat pituitaries. *Life Sciences*, **42**, 2259–68.

Mendoza, M.E. and Romano, M.C. (1989). Prepubertal rat thymus secretes a factor that modulates gonadotrophin secretion in cultured rat pituitary cells. *Thymus*, **14**, 232–42.

Michael, S.D. (1983). Interactions of the thymus and the ovary. In *Factors regulating ovarian function* (ed. G.S. Greenwald and P.F. Terranova), pp. 445–64. Raven Press, New York.

Michie, S.A. and Rouse, R.V. (1989). Traffic of mature lymphocytes into the mouse thymus. *Thymus*, **13**, 141–8.

Milenković, L., Lyson, K., Aguila, M.C., and McCann, S.M. (1992). Effect of thymosin α1 on hypothalamic hormone release. *Neuroendocrinology*, **56**, 674–9.

Miller, M.E. and Chatten, J. (1967). Ovarian changes in ataxia telangiectasia. *Acta Paediatrica Scandinavia*, **56**, 559–61.

Miller, K.G., Mills, S.P., and Bains, M.G. (1973). A study of the influence of pregnancy on the thymus gland of the mouse. *American Journal of Obstetrics and Gynaecology*, **117**, 913–18.

Mincheva-Nilsson, L., Hammarstrom, S., and Hammarstrom, M.L. (1992). Human decidual leukocytes from early pregnancy contain high numbers of gamma delta+ cells and show selective down-regulation of alloreactivity. *Journal of Immunology*, **149**, 2203–11.

Moberg, G.P. (1987). Influence of the adrenal axis upon the gonads. *Oxford Reviews of Reproductive Biology*, **49**, 456–96.

Mocchegiani, E. and Fabris, N. (1992). Interdependence of growth hormone and thyroid hormone action on thymulin release: clinical evidence. *Annals of the New York Academy of Sciences*, **650**, 91–3.

Moore, T.C. (1984). Modification of lymphocyte traffic by vasoactive neurotransmitter substances. *Immunology*, **52**, 511–18.

Moreno, A.M. and Zepata, A. (1991). *In situ* effects of estrogens on the stroma of rat thymus. In *Lymphatic tissues and in vivo immune responses* (ed. B.A. Imhof, S. Berrih-aknin, and S. Ezine), pp. 81–7. Marcel Dekker, New York.

Munroe, J.G. (1971). Progesteroids as immunosuppressive agents. *Journal of Reticuloendothelial Society*, **9**, 361–75.

Murgita, R.A., Goidl, E.A., Kontiainen, S., and Wigzell, H. (1977). α-fetoprotein induces suppressor T cells *in vitro*. *Nature*, **267**, 257–9.

Nieuwenhuis, P. (1990). Self-tolerance induction and the blood–thymus barrier. *Thymus Update*, **3**, 31–51.

Nishizuka, Y. and Sakakura, T. (1969). Thymus and reproduction: sex-linked dysgenesia of the gonad after neonatal thymectomy in mice. *Science*, **166**, 753–5.

Osoba, D. (1965). Immune reactivity in mice thymectomized soon after birth: normal response after pregnancy. *Science*, **147**, 298–9.

Ostensen, M., Aure, B., and Husby, G. (1983). Effect of pregnancy and hormonal changes on the activity of rheumatoid arthritis. *Scandinavian Journal of Rheumatology*, **12**, 69–72.

Padykula, H.A. and Tansey, T.R. (1979). The occurrence of uterine stromal and

210 Marion D. Kendall and Ann G. Clarke

intraepithelial monocytes and heterophils during normal late pregnancy in the rat. *Anatomical Record*, **193**, 329–55.

Pavia, C., Siiteri, P.K., Perlmann, J.D., and Stites, D.P. (1979). Suppression of murine allogeneic cell interactions by sex hormones. *Journal of Reproduction and Immunology*, **1**, 33–8.

Pearce, P.T., Khalid, B.A.K., and Funder, J.W. (1983). Progesterone receptors in rat thymus. *Endocrinology*, **113**, 1287–91.

Pepper, J.F. (1961). The effect of age, pregnancy and lactation on the thymus gland and lymph nodes of the mouse. *Journal of Endocrinology*, **22**, 335–48.

Persike, E.C. (1940). Involution of the thymus during pregnancy in young mice. *Proceedings of the Society for Experimental Biology and Medicine*, **45**, 315–17.

Persillin, R.H. (1981). Inhibitors of inflammatory and immune responses in pregnancy serum. *Clinical Rheumatic Disorders*, **7**, 769–80.

Philpott, K.I., Rastan, S., Brown, S., and Mellor, A.L. (1988). Expression of H-2 Class-I genes in murine extra-embryonic tissues. *Immunology*, **64**, 479–86.

Phuc, L.H., Papiernik, M., Berrih, S., and Duval, D. (1981*a*). Thymic involution in pregnant mice. 1. Characterization of the remaining thymocyte subpopulations. *Clinical and Experimental Immunology*, **44**, 247–52.

Phuc, L.H., Papiernik, M., and Dardenne, M. (1981*b*). Thymic involution in pregnant mice. II. Functional aspects of the remaining thymocytes. *Clinical and Experimental Immunology*, **44**, 253–61.

Pope, R.M. (1990). Immunoregulatory mechanisms present in the maternal circulation during pregnancy. *Baillière's Clinical Rheumatology*, **4**, 33–41.

Rebar, R.W., Morandini, I.C., Erikson, G.F., and Petze, J.E. (1981). The hormonal basis of reproductive defects in athymic mice: diminished concentrations in prepubertal females. *Endocrinology*, **108**, 120–6.

Rebar, R.W., Miyake, A., Erickson, G.F., Low, T.K.L., and Goldstein, A.L. (1983). The influence of the thymus gland: a hypothalamic site of action. In *Factors regulating ovarian function*. (ed. G.S. Greenwald and D.F. Terranova), pp. 465–9. Raven Press, New York.

Renz, H., Gentz, U., Schmidt, A., Dapper, T., Nain, M., and Gemsa, D. (1989). Activation of macrophages in an experimental rat model of arthritis induced by *Erysipelothrix rhusiopathial* infection. *Infective Immunology*, **57**, 3172–80.

Ritter, M.A. (1977). Embryonic mouse thymocyte development. Enhancing effect of corticosterone at physiological levels. *Immunology*, **33**, 241–6.

Ritter, M.A. and Crispe, I.N. (1992). The thymus. In *Focus* (ed. D. Male), pp 67–70. IRL Press, Oxford.

Roberts, J.L., Sharrow, S.O., and Singer, A. (1990). Clonal deletion and clonal anergy in the thymus induced by cellular elements with different radiation sensitivities. *Journal of Experimental Medicine*, **171**, 935–40.

Rozing, J., Coolen, C., Tielen, F.J., Weegenaar, J., Schuurman, H-J., Greiner, D.L., and Rossini, A.A. (1989). Defects in the thymic epithelial stroma of diabetes prone BB rats. *Thymus*, **14**, 125–35.

Russell, D.H., Kibler, R., Matrisian, L., Larson, D.F., Poulos, B., and Magun, B.E. (1985). Prolactin receptors on human T and B lymphocytes: antagonism of prolactin binding by cyclosporin. *Journal of Immunology*, **134**, 3027–31.

Saad, A.H. (1989). Pregnancy-related involution of the thymus in the viviparous lizard *Chalcides ocellatus*. *Thymus*, **14**, 223–32.

Safieh-Garabedian, B., Kendall, M.D., Khamashata, M.D., and Hughes, G. (1992). Thymulin and its role in immunomodulation. *Journal of Autoimmunity*, **16**, 158–63.

Savino, W. and Dardenne, M. (1984). Thymic hormone containing cells: VI. Immunohistologic evidence for the simultaneous presence of thymulin, thymopoietin and thymosin α1 in normal and pathological human thymuses. *European Journal of Immunology*, **14**, 987–91.

Savino, W., Bartoccioni, E., Homo-Delarche, F., Gagnerault, M. Cl., Iyoh, T., and Dardenne, M. (1988). Thymic hormone containing cells. IX. Steroids *in vitro* modulate thymulin secretion by human and murine thymic epithelial cells. *Journal of Steroid Biochemistry*, **30**, 479–84.

Schiff, R.I., Mercier, D., and Buckley, R.H. (1975). Inability of gestational hormones to account for the inhibitory effects of pregnancy plasma on lymphocyte responses *in vitro*. *Cellular Immunology*, **20**, 69–80.

Schlesinger, D.H., Goldstein, G., and Niall, H.D. (1975). The complete amino acid sequence of ubiquitin, an adenylate cyclase stimulating polypeptide probably universal in living cells. *Biochemistry*, **14**, 2214–18.

Schuurman, H-J., Nagelkerken, L., De Weger, R.A., and Rosing, J. (1991). Age-associated involution: significance of a physiological process. *Thymus*, **18**, 1–6.

Screpanti. I., Morrone, S., Meco, D., Santori, A., Gulino, A., Paolini, R., *et al*. (1989). Steroid sensitivity of thymocyte subpopulations during intrathymic differentiation. Effects of 17 β-estradiol and dexamethasone on subsets expressing T cell antigen receptor or IL-2 receptor. *Journal of Immunology*, **142**, 3378–83.

Selye, H. (1946). The general adaption syndrome and the diseases of adaption. *Journal of Clinical Endocrinology*, **6**, 117–230.

Shelesnyak, M.C., Kraicer, P.F., and Zeilmaker, G.H. (1963). Studies on the mechanism of decidualization. 1. The oestrogen surge of pseudopregnancy and progravidity and its role in the process of decidualization. *Acta Endocrinologica*, **42**, 225–32.

Shinomiya, N., Tsuru, S., Tsugita, M., Katsura, Y., Takemura, T., Rokutanda, M., and Nomoto, K. (1991). Thymic depletion in pregnancy: kinetics of thymocytes and immunological capacities of the hosts. *Journal of Clinical and Laboratory Immunology*, **34**, 11–22.

Smith, R.N. and Powell, A.E. (1977). The adoptive transfer of pregnancy-induced unresponsiveness to male skin grafts with thymus-dependent cells. *Journal of Experimental Medicine*, **146**, 899–904.

Smith, R. and Thomson, M. (1991). Neuroendocrinology of the hypothalamo-pituitary-adrenal axis in pregnancy and the puerperium. *Ballière's Clinical Endocrinology and Metabolism*, **5**, 167–86.

Spector, T.D. and Da Silva, J.A.P. (1992). Pregnancy and rheumatoid arthritis: an overview. *American Journal of Reproduction and Immunology*, **28**, 222–5.

Steinmann, G.G. (1986). Changes in the human thymus during aging. In *The human thymus* (ed. H.K. Müller-Hermelink), pp. 43–99. Springer-Verlag, Berlin.

Stimson, W.H. (1983). Are pregnancy-associated serum proteins responsible for the inhibition of lymphocyte transformation by pregnancy serum. *Clinical Experimental Immunology*, **40**, 157–60.

Stimson, W.H. and Hunter, I.C. (1980). Oestrogen-induced immunoregulation mediated through the thymus. *Journal of Clinical Laboratory Investigation*, **4**, 27–33.

Symonds, I.M. (1981). Steroid receptor concentrations in rat thymus during pregnancy. Thesis BMSc, Department of Obstetrics and Gynaecology, University of Nottingham.

Thomas, I.K. and Erickson, K.L. (1986). Gestational immunosuppression is mediated by specific Ly2+ T cells. *Immunology*, **57**, 201–6.

Trainin, N. (1974). Thymic hormones and the immune response. *Physiological Reviews*, **54**, 272–315.

Transey, C., Nash, S.R., David-Watine, B., Cochet, M., Hunt, S.W., Hood, L.E., and Kourilsky, P. (1987). A low polymorphic mouse H-2 Class I gene from the TLa complex is expressed in a broad variety of cell types. *Journal of Experimental Medicine*, **166**, 341–61.

Twigg, G.I. and Harris, S. (1982). Seasonal and age changes in the thymus gland of the Red fox, *Vulpes vulpes*. *Journal of Zoology*, **196**, 355–70.

Villa Verde, D.M.S., Defresne, M.P., Vannier-Dos-Santos, M.A., Dussault, J.H., Boniver, J., and Savino, W. (1992). Identification of nuclear triiodothyronine receptors in the thymic epithelium. *Endocrinology*, **131**, 1313–20.

Waites, G.T. and Whyte, A. (1987). Effect of pregnancy on collagen-induced arthritis in mice. *Clinical Experimental Immunology*, **67**, 467–76.

Webber, R.H., De Felice, R., and Ferguson, R.J. (1970). Bone marrow response to stimulation of the sympathetic trunks in rats. *Acta Anatomica*, **77**, 92–7.

Whyte, A., Hasan, Z., and Rutter, F. (1988). Anti-collagen antibody responses in DBA/1 (H-2q) mice associated with type II collagen-induced arthritis and the effects of pregnancy. *Journal of Reproduction and Immunology*, **13**, 277–86.

Whyte, A. and Williams, R.O. (1988). Bromocriptine suppresses post partum exacerbation of collagen-induced arthritis. *Arthritis and Rheumatism*, **31**, 927–8.

van de Wijngaert, F.P., Kendall, M.D., Schuurman, H-J., Rademakers, L.H.M.P., and Kater, L. (1984). Heterogeneity of human thymus epithelial cells at the ultrastructural level. *Cell and Tissue Research*, **237**, 227–37.

Wooley, P.H., Seibold, J.R., Whalen, J.D., and Chapdelaine, J.M. (1989). Pristane-induced arthritis: The immunological and genetic features of an experimental murine model of autoimmune disease. *Arthritis and Rheumatism*, **32**, 1022–30.

Yokoyama, M., Koga, Y., Taniguchi, K., Nakano, H., and Nomoto, K. (1988). T lymphocytes emigrating from the thymus to the spleen during postpartum regulate serum immunoglobulin levels in mice. *Immunology*, **63**, 151–6.

Yu, F-X., Lin, S-C., Morrison-Bogard, M., Atkinson, M.A.L., and Yin, H.L. (1993). Thymosin β10 and thymosin β4 are both actin monomer sequestering proteins. *Journal of Biological Chemistry*, **268**, 502–9.

Zaidi, S.A.A., Kendall, M.D., Gillham, B., and Jones, M.T. (1988). The release of luteinizing hormone from pituitaries perfused with thymic extracts. *Thymus*, **12**, 253–64.

Zhang, C.L., Cheng, L.R., Wang, H., Zhuang, L.Z., and Huang, W.Q. (1991). Neuropeptides and neurotransmitters in human placental villi. *Neuroendocrinology*, **53** (Suppl. 1), 77–83.

Zinkernagel, R.M. and Doherty, P.C. (1975). Immunological surveillance against altered self components by sensitised T lymphocytes in lymphocytic choriomeningitis. *Nature*, **251**, 547–8.

6 Reproductive Ageing and Neuroendocrine Function

JOSEPH MEITES and JOHN K.H. LU

I   INTRODUCTION

Ageing can be defined as a decline in body functions with time associated with reduced capacity to maintain homeostasis. Many theories have been developed to explain the cause(s) of ageing, but it is doubtful that any one theory is adequate to explain the entire ageing phenomenon. Most investigators agree that the genome has a major role in determining the length of the lifespan among different species, as well as the ageing process. Environmental factors also exert an important influence on ageing processes. It is our view that the genome and environment regulate ageing processes to a large degree via the neuroendocrine system, which controls or influences development, growth, and function of every organ and tissue in the body, including the reproductive system.

Early reports on the relation of hormones to ageing in animals and humans were based on examination of gross and microscopic changes in endocrine glands and their target organs and tissues, changes in size and weight of endocrine and target glands and tissues, bioassays and chemical assays (steroids) of hormones in the endocrine glands, blood, and urine, and on some clinical observations in old patients. The early studies were hampered by lack of knowledge of the important role of the hypothalamus in controlling hormone secretion by the endocrine glands, and unavailability of accurate methods for measuring the minute amounts of hormones present in the circulation, later provided by development of radioimmunoassays (RIAs). These early studies therefore offered only limited information on the role of hormones in ageing processes.

Research on the relation of the neuroendocrine system to ageing began in the 1960s and early 1970s with reports on the role of the hypothalamus on the reproductive decline in female rats. Aschheim (1976) showed that when ovaries from young rats were grafted into old ovariectomized rats, they failed to induce oestrous cycles. However, when ovaries from old non-cyclic rats were grafted into young ovariectomized rats, many of the young rats resumed cycles. Peng (1983) confirmed these results, and also demonstrated that when the pituitary was removed from old non-cyclic rats and grafted underneath the median eminence of hypophysectomized young rats, many of the young rats resumed cycles. These important studies demonstrated that neither the ovaries nor pituitary of old rats were mainly responsible for loss of oestrous cycles in old rats and suggested that the hypothalamus may be mainly at fault.

Direct proof of hypothalamic involvement in loss of cyclicity in rats was provided by Clemens et al. (1969), when it was shown that electrical stimulation of the preoptic area, which is involved in control of reproductive cycles, induced ovulation in old, constant oestrous rats.

This suggested that sufficient gonadotrophin-releasing hormone (GnRH) was present to induce ovulation, but the stimulus for its release was lacking. Most ageing rats undergo a prolonged constant oestrous state characterized by many well-developed follicles in the ovaries but fail to ovulate. Injection of adrenaline in oil also was shown to induce ovulation in the old constant oestrous rats, suggesting that a deficiency of catecholamines (CAs) in the hypothalamus might be responsible for failure of ovulation. Subsequently it was demonstrated that administration of L-dopa, the precursors of CAs, or iproniazid, an inhibitor of monoamine oxidase (MAO), the major enzyme responsible for catabolizing CAs, could reinitiate oestrous cycles in old constant oestrous rats (Quadri *et al.* 1973). Later it was demonstrated that the two CAs, dopamine (DA) and noradrenaline (NA), were both significantly reduced in the hypothalamus of old as compared with young rats (Simpkins *et al.* 1977), and this has been widely confirmed (Wise 1983, Simpkins 1984). It is now evident that the reduction in hypothalamic CAs in old rats is probably mainly responsible for loss of oestrous cycles in ageing female rats, and is also related to the decline in testosterone secretion and spermatogenesis in ageing male rats. Other neurotransmitters in the hypothalamus also may participate in the decline in reproductive functions in ageing rats, but little information is currently available on their possible role. In addition, as will be described elsewhere, the pituitary of old rats becomes less responsive to GnRH stimulation, and the gonads to gonadotrophin stimulation. These defects contribute to the reproductive decline, but are not believed to be as important as the faults that develop in the hypothalamus of the rat (Meites 1982, 1990, 1991).

Important differences as well as similarities have been demonstrated in the reproductive decline in primates as compared with that in ageing rats. Although reductions in hypothalamic functions have been reported in elderly humans, as will be noted elsewhere, defects that develop in the gonads rather than hypothalamic changes are mainly responsible for the reproductive decline in primates. The ovaries of premenopausal women became less responsive to gonadotrophic hormone stimulation and secretion of oestrogen is reduced, resulting in elevation of Follicle-stimulating hormone (FSH) and Luteinizing hormone (LH) secretion by the pituitary. Testes in ageing male primates also develop faults and secrete less testosterone and inhibin, resulting in elevated FSH and LH secretion. Ageing female rats approaching the end of cycling and women approaching the menopause both show irregularities in their cycles but the cycles become longer in rats and shorter in women. Ageing male rats and aging men maintain their fertility, if in good health, well into old age. These and other similarities and differences will be noted in the sections that follow.

II   CHANGES IN HYPOTHALAMIC AND PITUITARY FUNCTION WITH AGE

1   Hypothalamus

The hypothalamus is the ultimate regulator of reproduction since it secretes and releases GnRH which controls FSH and LH secretion by the pituitary, which in turn regulates ovarian and testicular functions. However, secretion of GnRH by the hypothalamus is not autonomous since its release is modulated by neurotransmitters, by environmental stimuli carried via the central nervous system, by negative or positive feedback from ovarian hormones and negative feedback by testicular androgens, and inhibition by high levels of pituitary gonadotrophins (short loop feedback) or GnRH (ultrashort loop feedback). Therefore, changes in reproductive functions with age may involve not only the hypothalamus, but also the pituitary, gonads, and environmental agents.

In the ageing rat, defects that develop in the hypothalamus are the primary cause for the decline in reproductive functions, as already mentioned. Several kinds of evidence indicate that the hypothalamus of old rats does not function normally. Thus, castration of old male or female rats results in a smaller rise in circulating LH and FSH than in young rats. The positive feedback effects of oestrogen or oestrogen followed by progesterone on LH release in old ovariectomized rats is also diminished as compared with that in young rats (Lu et al. 1977; Wise 1984). Although these studies do not clearly separate the role of the hypothalamus from that of the pituitary, more direct studies showed that the hypothalamus from old rats released less GnRH in vitro in response to stimulation by NA and naloxone than the hypothalamus from young rats (Jarjour et al. 1986). Rice et al. (1983) observed lower in vitro release of GnRH from hypothalamic granules from old than from young rats. When old rats were castrated, there was no fall in hypothalamic content of GnRH, in contrast to the large decrease seen in hypothalamic GnRH which normally occurs after castration in young rats, indicating that release of GnRH has occurred (Wise 1983). Also, the amplitude of the pro-oestrous surge of LH in old rats that are still cyclic is lower than in young cyclic rats (Cooper et al. 1980; Wise 1982; Nass et al. 1984). Although measurements of hypothalamic content of GnRH have not usually shown significant changes in old as compared with young rats, such assays provide no information on release of GnRH.

Sufficient GnRH is present in the hypothalamus of old rats to induce LH release and ovulation, since electrical stimulation of the preoptic area elicited LH release and ovulation in old constant oestrous rats (Clemens et al. 1969). The principal cause for the failure of LH release in old rats is believed to be due to a deficiency of NA, which normally rises just prior to the LH surge during prooestrus in young cyclic rats and is essential for

GnRH release (Simpkins 1983). It has been demonstrated that both DA and NA are significantly lower in the hypothalamus of old than in young rats (Simpkins *et al.* 1977). Lower DA concentrations have been observed in the median eminence, medial basal hypothalamus, and striatum of 25–26-month-old than in 3–4-month-old rats, whereas NA was lower in the preoptic–anterior hypothalamus and medial basal hypothalamus but not in the median eminence or striatum (Simpkins 1983). McInstosh and Westfall (1987) found a marked decrease in uptake and release of hypothalamic CAs *in vitro* in response to high frequency stimulation at 12 and 21–26 months of age in Fischer F344 rats as compared with rats 2–4 months old. Weiland and Wise (1990) observed a reduction in α_1 adrenergic receptors in rat median eminence and suprachiasmatic nucleus by middle age, and in all regions of the hypothalamus by old age.

Significant change in hypothalamic serotonin (5-hydroxytryptamine, 5-HT) levels has not been found in old rats (Simpkins 1983), but the ratio of CAs to 5-HT is altered and this may affect the normal rhythm of LH release. Brain opiates can depress gonadotrophin secretion by acting on the hypothalamus, but it is not known whether they have a role in the decrease in gonadotrophin secretion in rats during ageing. The possible influence of other neurotransmitters in the reproductive decline is unknown at present.

In contrast to the decrease seen in FSH and LH secretion in the ageing rat, prolactin (PRL) shows a progressive elevation in female rats with age which results in development of numerous mammary and PRL-secreting pituitary tumours (Meites 1982, 1991). The rise in PRL is believed to be due mainly to the reduction in hypothalamic DA which is the principal inhibitor of PRL secretion. The unopposed oestrogen secretion during the prolonged constant oestrous state may also contribute to the rise in PRL secretion. In ageing women, PRL secretion does not increase, and is usually decreased in the postmenopausal state due to loss of oestrogen. No significant changes in PRL secretion have been observed in old as compared with young men.

There are many potential causes for the decline in hypothalamic DA and NA in ageing rats. A significant loss of neurones has been observed in the medical preoptic area, arcuate nucleus, and ventromedial and lateral hypothalamic nuclei of old rats (Peng 1983). A decline in oestrogen receptors in the hypothalamus was also reported. Loss or damage to neurones may result from the oxidative metabolism of the CAs, leading to formation of hydrogen peroxide, superoxide anions, hydroxyl radicals, and highly reactive quinones (Davison 1987). MAO, the major enzyme that catabolizes CAs, was reported to be increased in the hypothalamus of old rats, whereas tyrosine hydroxylase, the rate-limiting enzyme for synthesis of CAs, was decreased (Davison 1987).

There is evidence that chronic exposure to oestrogen can damage or destroy hypothalamic neurones essential for reproductive functions. Thus, removal of the ovaries from rats and mice early in life, followed by grafting of fresh ovaries many months later, resulted in reinitiation of oestrous cycles that continued many months beyond the time when intact rats and mice ceased to cycle (Aschheim 1976; Finch *et al.* 1984). Administration of a long-acting oestrogen was shown to injure neurones in the arcuate nucleus and medial basal hypothalamus of young rats (Brawer *et al.* 1978). Sarkar *et al.* (1982) reported specific damage to DA neurones in young rats by chronic oestrogen action. Since oestrogen is well known to promote PRL secretion some of the damaging effects of oestrogen may be exerted via high PRL action on neurones. There may be other causes for the decline in hypothalamic CAs with age.

The role of hypothalamic CAs in the decline of reproductive performance in ageing rats receives additional support from studies on the effects of prolonged treatment with neuroleptic drugs (reserpinoids, phenathiazines, haloperidol, etc.) in animals and humans. Such drugs are known to decrease brain CAs as well as 5-HT, and chronic use can result in reduced gonadotrophin secretion and cessation of oestrous cycles in animals and menstrual cycles in women, increased PRL secretion and mammary growth, initiation of lactation, and decreased GH secretion (Martin and Reichlin 1987). These drugs also hasten development of mammary tumours in rats (Welsch and Nagasawa 1977). Can chronic use of neuroleptic drugs advance ageing processes?

Relatively little information is yet available on the role of the hypothalamus in ageing processes in human subjects. There are some indications that the sensitivity of the human hypothalamus to various stimuli, including hormones, is reduced with age. For example, it has been reported that physiologically adequate doses of oestrogen do not return the elevated levels of gonadotrophins in postmenopausal women to premenopausal levels, and older men require greater than normal sex steroid doses to inhibit gonadotrophin secretion (Muta *et al.* 1981). There is also evidence for a loss of circadian rhythm in testosterone secretion in older men, with a marked reduction in the morning peak (Bremner *et al.* 1983), suggesting hypothalamic involvement since it regulates rhythmic activity. Morley and Kaiser (1990) reported a decrease in LH amplitude without any change in pulse frequency in ageing men, which was attributed to a decline in hypothalamic function.

2   Pituitary

In addition to dysfunctions that develop in the hypothalamus with time, there is also evidence that ageing changes occur in the pituitary and its target glands. A decrease in LH and FSH release in response to multiple injections of GnRH has been demonstrated in old as compared with young male rats (Bruni *et al.* 1977). Similarly in old female rats, a smaller release of LH in response to a single injection of GnRH was noted (Watkins *et al.* 1975; Harman 1978). The decrease in LH release in response to GnRH administration in rats was found not to be caused by a reduction in pituitary GnRH receptors (Sonntag *et al.* 1984), but to a failure in calcium mobilization (Roth 1990). The pituitary of old rats has also been reported to be less responsive to administration of growth hormone–releasing hormone, thyrotrophin–releasing hormone, and corticotrophin-releasing factor than the pituitary of young rats (Meites 1990, 1991). By contrast, the ability of oestrogen to increase PRL secretion increases with age (Aschheim 1976; Wise 1983). Oestrogen has been shown to reduce hypothalamic CA activity, but can also directly stimulate the pituitary to increase PRL secretion (Meites 1982).

Basal FSH and LH secretion is lower in old than in young male rats (Bruni *et al.* 1977; Simpkins *et al.* 1977), reflecting both the reduced release of hypothalamic GnRH and decreased responsiveness of the pituitary to GnRH stimulation (Meites 1990, 1991). Basal serum FSH levels were reported to be higher and basal LH levels were about the same in old constant oestrous rats as in young rats on the afternoon of pro-oestrus or morning of oestrus (Huang *et al.* 1978). In old pseudopregnant rats, serum FSH values were lower but serum LH levels were about the same as in constant oestrous rats. However, the rise in circulating LH levels that normally occurs on the afternoon of pro-oestrus was observed to be lower in middle-aged cyclic rats than in young cyclic rats (Wise 1983), probably the result of decreased hypothalamic GnRH release as well as diminished responsiveness of the pituitary to GnRH stimulation. In old male rats, both LH and FSH levels as well as testosterone in the blood are significantly lower than in young male rats (Bruni *et al.* 1977, Simpkins *et al.* 1977).

III REPRODUCTIVE DECLINE IN AGEING FEMALE RODENTS

I Ovarian function and oestrous cycles

Puberty in the female rat occurs at about 35–40 days of age, followed by spontaneous and successive ovulatory cycles. Adult females exhibit regular oestrous cycles and normal reproductive function up to middle age. Beginning at 9–10 months, females show a progressive decrease in the incidence of regular cycles and a marked decline in fertility occurs (LaPolt *et al.* 1986; Matt *et al.* 1987*b*). By 14 months of age, most multiparous rats have ceased to undergo oestrous cycles and lost their ability to reproduce (Matt *et al.* 1986, 1987*a,b*). Reproductive declines with age occur much earlier in virgin than in multiparous female rats (LaPolt *et al.* 1986; Matt *et al.* 1987*b*).

During regular oestrous cycles in young adult rats, follicular development and growth result in increased oestradiol secretion, which reaches a peak on the early afternoon of pro-oestrus (LaPolt *et al.* 1986). It is believed that the gradual and large increase in circulating oestradiol levels during dioestrus day 2 and pro-oestrus stimulate both the hypothalamus and pituitary to elicit preovulatory LH and FSH surges on pro-oestrus (Smith *et al.* 1975), presumably due to an increase in hypothalamic discharge of GnRH by the positive feedback action of oestradiol (Sarker *et al.* 1976). In response to LH stimulation, the well developed preovulatory ovarian follicles secrete large amounts of progesterone during the late afternoon of pro-oestrus (LaPolt *et al.*, 1986), and follicle rupture occurs on the early morning of the next day. Following ovulation, the ruptured follicles are luteinized to form corpora lutea that produce substantial amounts of progesterone during dioestrus days 1 and 2. Concomitant with the LH surge, there is a significant increase in FSH secretion on pro-oestrus. A secondary rise in serum FSH levels occurs on the early morning of oestrus, which is thought to be important for new follicular recruitment (Schwartz 1974).

To study the age-related changes in reproductive and neuroendocrine functions, Lu *et al.* (1979) and Matt *et al.* (1987*b*) examined both virgin and multiparous Long-Evans female rats of various ages. Retired breeders 8–9 months of age were obtained from Charles River Laboratory, and over 95% of them were cycling regularly. By 12 months of age, about 65% of multiparous females displayed irregular, prolonged cycles, whereas only 30% of them were cycling regularly. Most often, the irregular oestrous cycles were characterized by a persistent cornification of vaginal epithelium and delayed ovulation for 2–6 days. Ovarian histology and hormone assays from these middle-aged and irregularly cycling rats revealed that the large, developed follicles produced substantial amounts of oestradiol, resembling those of preovulatory follicles in regularly cycling young females on prooestrus, but a preovulatory LH surge did not occur until several days after a rise in serum oestradiol levels (LaPolt *et al.* 1988*b*). These changes suggest that the hypothalamic GnRH response to oestradiol stimulation is probably delayed and/or attenuated in middle-aged rats. In fact, even among 9–10-month-old females that display consecutive regular cycles, some exhibit significantly attenuated and/or delayed preovulatory LH surges on pro-oestrus (Nass *et al.* 1984), while other middle-aged rats show a normal LH surge. Most interestingly, middle-aged females with attenuated LH surges subsequently show earlier cessation of regular cycles than middle-aged rats with normal LH surges (Nass *et al.* 1984). These findings indicate that prior to loss of regular cycles, ageing female rats experience a progressive decrease in hypothalamic GnRH and in the pituitary LH response to oestradiol stimulation. Such a decrease in neuroendocrine responsiveness is further manifested when the female no longer maintains regular cycles.

Following a short period of irregular cycling, most middle-aged females cease to display complete oestrous cycles and become persistently oestrous. Ageing persistently oestrous rats exhibit persistent cornification of the vaginal epithelium resulting from chronic anovulation and the presence of sustained, intermediate levels of serum oestradiol (Lu *et al.* 1979). This anovulatory phase may continue for 6 months or longer in individual female rats, and the persistently oestrous state is associated with the presence of large, well-developed follicles in the ovary, intermediate but sustained levels of serum oestrogens, and persistent low serum progesterone. Under these conditions, pituitary secretion of LH remains low, and there is a lack of an LH surge in response to oestrogen stimulation (Matt *et al.*, 1987*a*). It is evident that the absence of a spontaneous, preovulatory LH surge in these ageing persistently oestrous females is responsible for the chronic anovulation and the persistent oestrous state. It is interesting that administration of synthetic GnRH to persistently oestrous rats can readily elicit an increase in pituitary LH release followed by ovulation (Meites *et al.* 1978). Furthermore, caging/mating with fertile males causes LH surges in most middle-aged persistently oestrous rats (Matt *et al.* 1987*a*; Day *et al.* 1988). These findings demonstrate that the neuroendocrine mechanisms responsible for releasing hypothalamic GnRH and pituitary LH remain intact in ageing persistently oestrous female rats, and are responsive to male influences. However, unlike young rats, an increase in endogenous oestradiol secretion or injection of oestradiol to persistently oestrous females is unable to elicit the neuroendocrine signal essential for an increase in hypothalamic discharge of GnRH.

After a prolonged period of the anovulatory and persistently oestrous state, older females exhibit long periods of dioestrus with intermittent ovulatory activity at irregular intervals. The vaginal smears from these old rats show a prolonged dioestrous phase that is occasionally interrupted by 2–3 days of smears with nucleated and/or cornified epithelial cells (Lu *et al.* 1979; Lu 1983). Following each ovulation, corpora lutea are formed, and their function is maintained for 2–3 weeks or longer, presumably due to the presence of a chronically elevated PRL in the circulation (Lu *et al.* 1979; Damassa *et al.* 1980). Thus, after a prolonged period of anovulation, older females reinitiate spontaneous LH surges and ovulations. These ovulatory activities render these aged females repetitively pseudopregnant. However, a functional test has revealed that the hypothalamic-pituitary axis of these old pseudopregnant rats remains responsive to oestradiol and progesterone challenge and elicits an LH surge (Lu *et al.*, 1980), whereas that of persistently oestrous females is unresponsive (Lu *et al.*, 1981). The neuroendocrine mechanism underlying this functional reinstatement is not clear at present.

In the ovary of old pseudopregnant rats, ^{125}I-LH binds appreciably to the luteal cells, and high activity of 3β-hydroxysteroid dehydrogenase is also found in these cells (Steger *et al.* 1976). Microscopical examination has revealed that the large corpora lutea of old pseudopregnant rats consist of many enlarged luteal cells (Mandl 1959) that secrete copious amounts of both progesterone (Huang *et al.* 1978; Lu *et al.* 1979; Fayein and Aschheim 1980) and 20α-hydroxyprogesterone (Lu *et al.* 1979, 1980). It has been shown that PRL inhibits 20α-hydroxyprogesteroid dehydrogenase activity in granulosa and luteal cells of the rat ovary (Armstrong *et al.* 1969; Wang *et al.* 1979). In old pseudopregnant rats, however, plasma levels of both progesterone and 20α-hydroxyprogesterone are maintained at high levels while PRL secretion is also increased.

Following the repetitive pseudopregnant state, the oldest female rats exhibit constant anoestrus characterized by vaginal smears with a persistent dioestrous phase. The ovaries of these anoestrous females contain little follicular and luteal tissues (Meites and Huang 1976) and produce minimum amounts of steroids (Huang *et al.* 1978). Also, pituitary pathology and/or tumours are almost always found in these anoestrous rats.

2   Reproductive capacity

In young female rats during the oestrous cycle, a large increase in plasma oestradiol levels associated with follicular growth enhances sexual receptivity starting on the afternoon of proestrus and continues to the next morning. Since ovulation occurs on the early morning of oestrus, ova are readily seen in the oviducts by 0700 or 0800 h on oestrus. It is conceivable that fertile matings most likely occur between midnight of pro-oestrus and early morning of oestrus. Once every 4 days, each female rat ovulates 12–18 ova from both ovaries. If fertile mating occurs, day 1 of pregnancy refers to the day sperm are found in the vaginal lavage. Following fertilization, the zygotes remain in the oviducts to undergo early cleavage and embryogenesis. On day 5, developed blastocysts move down to the uterine horns, and implantation occurs between days 5 and 6. In the rat, it takes 3 weeks to complete a full term gestation, and parturition usually occurs on days 22 or 23 (Matt *et al.* 1986). In pregnant young rats, the number of pups delivered (that is, litter size) at the end of each gestation is in good agreement with the number of implantations found in the uterine horns, and the latter also correlates well with the ovulation rate (Matt *et al.* 1986). Thus, in young fertile female rats, almost all ova found in the oviducts are fertilized and develop into normal blastocysts for implantation, and no postimplantation failures are discerned (Matt *et al.* 1986).

In association with the cervical stimulation induced by mating, pregnant rats exhibit daily diurnal and nocturnal surges of PRL release, which continue for 11 consecutive days (Butcher *et al.* 1972). While the initial increases in plasma PRL rescue the corpora lutea of the oestrous cycle from involution, subsequent daily rises in PRL are essential for maintaining corpora luteal function. During pregnancy, plasma levels of progesterone increase markedly and progressively, reaching peak values at about days 16–19 of gestation (LaPolt *et al.* 1986; Day *et al.* 1991). In contrast, oestradiol secretion remains low for the first 2 weeks of gestation, followed by a gradual increase during the last few days (LaPolt *et al.* 1986; Matt *et al.* 1986). The rise in plasma oestradiol during late pregnancy is due to follicular development and growth. In fact, such an increase in plasma oestradiol in pregnant rats often elicits an LH surge, and these females readily show ovulation shortly after parturition.

In a longitudinal study of Long-Evans female rats, the chronological changes in oestrous cycle patterns between 4 and 12 months of age and their relation to fertility were examined in individual animals during five consecutive pregnancies (Matt *et al.* 1986). At 4 months, all females exhibited regular oestrous cycles and 94% of them produced fertile gestations after mating. At 4, 6, 8, 10, and 12 months of age, these same females (a colony of 100 rats) were mated with young fertile males and underwent full-term gestations. Beginning at 10 months of age, the incidence of regular cycles decreased considerably, while an increasing number of middle-aged rats became either irregular cyclic or pro-oestrus. On the other hand, beginning at 8 months of age the incidence of fertility among these females decreased significantly. By 10 months, 70% of these middle-aged females maintained regular cyclicity but only 24% of them produced fertility gestations following mating. By 1 year of age, the incidence of regular cycling dropped to 45%, whereas the fertility rate decreased to only 4%. For comparison with these results from multiparous rats, similar studies were performed in virgin females at 8, 10, and 12 months of age. Results from these longitudinal studies demonstrated that cessation of regular oestrous cyclicity during ageing occurs significantly earlier (at least 3–4 months) in virgin than in multiparous rats (LaPolt *et al.* 1986; Matt *et al.* 1987*b*). Fertility rate follows a similar but more dramatic pattern of decline than does the incidence of regular cyclicity in both virgin and multiparous females. Few irregular cycling and persistent-oestrous rats produce fertile gestations after mating, and increasing proportions of regularly cycling females fail to reproduce successfully at middle age. Thus, regular ovulatory cycles are apparently essential but not totally sufficient for fertile gestations in female rats, irrespective of chronologic age.

Beginning at 6 months of age, the litter sizes of multiparous rats decrease progressively, and these decreases are associated with a decline in the number of live but not dead pups delivered at parturition. Also, the percentage of dead pups/total number of pups delivered, increases steadily with advancing age in multiparous females (from 14 to 69%) but not in primiparous rats (Matt *et al.* 1986). The litter sizes of 8–10-month-old primiparous females are not different from those of multiparous rats. However, the litter sizes of irregular cycling females are consistently smaller than those of regular cycling rats. Thus, parity has little effect on fecundity in ageing female rats, whereas cessation of regular ovulatory cycles during ageing is associated with decreases both in incidence of fertility and litter size (Matt *et al.* 1987*b*).

In both multiparous and primiparous females, ovulation rates under the conditions of caging and mating with males remain relatively unchanged between 4 and 12 months of age. Whether spontaneous ovulation rates change with age is not clear at present, but preliminary observations suggest a relationship between ovulation rate and the pattern of the preovulatory LH surge in middle-aged rats (LaPolt *et al.* 1985, 1993). The percentage of morphologically normal blastocysts seen on day 5 of gestation decreases steadily, and this age-related decrease closely follows the decline in litter size. These findings demonstrate that cessation of regular cyclicity during ageing renders many female rats infertile, and the decline in litter size in middle-aged pregnant rats is directly related to a decrease in the number of normal blastocysts. Since many middle-aged rats also fail to reproduce successfully while displaying regular oestrous cycles, other functional defects in the neuroendocrine and/or reproductive system probably account for the age-related loss in fertility.

3   Embryonic development and fecundity

From the above recent studies it is clear that decreases in fertility and litter size precede cessation of regular oestrous cycles in middle-aged female rats, and that the decline in litter size is related to a decrease in the number of blastocysts that are present on day 5 of gestation, immediately prior to implantation. These results suggest that the pattern of embryonic development during the first 5 days of gestation may be altered in middle-aged females, resulting in fewer implanted blastocysts and smaller litter sizes. In a recent study ovulation rates, fertilization rates, and the patterns of embryonic development were examined during the first 5 days of pregnancy in regularly cycling, young (4 months) and middle-aged (10 and 13 months old) rats. The results demonstrated that the ovulation rate is significantly reduced in 13-month-old (9.0 ± 1.0 ova/rat) females, but not in 10-month-old rats (12.2 ± 0.8 ova/rat), as compared with that in young animals (12.8 ± 1.0 ova/rat). However, there is no decrease in fertilization rate in either the 10- or 13-month group. While the total numbers of embryos present on days 2–5 are similar among these three age groups, embryos from 10-month-old females show a delayed pattern of development and an increased incidence of morphological abnormalities. These abnormalities in embryo development are even more pronounced in 13-month-old females. By day 5 of gestation, there is a significant reduction in normal blastocysts in 10-month (7.3 ± 1.2 per rat) and 13-month (6.0 ± 1.6 per rat) -old rats, as compared with young females (10.6 ± 0.9 per rat). These data reveal that the decreased numbers of normal blastocysts in middle-aged rats result from delayed and abnormal patterns of embryonic development, possibly reflecting impaired oocyte quality and/or alteration in the oviductal environment during ageing (Day *et al.* 1989).

In a related study, both middle-aged (10 months) and young, pregnant rats produced similar numbers of total embryos on day 3 of gestation. In the young group, 95% of the embryos were morphologically normal and the remainder (5%) were abnormal. In contrast, middle-aged females show a high incidence (24%) of abnormal embryos. However, the reduced numbers of normal embryos were not associated with any discernible alteration in plasma pattern of PRL, progesterone, or oestradiol during days 1–3 of pregnancy. Since embryonic abnormalities become apparent prior to any significant changes in plasma progesterone and oestradiol levels, these findings indicate that reduced numbers of normal embryos in middle-aged females are unlikely to be caused by defect(s) in luteal function early in pregnancy. Taken together, these findings suggest that the quality of the ova produced from the ovaries of middle-aged rats during regular cycles may not be uniform at ovulation (Day *et al.* 1991).

4 Influence of oestradiol and progesterone in young female rats on subsequent age-related changes in neuroendocrine and reproductive functions

As already mentioned, reproductive declines during ageing occur significantly earlier in virgin than in multiparous female rats (LaPolt *et al.* 1986; Matt *et al.* 1987b), and pregnancy is associated with a persistent rise in plasma progesterone and a decrease in oestradiol (LaPolt *et al.* 1986; Matt *et al.* 1986). Since cessation of oestrous cyclicity at middle age is directly related to loss of the LH surge response to oestradiol stimulation (Lu *et al.* 1980; Matt *et al.* 1987a), it was proposed that repetitive rises in plasma oestradiol during oestrous cycles eventually render the hypothalamus unresponsive to further oestradiol stimulation of GnRH release. The effects of a chronic decrease in circulating oestradiol, associated with a persistent rise in progesterone on age-related reproductive declines were examined in longitudinal studies (Lu *et al.* 1985; LaPolt *et al.* 1986, 1988a). Between 4 and 6 months of age, virgin rats received three consecutive Silastic implants of progesterone or blank implants, subcutaneously. While the virgin controls exhibited oestrous cycles associated with cyclic increases in oestradiol secretion and ovulations, progesterone-implanted female rats showed high levels of plasma progesterone and low oestradiol constantly without follicular maturation or ovulation. Following a total of 9 weeks of treatments with progesterone, 85% of these virgin females maintained regular cycles up to 8 months of age. By contrast, only 51% of virgin controls displayed regular cycles at 8 months. By 10 months, less than one-third of virgin controls were regularly cycling, while over 50% of females previously given progesterone continued to show regular cycles. At 10 and 12 months of age, the virgin rats previously given progesterone not only showed a higher incidence of regular oestrous cycles, but also exhibited a two-to three fold greater fertility rate after mating (LaPolt et al. 1986). These results demonstrate that successive treatments with progesterone implants in young virgin rats inhibit ovarian follicular development, reduce ovarian production of oestradiol, and result in maintenance of regular cyclicity for an extended period of time. They significantly improved reproductive function at middle age.

In a separate study, retired breeder female rats each received five consecutive Silastic implants (subcutaneously) of progesterone at 8–13 months of age, while multiparous controls were grouped into five rats per cage. Following a total of 15 weeks of treatment with progesterone, over 50% of retired breeders continued to exhibit regular cyclicity at least up to 17 months, whereas only 20% of multiparous controls were cycling regularly (Lu et al. 1985). Even at 19 months, these retired breeders previously treated with progesterone showed a normal pattern of pro-oestrous LH surges during regular cycles. These results demonstrate that a chronic decrease in plasma oestradiol, in association with a persistent increase in progesterone, contributes to maintaining a normal hypothalamic GnRH response to oestradiol stimulation.

In female rats and mice, reproductive declines during ageing are associated with significant decreases in the numbers of resting follicles present in the ovary (Mandl and Shelton 1959; Gosden et al. 1983). Reports have also shown that in middle-aged multiparous rats, an experimental reduction in the size of the follicular pool increases the incidence of irregular cycles and reproductive dysfunctions (Sopelak and Butcher 1982; Butcher 1985). While oestradiol is known to stimulate follicular development (Goldenberg et al. 1972), high levels of plasma progesterone suppress follicular recruitment (Pederson and Peters 1971), decrease oestradiol production by granulosa cells (Fortune and Vincent 1983) and inhibit follicular growth and ovulation (Buffler and Rosen 1974; Wilks et al. 1983). In view of these observations, we have examined the effects of a chronic increase in circulating oestradiol and/or progesterone on follicular pool size, and the relationship between the latter and reproductive declines (LaPolt and Lu 1987). Between 3 and 8 months of age, virgin rats received five consecutive Silastic implants of progesterone or oestradiol plus progesterone, subcutaneously, while control females received blank implants. During these 5 months of steroid implant treatments, animals in the progesterone only group exhibited high levels of plasma progesterone and low oestradiol constantly, whereas females in the progesterone plus oestradiol group showed sustained, intermediate levels of plasma oestradiol and high progesterone (LaPolt et al. 1988a). At 8 months of age, the numbers of resting follicles in the ovaries of the progesterone only group females were at least two times higher than those found in the ovaries of virgin controls. The size of the follicular pool in females given both oestradiol and progesterone were intermediate between the other two groups. At 8–10 months of age, the percent incidence of regular cyclicity in these three groups of virgin female rats appeared to be inversely related to the overall amounts of circulating oestradiol during the entire period (5 months) of treatment. Thus, at 8 months, 76 and 55% of animals previously given progesterone and oestradiol plus progesterone, respectively, displayed regular oestrous cycles, whereas only 26% of virgin controls exhibited regular cycles. Moreover, linear regression analyses revealed a significant correlation between the numbers of resting follicles present in the ovaries and the per cent incidence of regular cyclicity in all three groups of virgin rats. At 10 months, fertility rates among these regularly cyclic virgins were 73 and 44% for the progesterone only group and progesterone plus oestradiol group, respectively, but were only 20% for virgin controls. These findings indicate that plasma levels of oestradiol rather than progesterone are more influential on the age-related reproductive declines, and further suggest that a long-term reduction in circulating oestradiol may contribute to a delay in the reproductive decline through a conservation of ovarian follicles, in addition to potentially deleterious effects of oestradiol on hypothalamic function (LaPolt et al. 1988a). Since litter sizes produced by 10–12-month-old females previously given progesterone were significantly greater than those of virgin controls, it is possible that a persistent decrease in oestradiol production by the ovaries may also contribute to a better quality of oocytes.

IV   REPRODUCTIVE DECLINE IN AGEING WOMEN

1   Endocrine physiology of the normal menstrual cycle

In women, cyclic and orderly changes in ovarian morphology and function during the normal menstrual cycle are timed and closely regulated by the actions of pituitary gonadotrophins. In turn, steroid hormones produced by the ovary modulate hypothalamic function and pituitary gonadotrophin secretion through feedback mechanisms (Lu 1986; Lu and Judd 1991). During the follicular phase of the cycle, gonadotrophins stimulate follicle development and growth and prepare one mature follicle for ovulation. At mid-cycle, the hypothalamic–pituitary axis exhibits a preovulatory surge of gonadotrophins, followed by ovulation. Under LH action, the ruptured follicle is luteinized to form a corpus luteum which secretes hormones preparing the uterine endometrium for implantation of the blastocyst. Following implantation, human chorionic gonadotrophin (hCG) produced from early placental tissue prevents the corpus luteum from regression. However, if implantation is not ensured, menstruation occurs as the result of cessation of corpus luteal function.

During menstruation and early follicular phase of the cycle, ovarian secretion of oestradiol is low. Beginning at 7–8 days prior to ovulation, there is a steady and marked increase in circulating oestradiol, reaching a maximum 1 day before the mid-cycle LH surge. After this peak, and before ovulation, there is a transient but significant fall in oestradiol concentrations, resulting from a partial luteinization of the preovulatory follicle. During the luteal phase of the cycle, the functioning corpus luteum produces both oestradiol and progesterone which reach a maximum approximately 5–7 days after ovulation. Before the next menstruation, plasma oestradiol levels return to baseline values as a result of corpus luteum regression. The ovary produces less oestrone than oestradiol, and peripheral conversion of androstenedione to oestrone accounts for most oestrone in the circulation.

Both the ovary and the adrenal secrete androstenedione, and most testosterone in the circulation comes from conversion of androstenedione. The ovary secretes only small amounts of dihydrotestosterone (DHT), and enzymatic conversions of testosterone and androstenedione contribute to most DHT in the circulation. Great portions of both dehydroepiandrosterone (DHEA) and DHEA sulphate come from the adrenals, while small amounts are secreted by the ovary. During the follicular phase of the cycle, progesterone secretion by the developing follicles is minimal. During the mid-cycle LH surge, a partial luteinization of the preovulatory follicle produces a small rise in plasma progesterone and a major increase in 17α-hydroxyprogesterone (17α-OH-P). Following ovulation, the functioning corpus luteum produces large amounts of both progesterone and oestradiol and some 17α-OH-P, reaching a maximum about 5–7 days after ovulation.

2   Neuroendocrine control of gonadotrophin secretion

Pituitary secretion of gonadotrophins is mainly under direct control by the hypothalamus. Hypothalamic GnRH is discharged in a pulsatile fashion into the pituitary portal blood, resulting in pulsatile patterns of LH and FSH release from the pituitary. Also, oestradiol and progesterone can modulate gonadotrophin secretion by changing the amplitude and, to a lesser extent, frequency of hypothalamic GnRH release. Toward the end of the luteal phase, decreasing oestradiol and progesterone in the circulation lead to a preferential rise in FSH secretion which stimulates follicular development and growth. At mid-cycle, the sustained rise in circulating oestradiol initiates the preovulatory LH and FSH surges through a positive feedback mechanism on the pituitary (Knobil and Plant 1978). In women and other primates, whether the mid-cycle gonadotrophin surge is mediated by a concomitant increase in hypothalamic discharge of GnRH is still a controversial issue. The LH surge is accompanied by a small increase in progesterone produced by the preovulatory follicle, and this small rise in progesterone at mid-cycle may enhance and prolong the LH and FSH surges. The LH surge-mediated increase in progesterone production by the preovulatory follicles is much greater in female rats than in women (LaPolt *et al.*, 1986). Following ovulation, high levels of circulating oestradiol and progesterone produced by the functioning corpus luteum exert a potent inhibition on LH and FSH secretion. This inhibition of gonadotrophins persists until the end of the luteal phase, when the secretion of oestradiol and progesterone decreases as result of corpus luteum regression. At that time, low levels of both oestradiol and progesterone allow a small rise in FSH secretion, which stimulates new follicular development and growth for the next cycle.

3   Perimenopause: the ageing transition

The average women goes through the menopause at 49–52 years of age (Judd 1983), and the most dramatic event associated with the meno-pause is the loss of cyclic ovarian function and cessation of menstrual cycles. The total number of follicles in the ovaries of a healthy young women is estimated to be 400000, and the follicular pool decreases with age (Block 1951). Among these many primordial follicles, only 0.1% are actually involved in ovulatory function, while the great major-ity of the ovarian follicles disappear by atresia during the reproduc-tive years. It is interesting to note that after the age of 40–42 in women, the rate of follicle depletion from the ovarian pool is accelerated (Richardson *et al.* 1987). After menopause, a few follicles remain in the ovary but they are usually non-responsive to gonadotrophins and eventually disappear.

The perimenopause represents the transition from regular menstrual cycles to the permanent amenorrhoea of the menopause and is charac-terized by menstrual irregularity. During this transitional period, changes in steroid hormone secretion resulting from altered function of the ageing ovary becomes obvious. It has been shown that, in regularly menstruating women over the age of 45, the length of the follicular but not luteal phase is decreased as compared with that of the normal cycle in young adults. In this group of older women, oestradiol levels in the follicular, periovulatory, and luteal phases are lower, while no differences in progesterone are found. The consistent reduction in circulating oestradiol throughout the menstrual cycle is associated with a significant rise in circulating FSH but not LH, sug-gesting a reduction in the number of functional ovarian follicles (Sherman *et al.* 1976). The marked menstrual irregularity during the perimenopause suggests an irregular pattern of follicular maturation which may or may not lead to ovulation. It is conceivable that the residual follicles remaining in the perimenopausal ovary are responsible for the limited amount of oestradiol secretion and the irregular pattern of follicular maturation as well.

4   Menopause

The transitional period of the perimenopause is followed by a cessation of menstrual cycles, and the menopause is associated with major changes in ovarian steroid hormone and pituitary gonadotrophin secretion. Following the menopause, oestrogen production decreases and is mostly due to the reduction in circulating oestradiol. The low oestradiol levels in postmenopausal women are clearly less than those found in any phase of the menstrual cycle and are comparable with those seen in ovariectomized young women (Judd *et al.* 1974). After menopause, plasma concentrations of oestrone are usually higher than those of oestradiol, and neither the adrenal nor the ovary secretes oestrone. Most oestrone in the circulation comes from peripheral conversion of androstenedione, while most of the oestradiol in postmenopausal women is derived from conversion of oestrone and testosterone.

In young women, androstenedione is the principal androgen secreted by the developing follicles. With the menopause, plasma levels of androstenedione are reduced to about half of that in premenopausal women, with a major source from the adrenal. In contrast, the postmenopausal ovary continues to secrete testosterone, probably from the hilar cells and the luteinized stromal cells, so that the mean levels of circulating testosterone are only slightly lower in postmenopausal women than in young individuals. DHEA and DHEA sulphate are major androgens secreted by the adrenals, with the ovarian contribution less than 25%. Beginning in the fourth decade, production of DHEA and its sulphate declines with advancing age, independent of ovarian function. After menopause, secretion of progesterone decreases markedly in the absence of follicular activity and the corpus luteum.

In premenopausal women, pituitary secretion of gonadotrophins is under direct control by episodic release of hypothalamic GnRH and by ovarian hormones, although the magnitude of these LH and FSH pulses are relatively small. After menopause, the minimal amounts of both oestradiol and progesterone in the circulation are insufficient to inhibit gonadotrophin release through a normal negative feedback mechanism. Under these conditions, the pituitary secretion of both LH and FSH is markedly enhanced, and large pulses of both LH and FSH discharge occur once every 60–90 min in postmenopausal women, presumably reflecting augmented GnRH output by the hypothalamus in the face of diminished oestradiol and progesterone in the circulation. Later in the postmenopausal period, there is a decline in FSH and LH secretion which still remains above normal.

V   CHANGES IN SEXUAL BEHAVIOUR IN AGEING FEMALE RODENTS AND WOMEN

1   Rodents

Ageing female rats exhibit a series of irregular, lengthened oestrous cycles beginning about 7–8 months of age, associated with extended periods of vaginal cornification and delayed ovulation. At about 10 months of age they enter a persistent oestrous state characterized by many well-developed ovarian follicles, constant vaginal cornification, no ovulation, and relatively high oestrogen secretion. During this state, sexual receptivity (lordosis response) is continuous in both in middle-aged rats (12–14 months old) and older (18–20 months old) rats. The persistent oestrous state is often followed by occurrence of a series of prolonged periods of pseudopregnancy, characterized by ovaries with many corpora lutea that secrete progesterone. These rats show no sexual receptivity. The oldest rats, 2–3 years of age, become anoestrus, have shrunken ovaries and uteri, and do not exhibit sexual receptivity (Meites 1982; Davidson *et al*. 1983). Nass *et al*. (1982) observed that when ageing female rats were placed in the vicinity of males, oestrous cycles continued beyond the time when control rats not in the vicinity of males ceased to cycle. Thus the sight or smell of male rats can influence neuroendocrine mechanisms that regulate oestrous cycles.

When middle-aged female rats (14 months old) were ovariectomized and treated with oestrogen or oestrogen plus progesterone, and placed with vigorous males, sexual receptivity remained the same as in young rats (Borchardt *et al*. 1980). Similarly, Peng *et al*. (1980) tested sexual receptivity in ovariectomized female rats 5–8, 15–23, and over 24 months of age given high doses of oestradiol benzoate and found no differences. However, in spayed oestrogen-treated female hamsters, a gradual decline in sexual receptivity was reported, which became marked in the oldest hamsters (Farrel *et al*. 1977).

2   Women

According to Davidson *et al*. (1983), libido and potency define the major dimensions of sexuality both in men and women. An interesting question is whether the menopause alters sexual interest and activity by women. According to Masters and Johnson (1966), the menopause, which occurs at about the age of 50–52 years, results in decreased clitoral tumescence, thinning of the walls and loss of lubrication in the vagina, decreased vasocongestion in the vagina, decreased lubrication of the labia, involution of the uterus, and loss of dilation of the cervical os. There are also major vasomotor disorders, including development of 'hot flushes' (Tataryn *et al.* 1979), palpitations, and headaches, which may last for several years after the onset of menopause. Despite these consequences of the menopause, most investigators have reported no significant decrease in libido until about the sixth decade of life. Kinsey *et al.* (1953) recorded a definite decline in the frequency of sexual acts by the sixth decade. The decline was stated by some women to occur because of a decrease in sexual interest by the male partner. The view that the menopause was not necessarily a barrier to continued sexual activity was also supported by Masters and Johnson (1966) and by Verwoerdt *et al.* (1969), even though many women stated that the menopause resulted in a definite fall in sexual interest. In a study of 800 Swedish women 46–54 years of age, most of the women reported reduced sexual interest and decreased sexual and orgasmic activity which they attributed to the menopause (Hallstrom 1977). Many of these women reported that their husbands developed a greater sex drive after the menopause. Declining sexual interest was found to be more prevalent in lower than in higher socio-economic groups. It should be appreciated that most of the above information on women was obtained by questionnaires which may not be free of error.

The effects of oestrogen on the brain are believed to be profound, and loss of this hormone after the menopause may have consequences on behaviour that cannot be adequately assessed by questionnaires. No dose–response relationship has been demonstrated between oestrogen levels and sex desire in women.

VI   DECLINE IN TESTICULAR FUNCTION AND SEXUAL INTEREST IN AGEING MALE RODENTS AND MEN

1   Rodents

In the ageing male rat, testosterone secretion declines (Bruni *et al.* 1977; Gray 1978; Kaler and Neaves 1981), mainly as the result of diminished secretion of FSH and LH (Simpkins *et al.* 1977; Harman and Talbert 1985), in turn due to decreased release of GnRH from the hypothalamus associated with deficient NA activity (Meites 1982, 1990, 1991). Reduced FSH and LH secretion can also be attributed in part to the diminished responsiveness of the pituitary to GnRH stimulation (Bruni *et al.* 1977). Reduced testosterone secretion is due in part to decreased responsiveness of the testes to gonadotrophic hormone stimulation (Riegle and Miller 1978), perhaps the result of increased loss of Leydig cells in the testes.

A decline in sperm production as well as an increase in sperm abnormalities may also occur in old male rats and mice. Talbert (1977) reported there were no differences between old and young rats in the length of time required for sperm to develop from the spermatogonium to mature male gametes. Saksena (1979) observed no differences in the number of spermatozoa in the rat caput epididymis between 72 and 500 days of age, but the number of pups produced per litter decreased significantly with time (Matt *et al.* 1986). The reduction in pups per litter may be due to reduced sperm quality, altered sperm transport, or other factors (Harman and Talbert 1985). Bronson and Desjardins (1977) observed a reduction in sperm in 24- and 30-month-old CBF-1 male mice as compared with mice 6 months old. In ageing rats, Humphreys (1977) found tubular degeneration, spermatogenic arrest, and thickening and folding of the basement membrane.

There is a decline in sexual interest in ageing male rats. Old males exhibit a decreased interest in copulation when placed with receptive females, and there are fewer mounts, intromissions, and ejaculations (Larson 1958). Davidson *et al.* (1983) studied the sexual behaviour of middle-aged male rats (13–15 months old) as compared with young mature rats, since testosterone secretion significantly declines by middle age. They found that middle-aged rats exhibited decreased mounts in the presence of receptive females, and showed an increased time to intromission and ejaculation. There was also a quantitative reduction in erection frequency (Gray *et al.* 1981). Whereas 100% of male rats showed complete copulatory behaviour (mounting, intromission, and ejaculation) at 7 months of age, only 39% of the animals that survived to 27 months of age exhibited complete copulatory behaviour. Although testosterone secretion also shows a decline by middle age in male rats, only a partial relationship could be demonstrated between testosterone levels and sexual behaviour. When middle-aged rats were castrated and given testosterone at low or supraphysiological levels, the decline in the erectile reflex response was corrected, but not the decrease in sexual motivation or copulatory behaviour. Davidson *et al.* (1983) therefore concluded that the decline in testosterone secretion does not completely explain the decrease in sexual motivation and ejaculatory behaviour, and that other factors need to be considered. Behavioural changes in ageing laboratory mice appear to be similar to those in ageing rats (Huber *et al.* 1980).

2   Men

i   Hypothalamic–pituitary function

The decrease in testosterone and inhibin secretion by the testes of ageing human males leads to elevation of FSH and LH secretion by the anterior pituitary beginning at about 45–50 years of age (Pedersen-Bjergaard and Jonnesen 1948; Albert 1956). The rise in these two hormones is not nearly as great as in pre- and postmenopausal women. The greater rise in FSH than in LH probably is due to the reduction in inhibin secretion by the testes. As with testosterone secretion, there is considerable individual variation in gonadotrophin secretion, and in some healthy older men the increase in gonadotrophins is relatively small. Despite the rise in gonadotrophin secretion in older men, the pituitary exhibits less LH and FSH release in response to administration of GnRH than the pituitary of young men (Harman 1978; Winters and Troen 1982).

The absence of a testosterone circadian rhythm, and of a pulsatile pattern of LH secretion in older men (McFayden *et al.* 1980) suggests a change in the nature of GnRH release by the hypothalamus. Morley and Kaiser (1990) found an increase in occurrence of eugonadotrophic hypogonadism with age in men, suggesting a decrease in responsiveness of the hypothalamic–pituitary system to circulating levels of bioactive testosterone. NA and DA were found to be reduced in the hypothalamus of old men and women (Robinson *et al.* 1972; Hornykiewicz 1987; O'Neil *et al.* 1987) as mentioned previously, but it is unknown whether this influences release of GnRH.

ii   Testicular function

Most investigators have reported a progressive decline in testosterone secretion in men beginning at about 40–50 years of age, with considerable differences present among individuals (Vermeulen *et al.* 1972; Bremner *et al.* 1983; Harman and Talbert 1985; Morley and Kaiser 1990). There is also a loss of circadian rhythm in testosterone secretion and blunting of the morning peak of testosterone secretion. Dihydrotestosterone, which is converted from testosterone in target tissues, may be unchanged in older men. Most workers have found no significant changes in oestrogen levels in ageing men. However, inhibin secretion by the testes is reduced, probably due to loss of Sertoli cells.

The decrease in testosterone secretion in older men is believed to be caused mainly by the reduction in Leyding cell number and perhaps is also a result of decreased sensitivity of the Leydig cells to gonadotrophic hormone stimulation (Sarjent and McDonald 1948; Harman and Talbert 1985; Morley and Kaiser 1990). The loss of Leydig cells appears to be due to atherosclerotic changes found in blood vessels in the testes. In a study of testes obtained at autopsy from 121 men aged 14–73 years of age, an increase in degenerative changes was observed which was closely related to the distribution of the regional blood supply (Khoury and Sowers 1988). A decrease in response of the testes of older men to hCG has also been reported, and is believed to be due mainly to loss of Leydig cells. A similar reduction in release of testosterone in response to hCG administration has been reported in old male rats (Riegle and Miller 1978).

In addition to the decrease in testosterone secretion in ageing men, a decrease in spermatogenesis may occur (Harman and Talbert 1985). Sperm were observed to decrease from 68.5% in the sixth decade to

48% in the eighth decade. Nieschlag *et al.* (1982) observed that sperm number and morphology were unchanged in healthy older men, but degenerative changes were observed in the germinal tubular membrane, as well as peritubular fibrosis and failure of spermatogenic maturation due to sclerosis and closing of the tubular lumen. Autopsy studies have shown that areas of normal spermatogenesis are present in more than 50% of men over 70 years of age (Engle 1952). Despite these changes in testicular function, healthy old human males retain sufficient function of the seminiferous tubules to reproduce even into old age and cases have been reported of men even in their nineties who have fathered children (Harman 1978).

iii   Sexual behaviour

Male sexual interest declines with age but is not lost in healthy older individuals. According to early studies by Kinsey *et al.* (1948) peak monthly seminal ejaculations occur in the mid-teens and then gradually decrease to very low levels by 80–90 years of age. Sexual events were reported to decline after age 30 years from about three per week to about 0.5 per week by age 70 years, and to near zero by age 80 years (Martin 1977). Impotence (inability to sustain an erection adequate for sexual intercourse) occurs with increasing frequency with advancing age (Morley and Kaiser 1990). It was estimated that by age 65 years, 25% of men are impotent, and this rises to 55% by 75 years of age. The frequency in ability to ejaculate more than once within a short time span also declines with age. The four most common causes of impotence with advancing age were stated to be vascular disease, medications, diabetes mellitus, and hormonal dysfunctions (Morley and Kaiser 1990).

The relationship of testosterone secretion with the decline in male sexual behaviour has been investigated by Davidson *et al.* (1983). They reported that declining testosterone levels and decreased sexual activity and interest (libido) were only partly related. No positive correlation was observed between testosterone levels and sexual function in a normal population of young men. However, Harman and Talbert (1985) reported that when the relation of sexual activity (total number of events leading to orgasm) to free testosterone levels was tested, a significant correlation was found in the oldest men. Those who maintained the greatest level of sexual activity tended to exhibit higher testosterone values. Vermeulen (1979) also published data suggesting a correlation between the decline in free testosterone and sexual activity in ageing men. It is clear that many factors other than testosterone levels can influence sexual behaviour. A correlation has been observed between high sexual performance in youth and retention of performance in older age. Other factors that can influence sexual behaviour include health, nutrition, intake of alcohol, stress, psychological disturbances, etc.

238 Joseph Meites and John K.H. Lu

VII CONCLUSIONS

Faults that develop in the hypothalamus are mainly responsible for the reproductive decline with age in female and male rats, although defects also appear in the pituitary and gonads. The major fault appears to be the decrease in hypothalamic CAs, particularly NA, which is normally required to evoke increased release of GnRH to induce pituitary LH release and ovulation. The decrease in NA does not permit an adequate response by the hypothalamus to hormone stimulation from the ovaries and to other stimuli that normally elicit LH release. There are many potential causes for the decline in hypothalamic CAs, including: (a) damage to neurones that secrete CAs as a result of the catabolism of CAs by MAO, leading to the production of 'free radicals'; (b) the decrease in tyrosine hydroxylase, the rate-limiting enzyme required to synthesize CAs; (c) the increase in MAO, the major enzyme responsible for catabolism of CAs; (d) damage to neurones in the hypothalamus by oestrogen action; and (e) other factors.

The effects of oestrogen on the maintenance of oestrous cycles are of particular interest. Administration of a long-acting oestrogen has been shown to damage neurones in the arcuate nucleus and medial basal hypothalamus (Brawer *et al.* 1978; Sarkar *et al.* 1982), whereas removal of the ovaries in cyclic rats followed by replacement with fresh ovaries many months later extends oestrous cycles many months beyond the normal period when intact rats cease to cycle (Aschheim 1976; Finch *et al.* 1984). Furthermore, Lu and his colleagues (1983, 1985, 1990) have demonstrated that: (a) oestrous cycles cease earlier in virgin than in multiparous rats due to the inhibitory effects of progesterone on oestrogen secretion during gestations in the multiparous rats; (b) when progesterone was given to virgin rats, it maintained oestrous cycles for a much longer period than in virgin rats not given progesterone; (c) administration of progesterone to middle-aged retired breeders also resulted in prolonged maintenance of oestrous cycles in these rats, and (d) progesterone was shown to inhibit follicular development and reduce ovarian secretion of oestrogen, thereby inhibiting the damaging effects of oestrogen on the hypothalamus. In addition, by increasing the number of resting follicles in the ovaries and reducing oestrogen secretion, the quality of the ova remaining in the ageing rat is improved (LaPolt *et al.* 1986). It has been demonstrated that oestrogen can decrease hypothalamic CAs in the rat (Meites 1982), thereby inhibiting gonadotrophin secretion. Oestrogen may adversely affect other hypothalamic mechanisms as well.

Administration of L-dopa, the precursor of the CAs, to ageing cycling rats, extends the period of oestrous cycles in these rats, and can also reinitiate cycles in old constant oestrous rats after they have ceased to cycle (Meites 1990, 1991). Injections of iproniazid, deprenyl, and progesterone are also effective in restoring oestrous cycles in constant oestrous rats. In addition, the presence of male rats extends maintenance of oestrous cycles in ageing female rats, although the mechanisms involved are not understood (Nass *et al.* 1982).

It is of interest that in old rats some small follicles and ova remain in the ovaries practically to the end of life, and can be activated by removing and grafting the ovaries into young ovariectomized rats or by administration of gonadotrophic hormones (Meites 1982, 1991). This is in contrast to the ovaries of postmenopausal women whose follicles and ova disappear

As previously indicated, faults that develop in the gonads are primarily responsible for the decline in reproductive functions in ageing men and women, although defects have also been reported in hypothalamic and pituitary function. The ovaries of women approaching the menopause show a decrease in ability to secrete oestrogen primarily due to loss of follicles by atresia, and this results in elevated secretion of FSH and LH. Stromal tissues are also lost, which combined with loss of follicles result in shrinkage of the ovaries. In the postmenopausal period, the ovaries gradually become more shrunken and fibrotic, lose their few remaining follicles and ova, and become less responsive to the increased circulating levels of FSH and LH.

In ageing men, there is a decrease in testicular weight and in the number of androgen-secreting Leydig cells, resulting in reduced testosterone bio-synthesis, which leads to elevated gonadotrophin secretion. There may be some loss of seminiferous tubules in selected areas of the testes, but spermatogenesis is maintained. A decrease in inhibin secretion by the Sertoli cells is believed to be partly responsible for the rise in FSH secretion. In the ageing male rat, there is also a decrease in testosterone secretion, but this is due mainly to the decline in FSH and LH secretion caused mainly by inadequate GnRH release from the hypothalamus. The pituitary of ageing men also shows a diminished response to GnRH, and the testes to gonadotrophin stimulation (Harman 1978; Harman and Talbert 1985).

Why major differences exist in the regulation of reproductive ageing in rodents as compared with primates is not entirely clear at present, particularly since the hypothalamus is the ultimate regulator of reproductive functions. However, cyclic reproductive activities in rodents (oestrous cycles) are controlled mainly by the hypothalamus, whereas in primates menstrual cycles are regulated mainly by the ovaries (Knobil and Plant 1978). The major action of the primate ovaries is on the pituitary rather than on the hypothalamus. Loss of oestrous cycles can be delayed or reversed in ageing rats by correcting hypothalamic dysfunctions, but in primates losses in ovarian follicles with age prevent such restoration. It is possible that grafting of ovaries from young into postmenopausal primates may permit some restoration of reproductive cycles.

REFERENCES

Albert, A. (1956). Human urinary gonadotrophins. *Recent Progress in Hormone Research*, **12**, 227–301.

Armstrong, D.T., Miller, L.S., and Knudsen, K.A. (1969). Regulation of lipid metabolism and progesterone production in rat corpora lutea and ovarian interstitial elements by prolactin and luteinizing hormone. *Endocrinology*, **85**, 393–401.

Aschheim, P. (1976). Aging in the hypothalamic–hypophyseal ovarian axis in the rat. In *Hypothalamus, pituitary, and aging* (ed. A.V. Everitt and J.A. Burgess), pp. 376–418. Charles C. Thomas, Springfield, IL.

Block, E. (1951). Quantitative morphological investigations of the follicular system in women: Variations in different phases of the sexual cycle. *Acta Endocrinologica*, **8**, 33–8.

Borchardt, C.M., Lehman, J.R., and Hendricks, S.E. (1980). Sexual behaviour and some of its physiological consequences in persistently estrous aged female rats. *Ageing*, **3**, 59–62.

Brawer, J.R., Naftolin, F., Martin, J., and Sonnenschein, C. (1978). Effects of a single injection of estradiol valerate on the hypothalamic arcurate nucleus and on reproductive function in the female rat. *Endocrinology*, **103**, 501–12.

Bremner, W.J., Vitiello, M.V., and Prinz, P.N. (1983). Loss of circadian rhythmicity in blood testosterone levels with aging in normal men. *Journal of Clinical Endocrinology and Metabolism*, **56**, 1278–81.

Bronson, F.H. and Desjardins, C. (1977). Reproductive failure in aged CBF male mice: interrelationships between pituitary gonadotrophic hormones, testicular function, and mating success. *Endocrinology*, **101**, 939–45.

Bruni, J.F., Huang, H.H., Marshall, S., and Meites, J. (1977). Effects of single and multiple injections of synthetic GnRH on serum LH, FSH and testosterone in young and old male rats. *Biology of Reproduction*, **17**, 309–12.

Buffler, G. and Rosen, S. (1974). New data concerning the role played by progesterone in the control of follicular growth in the rat. *Acta Endocrinologica*, **75**, 569–78.

Butcher, R.L. (1985). Effect of reduced ovarian tissue on cyclicity, basal hormone levels and follicular development in old rats. *Biology of Reproduction*, **32**, 315–21.

Butcher, R.L., Fugo, N.W., and Collins, W.E. (1972). Semicircadian rhythm in plasma levels of prolactin during early gestation in the rat. *Endocrinology*, **90**, 1125–7.

Clemens, J.A., Amenomori, Y., Jenkins, T., and Meites, J. (1969). Effects of hypothalamic stimulation, hormones, and drugs on ovarian function in old female rats. *Proceedings of the Society for Experimental Biology and Medicine*, **132**, 561–63.

Cooper, R.L., Conn, P.M., and Walker, R.F. (1980). Characterization of the LH surge in middle-aged female rats. *Biology of Reproduction*, **23**, 611–15.

Damassa, D. A., Gilman, D.P., Lu, K.H., Judd, H.L., and Sawyer, C.H. (1980). The 24-hour pattern of prolactin secretion in aging female rats. *Biology of Reproduction*, **22**, 571–5.

Davidson, J.M., Gray, G.D., and Smith, E.R. (1983). The sexual psychoneuro-endocrinology of aging. In *Neuroendocrinology of aging* (ed. J. Meites), pp. 221–58. Plenum Press, New York.

Davison, A.N. (1987). Functional morphology of neurons during normal and pathological aging. In *Modifications of cell to cell signals during normal and pathological aging* (ed. S. Govoni and S. Battaini), pp. 1–16. Springer-Verlag, Berlin.

Day, J.R., Morales, T.H., and Lu, J.K.H. (1988). Male stimulation of luteinizing hormone surge, progesterone secretion and ovulation in spontaneously persistent estrous aging rats. *Biology of Reproduction*, **38**, 1019–26.

Day, J.R., LaPolt, P.S., Morales, T.H., and Lu, J.K.H. (1989). An abnormal pattern of embryonic development during early pregnancy in aging rats. *Biology of Reproduction*, **41**, 933–9.

Day, J.R., LaPolt, P.S., and Lu, J.K.H. (1991). Plasma patterns of prolactin, progesterone and estradiol during early pregnancy in aging rats. Relation to embryonic development. *Biology of Reproduction*, **44**, 786–90.

Engle, E.T. (1952). The male reproductive system. In *Cowdry's problems of aging*, 3rd ed. (ed. A.I. Lansing), pp. 708–29. Williams and Wilkins, Baltimore.

Farrell, A., Gerall, A.A., and Alexander, M.J. (1977). Age-related decline in receptivity in normal, neonatally androgenized female and male hamsters. *Experimental Aging Research*, **103**, 117–28.

Fayein, N.A. and Aschheim, P. (1980). Age-related temporal changes of levels of circulating progesterone in repeatedly pseudopregnant rats. *Biology of Reproduction*, **23**, 616–20.

Finch, C.E., Felicio, L.S., Mobbs, C.V., and Nelson, J.F. (1984). Ovarian and steroidal influences on neuroendocrine aging processes in female rodents. *Endocrine Reviews*, **5**, 467–97.

Fortune, J.E. and Vincent, S.E. (1983). Progesterone inhibits the induction of aromatase activity in rat granulosa cells *in vitro*. *Biology of Reproduction*, **28**, 1078–89.

Goldenberg, R.L., Vaitudaitis, J.L., and Ross, G.T. (1972). Estrogen and follicle-stimulating hormone interactions on follicle growth in rats. *Endocrinology*, **90**, 1492–8.

Gosden, R.G., Laing, S.C., Felicio, L.S., Nelson, J.F., and Finch, C.E. (1983). Imminent oocyte exhaustion and reduced follicular recruitment mark the transition to acyclicity in aging C57BL/6J mice. *Biology of Reproduction*, **28**, 255–60.

Gray, G.D. (1978). Changes in the levels of luteinizing hormone and testosterone in the circulation of aging male rats. *Journal of Endocrinology*, **76**, 551–2.

Gray, G.D., Smith, E.R., Dorsa, D.M., and Davidson, J.M. (1981). Sexual behavior and testosterone in middle-aged male rats. *Endocrinology*, **109**, 1597–604.

Hallstrom, T. (1977). Sexuality in the climacteric. *Clinical Obstetrics and Gynecology*, **4**, 227–39.

Harman, S.M. (1978). Clinical aspects of aging of the male reproductive system. In *The aging reproductive system* (ed. E.L. Schneider), pp. 29–58. Raven Press, New York.

Harman, S.M. and Talbert, G.B. (1985). Reproductive aging. In *Handbook of the biology of aging*, 2nd ed (ed. C.E. Finch and E.L. Schneider), pp. 457–510. Van Nostrand Reinhold Co., New York.

Hornykiewicz, O. (1987). Neurotransmitter changes in human brain during aging. In *Modification of cell to cell signals during normal and pathological aging* (ed. S. Stefano and F. Battaini), pp. 169–82. Springer-Verlag, Berlin.

Huang, H.H., Steger, R.W., Bruni, G.F., and Meites, J. (1978). Patterns of sex steroid and gonadotropin secretion in aging female rats. *Endocrinology*, **103**, 1855–9.

Huber, M.H.R., Bronson, F.H., and Desjardins, C. (1980). Sexual activity of aged male mice: Correlation with level of arousal, physical endurance, pathological status and ejaculatory capacity. *Biology of Reproduction*, **23**, 3095–116.

Humphreys, P.N. (1977). The histology of the testis in aging and senile rats. *Experimental Gerontology*, **12**, 27–34.

Jarjour, L.T., Handelsman, D.J., and Swerdloff, R.S. (1986). Effects of aging on the *in vitro*. release of gonadotrophin releasing hormone. *Endocrinology*, **119**, 1113–17.

Judd, H.L. (1983). Pathophysiology of menopausal hot flushes. In *Neuroendocrinolog of aging* (ed. J. Meites), pp. 173–202. Plenum Press, New York.

Judd, H.L., Judd, G.E., Lucas, W.E., and Yen, S.S.C. (1974). Endocrine function of the postmenopausal ovary: concentration of androgens and estrogens in ovarian and peripheral vein blood. *Journal of Clinical Endocrinology and Metabolism*, **39**, 1020–4.

Kaler, L.W. and Neaves, W.B. (1981). The androgen status of aging male rats. *Endocrinology*, **108**, 712–19.

Khoury, S.A., and Sowers, G.R. (1988). Age-related changes in male sexual function. In *The endocrinology of aging* (ed. J.R. Sowers and J.V. Felicetta), pp. 113–34. Raven Press, New York.

Kinsey, A.C., Pomeroy, W.E., and Martin, C.E. (1948). *Sexual behaviour in the human male*. W.B. Saunders, Philadelphia.

Kinsey, A.C., Pomeroy, W.E., and Martin, C.B., and Behbard, P.H. (1953). *Sexual behaviour in the human female*. W.B. Saunders, Philadelphia.

Knobil, E. and Plant, T.M. (1978). Neuroendocrine control of gonadotropin secretion in the female rhesus monkey. In *Frontiers in neuroendocrinology*, Volume 5 (ed. W.F. Ganong and L. Martini), pp. 249–64. Raven Press, New York.

LaPolt, P.S. and Lu, J.K.H. (1987). Effects of increased circulating progesterone on ovarian follicular loss and reproductive aging. *Biology of Reproduction*, **36**, (Supplement 1), 173.

LaPolt, P.S., Matt, D.W., Judd, H.L., and Lu, J.K.H. (1985). Relation of the proestrous LH surge to estrous cyclicity, ovulation rate and corpora luteal function in aging rats. *Federation Proceedings*, **44**, 1359.

LaPolt, P.S., Matt, D.W., Judd, H.L., and Lu, J.K.H. (1986). The relation of ovarian steroid levels in young female rats to subsequent estrous cyclicity and reproductive function during aging. *Biology of Reproduction*, **35**, 1131–9.

LaPolt, P.S., Yu, S.M., and Lu, J.K.H. (1988*a*). Early treatment of young female rats with progesterone decelerates the aging-associated reproductive decline; A counteraction by estradiol. *Biology of Reproduction*, **38**, 987–95.

LaPolt, P.S., Day, J.R., Brooks, E., Morales, T., and Lu, J.K.H. (1988*b*).

Alterations in ovarian steroid secretion and positive feedback on LH release both contribute to a loss of regular ovulatory cycles in aging female rats. *35th Annual Meeting of the Society for Gynecologic Investigation.* Scientific Programme and Abstracts p. 108 (Abstract number 97).

LaPolt, P.S., Lu, J.K.H., and Pu, S.F. (1993). Relation of proestrous LH surge magnitude to ovulation rate and ovarian tissue-type plasminogen activator (tPA) mRNA in aging female rats. *75th Annual Meeting of the Endocrine Society.* Programme and Abstracts p. 369 (Abstract number 1273).

Larson, K. (1958). Sexual activity in senile male rats. *Journal of Geronotology*, **13**, 136–9.

Lu, J.K.H. (1983). Changes in ovarian function and gonadotropin and prolactin secretion in aging female rats. In *Neuroendocrinology of aging* (ed. J. Meites), pp. 103–22. Plenum Press, New York.

Lu, J.K.H. (1986). Physiology and biochemistry of the normal reproductive cycle. In *Essentials of obstetrics and gynecology* (ed. N.F. Hacker and J.G. Moore), pp. 395–402, W.B. Saunders, Philadelphia.

Lu, J.K.H. and Judd, H.L. (1991). The neuroendocrine aspects of menopausal hot flushes. In *Progress in basic and clinical pharmacology*, Vol. 6 (ed. P. Lomax and E.S. Vessell), pp. 83–99. pp. S. Karger AG, Basel.

Lu, J.K.H., Huang, H.H., Chen, H.T., Kurcz, M., Mioduzewski R., and Meites, J. (1977). Positive feedback by estrogen and progesterone on LH release in old and young rats. *Proceedings of the Society for Experimental Biology and Medicine*, **154**, 82–5.

Lu, J.K.H., Hopper, B.R., Vargo, T.M., and Yen, S.S.C. (1979). Chronological changes in sex steroid, gonadotropin and prolactin secretion in aging female rats displaying different reproductive states. *Biology of Reproduction*, **21**, 193–203.

Lu, J.K.H., Damassa, D.A., Gilman, D.P., Judd, H.L., and Sawyer, C.H. (1980). Differential patterns of gonadotropin responses to ovarian steroids and to LH-releasing hormone between constant-estrous and pseudopregnant states in aging rats. *Biology of Reproduction*, **23**, 345–51.

Lu, J.K.H., Gilman, D.P., Meldrum, D.R., Judd, H.L., and Sawyer, C.H. (1981). Relationship between circulating estrogens and the central mechanisms by which ovarian steroids stimulate luteinizing hormone secretion in aged and young female rats. *Endocrinology*, **108**, 836–41.

Lu, J.K.H., LaPolt, P.S., Nass, T.E., Matt, D.W., and Judd, H.L. (1985). Relation of circulating estradiol and progesterone to gonadotropin secretion and estrous cyclicity in aging female rats. *Endocrinology*, **116**, 1953–9.

Lu, J.K.H., Matt, D.W., and LaPolt, P.S. (1990). Modulatory effects of estrogens and progestins on female reproductive aging. In *Ovarian secretion and cardiovascular and neurological function* (ed. F. Naftolin, J.N. Gutmann, and A.H. DeCherney), pp. 297–306. Raven Press, New York.

Mandl, A.M. (1959). Corpora lutea in senile virgin laboratory rats. *Journal of Reproductive Endocrinology*, **18**, 438–43.

Mandl, A.M. and Shelton, M. (1959). A quantitative study of oocytes in young and old nulliparous laboratory rats. *Journal of Endocrinology*, **18**, 444–50.

Martin, C.E. (1977). Sexual activity in the aging male. In *Handbook of Sexology* (ed. J. Money and H. Musaph), pp. 813–24. ASP Biology and Medicine Press, Amsterdam.

Martin, J.B. and Reichlin, S. (1987). Neuropharmacology of anterior pituitary regulation. In *Clinical Neuroendoctinology*, 2nd ed. (ed. J.B. Martin and S. Reichin)pp 45–63. F.A. Davis Co., Philadelphia

Masters, W.H. and Johnson, V.E. (1966). *Human sexual responses*. Little Brown, Boston.

Matt, D.W., Lee, J., Sarver, P.L., Judd, H.L, and Lu, J.K.H. (1986). Chronological changes in fertility, fecundity and steroid hormone secretion during consecutive pregnancies in aging rats. *Biology of Reproduction*, **34**, 478–87.

Matt, D.W., Coquelin, A., and Lu, J.K.H. (1987a). Neuroendocrine control of luteinizing hormone secretion and reproductive function in spontaneously persistent estrous rats. *Biology of Reproduction*, **37**, 1198–206.

Matt, D.W., Sarver, P.L., and Lu, J.K.H. (1987b). Relation of parity and estrous cyclicity to the biology of pregnancy in aging female rats. *Biology of Reproduction*, **37**, 421–30.

McFadyen, I.J., Bolton, A.E., Cameron, E.H.D., Hunter, W.M., Rabb, W.M., and Forrest, A.P.M. (1980). Gonadal-pituitary hormone levels in gynaecomastia. *Clinical Endocrinology*, **13**, 77.

McIntosh, H.H. and Westfall, T.C. (1987). Influence of aging on catecholamine levels, accumulation, and release in F-344 rats. *Neurobiology of Aging*, **8**, 233–9.

Meites, J. (1982). Changes in neuroendocrine control of anterior pituitary function during aging. *Neuroendocrinology*, **34**, 151–6.

Meites, J. (1990). Aging: hypothalamic catecholamines, neuroendocrine-immune interactions, and dietary restriction. *Proceedings of the Society for Experimental Biology and Medicine*, **195**, 304–11.

Meites, J. (1991). Aging of the endocrine brain, basic and clinical aspects. In *Brain endocrinology* 2nd ed. (ed. M. Motta), pp. 449–60. Raven Press, New York.

Meites, J. and Huang, H.H. (1976). Relation of neuroendocrine system to loss of reproductive function in aging rats. In *Neuroendocrine regulation of fertility* (ed. A. Kumar), pp. 246–58. Karger, Basel.

Meites, J., Huang, H.H. and Simpkins, J.W. (1978). Recent studies on neuroendocrine control of reproductive senescence in rats. In *The aging reproductive system* (ed. E.L. Schneider), pp. 213–35. Raven Press, New York.

Morley, J.E. and Kaiser, F.E. (1990). Testicular function in the aging male. In *Endocrine function and aging* (eds. H.J. Armbrecht, R.M. Coe, and N. Wongsurawat), pp. 99–114, Springer-Verlag, New York.

Muta, K., Kato, K., Akamine, Y., and *Ibayashi* (1981). Age-related changes in the feedback regulation of gonadotropin secretion by sex steroids in men. *Acta Endocrinologica*, **96**, 154–62.

Nass, T.E., LaPolt, P.S., and Lu, J.K.H. (1982). The effects of prolonged caging with fertile males on reproductive functions in aging female rats. *Biology of Reproduction*, **27**, 609–15.

Nass, T.E., LaPolt, P.S., Judd, H.L., and Lu, J.K.H. (1984). Alterations in ovarian steroids and gonadotrophin secretion preceding the cessation of regular oestrous cycles in ageing female rats. *Journal of Endocrinolgy*, **100**, 43–50.

Nieschlag, E., Lammers, U., Freischem, C.W., and Wickings, E.J. (1982). Reproductive functions in young fathers and grandfathers. *Journal of Clinical Endocrinology and Metabolism*, **55**, 676–81.

O'Neil, C., Marcusson, J., Norberg, A., and Winblad, B. (1987). The influence of age on neurotransmitters in the human brain. In *Modification of cell to cell signals during normal and pathological aging* (ed. S. Stefano and F. Battaini), pp. 183–98. Springer-Verlag, Berlin.

Pedersen-Bjergaard, K., and Jonnesen, M. (1948). Sex hormone analysis: excretion of sexual hormones by normal males, polyarthritics, and prostatics. *Acta Medica Scandinavica and* **213**, (Suppl.) 284–91.

Pederson, T. and Peters, H. (1971). Follicle growth and cell dynamics in the mouse ovary during pregnancy. *Fertility and Sterility*, **22**, 42–52.

Peng, M.T. (1983) Changes in hormone uptake and receptors in the hypothalamus during aging. In *Neuroendocrinology of aging* (ed. J. Meites), pp. 61–72. Plenum Press, New York.

Peng, M.T., Yao, C.T., and Wan, W.C.M. (1980). Dissociation between female sexual behavior and luteinizing hormone release in old female rats. *Physiological Behavior*, **25**, 633–6.

Quadri, S.K., Kledzik, G.S., and Meites, J. (1973). Reinitiation of estrous cycles in old constant-estrous rats by central acting drugs. *Neuroendocrinology*, **11**, 148–55.

Rice, G.E., Cho, G., and Barnea, A. (1983). Age related reduced release of LH-releasing hormone from hypothalamic granules. *Neurobiology of Aging*, **4**, 217–22.

Richardson, S.J., Senikas, V., and Nelson, J.F. (1987). Follicular depletion during the menopausal transition: Evidence for accelerated loss and ultimate exhaustion. *Journal of Clinical Endocrinology and Metabolism*, **65**, 1231–7.

Riegle, G.D. and Miller, A.E. (1978). Aging effects on the hypothalamic-hypophyseal-gonadal control system in the rat. In *The aging reproductive system*, (ed. E.L. Schneider), pp. 159–92. Raven Press, New York.

Robinson, D.S., Nies, A., Davis, J., *et al.* (1972). Aging, monoamines and monoamine oxidase levels. *Lancet*, **i**, 290–1.

Roth, G.S. (1990). Changes in hormone action with age: altered calcium mobilization and/or responsiveness impairs signal transduction. In *Endocrine function and aging* (ed. H.J. Armbrecht, R.M. Coe, and N. Wongsurawat), pp. 26–34. Springer-Verlag, New York.

Saksena, S.K., Lau, I.F., and Chang, M.C. (1979). Age dependent changes in sperm population and fertility in the male rat. *Experimental Aging Research*, **5**, 373–81.

Sarjent, J.W. and McDonald, J.R. (1948). A method for quantitative measurement of Leydig cells in the human testis. *Mayo Clinic Proceedings*, **23**, 249–54.

Sarker, D.K., Chiappa, S.A., and Fink, G. (1976). Gonadotrophin-releasing hormone surge in proestrous rats. *Nature*, **264**, 461–3.

Sarkar, D.K., Gottschall, P.E., and Meites, J. (1982). Damage to hypothalamic dopaminergic neurons is associated with development of prolactin-secreting tumors. *Science*, **218**, 684–6.

Schwartz, N.B. (1974). The role of FSH and LH and of their antibodies on follicle growth and on ovulation. *Biology of Reproduction*, **10**, 236–72.

Sherman, B.M., West, J.H., and Korenman, S.G. (1976). The menopausal transition: Analysis of LH. FSH, estradiol and progesterone concentrations during menstrual cycles of older women. *Journal of Clinical Endocrinology and Metabolism*, **42**, 629–36.

Simpkins, J.W. (1983). Changes in hypothalamic hypophysiotropic hormones and neurotransmitters during aging. In *Neuroendocrinology of aging.* (ed. J. Meites), pp. 41–59. Plenum Press, New York.

Simpkins, J.W. (1984). Regional changes in monoamine metabolism in aging constant estrous rats. *Neurobiology of Aging*, **4**, 3309–14.

Simpkins, J.W., Mueller, G.P., Huang, H.H., and Meites, J. (1977). Evidence for depressed catecholamine and enhanced serotonin metabolism in aging male rats. *Endocrinology*, **100**, 1672–8.

Smith, M.S., Freeman, M.E., and Neill, J.D. (1975). The control of progesterone secretion during the estrous cycle and early pseudopregnancy in the rat: prolactin, gonadotropin and estradiol levels associated with rescue of the corpus luteum of pseudopregnancy. *Endocrinology*, **96**, 219–26.

Sonntag, W.E., Forman, L.J., Fiori, J.M., Hylka, V.W., and Meites, J. (1984). Decreased ability of old male rats to secrete luteinizing hormone (LH) is not due to alterations in pituitary LH-releasing hormone receptors. *Endocrinology*, **114**, 1657–64.

Sopelak, V.M. and Butcher, R.L. (1982). Decreased amount of ovarian tissue and maternal age affect embryonic development in old rats. *Biology of Reproduction*, **27**, 449–55.

Steger, R.W., Peluso, J.J., Huang, H.H., Hafez, E.S., and Meites, J. (1976). Gonadotropin binding sites in the ovary of aged rats. *Journal of Reproduction and Fertility*, **48**, 205–7.

Talbert, G.B. (1977). Aging of the reproductive system. In *Handbook of the biology of aging* (ed. C.E. Finch and L. Hayflick), pp. 318–56. Von Nostrand Reinhold Co., New York.

Tataryn, I.V., Meldrum, D.R, Lu, K.H., Frumar A.M., and Judd, H.L. (1979). LH, FSH and skin temperature during the menopausal hot flash. *Journal of Clinical Endocrinology and Metabolism*, **49**, 152–4.

Vermeulen, A. (1979). Decline in sexual activity in aging men: correlation with sex hormone levels and testicular changes. *Journal of Biosocial Science*, **6** (Suppl.), 5–18.

Vermeulen, A., Rubens, R., and Verdonch, L. (1972). Testosterone secretion and metabolism in male senescence. *Journal of Clinical Endocrinology and Metabolism*, **34**, 730–5.

Verwoerdt, A., Pfeiffer, E., and Wang, H.S. (1969). Changes in sexual activity and interest of aging men and women. *Journal of Geriatric Psychiatry*, **2**, 163–80.

Wang, C., Hsueh, A.J.W., and Erickson, G.F. (1979). Induction of functional prolactin receptors by follicle-stimulating hormone in rat granulosa cells *in vivo* and *in vitro*. *Journal of Biological Chemistry*, **254**, 11330–6.

Watkins, B.E., Meites, J., and Riegle, G.D. (1975). Age-related changes in pituitary responsiveness to LHRH in the female rat. *Endocrinology*, **97**, 543–8.

Welsch, C.W. and Nagasawa, H. (1977). Prolactin and mammary tumorigenesis: a review. *Cancer Research*, **37**, 951–63.

Weiland N.G. and Wise, P.M. (1990). Aging progressively decreases the densities and alters the diurnal rhythm of α-1 adrenergic receptors in selected hypothalamic regions. *Endocrinology*, **126**, 2392–7.

Wilks, J.W., Spilman, C.H., and Campbell, J.A. (1983). Arrest of folliculogenesis and inhibition of ovulation in the monkey following weekly administration of progestins. *Fertility and Sterility*, **40**, 688–92.

Winters, S.J. and Troen, P. (1982). Episodic luteinizing hormone (LH) secretion and the response of LH and follicle stimulating hormone to LH-releasing hormones in aged men: Evidence for coexistent primary testicular insufficiency and an impairment in gonadotropin secretion. *Journal of Clinical Endocrinology and Metabolism*, **55**, 560–5.

Wise, P.M. (1982). Alterations in proestrous LH, FSH, and prolcatin surges in middle-aged rats. *Proceedings of the Society for Experimental Biology and Medicine*, **169**, 348–54.

Wise, P.M. (1983). Aging of the female reproductive system. *Review of Biological Research in Aging*, **1**, 195–222.

Wise, P.M. (1984). Estradiol-induced daily luteinizing hormone and prolactin surges in young and middle-aged rats: correlations with age-related changes in pituitary responsiveness and catecholamine turnover rates in microdissected brain areas. *Endocrinology*, **115**, 801–9.

7 Prolactin, a hormone for all seasons: endocrine regulation of seasonal breeding in the Macropodidae

LYN A. HINDS

I   INTRODUCTION

In this chapter the endocrine control of reproduction in the more advanced marsupials, the superfamily Macropodoidea (kangaroos, wallabies, and rat-kangaroos), will be presented, with a specific outline of the endocrine mechanisms which have evolved to regulate breeding in two obligate seasonal breeders in the sub-Family Macropodidae, the tammar wallaby, *Macropus eugenii*, and the Bennett's wallaby, *Macropus rufogriseus rufogriseus*. As the chapter progresses it will become apparent that prolactin is central to the regulation of their annual reproductive cycles, in contrast to the major role played by the hypothalamic pulse generator and gonadotrophin secretion as described for several eutherian species. Nevertheless, while the mechanisms which have evolved in the two macropodid species both involve prolactin, they are not identical and are likely to have arisen independently through convergent adaptation of reproductive processes. Of the two species, most is known of the endocrine mechanisms in the tammar. Although many other marsupials are seasonal breeders (Tyndale-Biscoe and Renfree 1987), only limited experimental studies have been undertaken to determine the factors regulating their seasonality. The reader is referred to Tyndale-Biscoe and Renfree (1987) for an overview of the reproductive physiology of marsupials, to Godfrey (1969) and Smith *et al.* (1978) for studies on *Sminthopsis crassicaudata*, and to Gemmell (1987; 1990) for work on the brush tail possum *Trichosurus vulpecula*.

The earliest fossil record of mammals appeared in the late Triassic, with the first recognizable forms of eutherian and metatherian mammals appearing in the Cretaceous period more than 100 million years ago. While there is a vast spectrum of reproductive patterns among these two subclasses of mammal, in individual species the pressures of natural selection have led to convergent adaptations, such as the adoption of seasonality, which ensure continued success in the reproductive process. In non-equatorial climates some seasons are obviously more favourable for the rearing of young than others. Thus for most eutherian species the time of birth coincides with spring, a time of food abundance and moderate environmental conditions. Because different species have different lengths of gestation it is important that each can accurately time conception so that parturition occurs at the most optimal time for survival of young and adults (Lincoln and Short 1980; Karsch *et al.* 1984; Nicholls *et al.* 1988). In theory, any environmental cue, such as light, rainfall, or temperature could be used, but of the factors which change from season to season the most predictable is photoperiod, particularly with increasing distance from the equator. Thus many species use the annual changes in daylength to ensure appropriate timing of reproductive processes such as gonadal development, gamete production, establishment of territories, conception, and pregnancy, so that parturition occurs in spring.

A major factor influencing the timing of the onset of breeding is the length of gestation. Species such as birds and most rodents, which have relatively short incubation or gestation periods, respond to the increasing daylengths after the winter solstice (long day breeders), while the larger species (sheep, goats, deer, rhesus monkeys) with longer gestation periods begin their breeding season after the summer solstice as the daylength is decreasing (short day breeders) (Lincoln and Short 1980; Karsch *et al*. 1984; Nicholls *et al*. 1988). Both patterns result in births in Spring.

Many marsupials are also seasonal breeders, although there are major differences between the Eutheria and Metatheria in their overall strategy of reproduction (Russell 1982; Renfree 1983; Lee and Cockburn 1985; Tyndale-Biscoe and Renfree 1987). In eutherians, maternal investment in the young largely occurs during pregnancy while in marsupials the major investment in the young occurs during lactation. Thus, in contrast to eutherians, the gestation period of marsupials is relatively short and the young are born at an immature stage of development (Tyndale-Biscoe and Janssens 1988). Over a period of weeks or months the young grows and differentiates. As it becomes physiologically mature it emerges from the pouch or nest and begins to be weaned. This period of weaning from a milk to herbivorous or carnivorous diet is a crucial time for the survival of the young, and generally occurs in spring and summer. Thus conception must occur at an appropriate time sufficiently in advance of spring so that both pregnancy and lactation can be accommodated (Fig. 7.1).

Several endocrine axes are critical to the success of breeding. Common to all species is the axis involving the hypothalamus, the pituitary, and gonads. In seasonally breeding mammals there is a fourth functional component in the axis, the pineal gland, which produces the neurohormone melatonin. For photoperiod information to be perceived there must be a photoreceptor with an intact sympathetic innervation to transmit this information to the pineal gland. The presumptive pathway is via the suprachiasmatic nuclei and the superior cervical ganglia (Arendt 1986). The neural information is transduced to an endocrine signal by the pineal gland, translated into a response by the hypothalamic–pituitary axis and, finally, the gonads. To date the transmission of the neural signal from the eye to the pineal gland has not been measured for any species, and the mode of transduction of the melatonin signal to the hypothalamus remains unknown. However, the nature of the pineal signal and the hypothalamic–pituitary response it induces have been examined in various detail for a few species (Karsch *et al*. 1984; Goodman 1988; Goldman and Elliot 1988; Bartke and Steger 1992). For example, in the ewe, responses to photoperiod manipulation are reflected in changes in the hypothalamic pulse generator controlling gonadotrophin-releasing hormone GnRH) secretion and differences in the negative feedback actions of oestradiol on tonic luteinizing hormone (LH) release; both LH pulse frequency and magnitude vary depending on the stage of the annual cycle (Legan *et al*. 1977; Karsch *et al*. 1984; Robinson *et al*. 1985*a*). The steroid-dependent effects are manifested via changes in the generation of pulses of GnRH from the hypothalamus, but there are also steroid-independent effects on the LH pulse generator that are

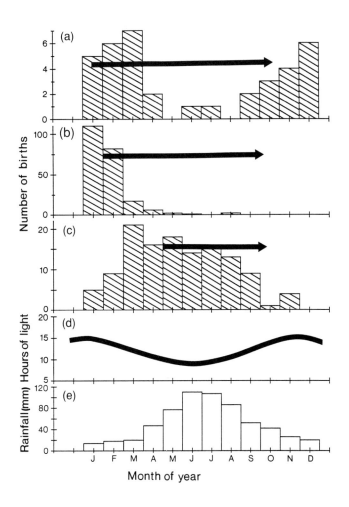

Fig. 7.1 Distribution of births of (a) grey kangaroos, *Macropus fuliginosus fuliginosus*, (b) tammar wallabies, *Macropus eugenii* on Kangaroo Island, South Australia (latitude 36°S), and (c) brush tail possums, *Trichosurus vulpecula* in Adelaide, South Australia, and relation to (d) the photoperiod and (e) the average rainfall for Cape Borda, Kangaroo Island for 1865–1965). Length of lactation indicated by horizontal arrow. Data for grey kangaroos from Poole (1976), tammars from Andrewartha and Barker (1969), and brush tail possum from Pilton and Sharman (1962), other data from Bureau of Meteorology, Canberra.

driven by photoperiod and the changing pattern of melatonin (Bittman and Karsch 1984). There is some evidence (Goodman 1988) that the long day pattern of melatonin affects two inhibitory neural systems, a set of catecholaminergic neurones and a set of serotonergic neurones, which affect the LH pulse generator, and this leads to seasonal alterations in LH pulse frequency. However, Goodman's (1988) model does not account for the mechanisms underlying both long and short day photorefractoriness in the ewe (Robinson and Karsch 1984; Robinson et al. 1985b) or the existence of endogenous circannual rhythms (Karsch et al. 1989).

Given that the strategy of reproduction differs between eutherian and metatherian mammals, do the endocrine mechanisms controlling pregnancy and seasonality also differ? Later in this review it will be clear that the mechanisms regulating seasonal breeding are different in marsupials, at least for those in which there has been sufficiently detailed examination. In the sheep (and probably most photoperiodic eutherian species and birds; Nicholls et al. 1988) the LH pulse generator is a central player in the hypothalamic-hypophyseal response to photoperiod and no definitive role can be attributed to prolactin, although this hormone shows distinct seasonal changes in the peripheral circulation (Walton et al. 1977; McNeilly and Land 1979; Karsch et al. 1984; 1989). By contrast, in the two obligate seasonally breeding marsupials which have been examined, prolactin dominates the control of the annual reproductive cycle.

To begin, some of the general characteristics of reproduction and its regulation in marsupials will be presented, before the roles of the various endocrine axes (hypothalamic-pituitary–follicular axis; CL–follicular axis and pituitary–CL axis) involved in the hormonal control of reproduction in the tammar are assessed in relation to their importance in regulating seasonality. The endocrine responses to photoperiod manipulation, the influence of the pineal gland on the pituitary–CL axis and the role of photorefractoriness in the annual reproductive cycle will then be assessed. Finally, a model of the mechanisms regulating the annual cycle will be proposed.

II   PATTERNS OF REPRODUCTION IN MARSUPIALS

Some features of reproduction are common to all marsupials. So far as is known most are polyoestrus (Tyndale-Biscoe and Renfree 1987), and with two exceptions (*Monodelphis domestica*, Fadem et al. 1982; *Bettongia penicillata*: Hinds and Smith 1992; Smith 1992), ovulation is spontaneous. After ovulation the new CL grows and secretes progesterone which induces an obvious secretory phase in the uterus (Sharman 1955a) independent of the presence of an embryo. If conception occurs, pregnancy

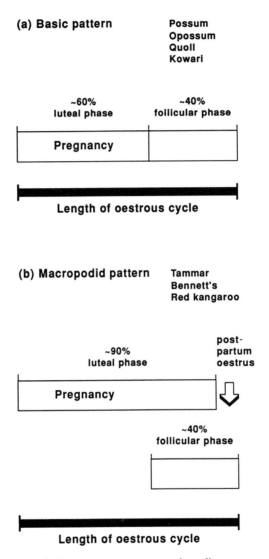

Fig. 7.2 Diagram of the two most common breeding patterns in marsupials, illustrating the different relationships of the oestrous cycle and pregnancy. (a) Basic or Group 1 pattern and (b) Macropodid or Group 3 type pattern.

generally occupies a period less than the length of the oestrous cycle, and after parturition subsequent ovarian activity is suppressed by lactation. Nevertheless four patterns of reproduction are recognized (Sharman *et al*. 1966; Tyndale-Biscoe and Renfree 1987). The two major patterns will be considered here.

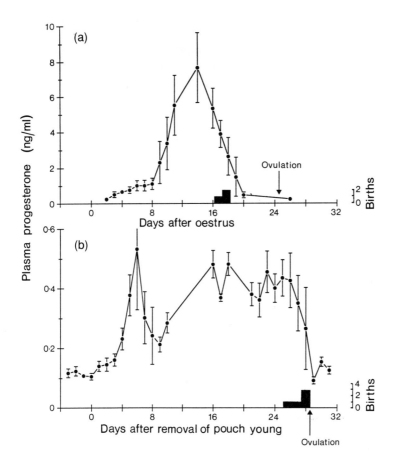

Fig. 7.3   Comparison of plasma progesterone profiles through pregnancy for (a) the brush tail possum, *Trichosurus vulpecula* (Group 1 breeding pattern) and (b) the tammar, *Macropus eugenii* (Group 3 breeding pattern). From Hinds (1983) and Hinds and Tyndale-Biscoe (1982a).

The most primitive or basic reproductive pattern (Group 1; Tyndale-Biscoe and Renfree 1987) is seen in such species as the monotocous brush tail possum, *T. vulpecula* (Phalangeridae; Pilton and Sharman 1962), the polytocous American opossum, *Didelphis virginiana* (Didelphidae; Hartman 1923) and the polytocous eastern quoll, *Dasyurus viverrinus* (Dasyuridae; Hinds 1989a). Pregnancy occupies approximately 60% of the length of the oestrous cycle, and parturition coincides with the end of the luteal phase when progesterone secretion by the CL declines (Figs. 7.2a and 7.3a). The next follicular phase, pro-oestrus, oestrus, and ovulation are suppressed during lactation by the sucking stimulus of the young, but if the young is lost or removed follicular growth resumes.

The second major pattern of reproduction (Group 3; Tyndale-Biscoe and Renfree 1987) is derived from this basic pattern, and is seen in the more advanced super-Family Macropodoidea (kangaroos, wallabies, and rat-kangaroos). In these species, the luteal phase has extended until it occupies approximately 90% of the oestrous cycle (Figs. 7.2b and 7.3b). Parturition is followed by a post-partum oestrus and ovulation, and the development of the newly formed CL is suppressed by the sucking stimulus of the new young. If conception occurs, the embryo develops to a unilaminar blastocyst and then enters diapause. This period, when the sucking stimulus of the young maintains the CL in a quiescent state and the blastocyst is in diapause, is known as lactational quiescence, and is a condition common to the great majority of Macropodidae (Tyndale-Biscoe et al. 1974; Tyndale-Biscoe 1989). Delayed implantation or embryonic diapause was first discovered in the quokka, *Setonix brachyurus* (Sharman 1954, 1955b) and then confirmed in the tammar wallaby, *M. eugenii* (Sharman 1955c). Subsequently, the phenomenon has been demonstrated, with one exception (*Macropus fuliginosus*, western grey kangaroo; Poole and Catling 1974), in all species of macropodids which have been examined. The proximate stimulus is the sucking of the young because loss or removal of the pouch young allows reactivation or renewed development of the quiescent CL and the blastocyst. In the absence of a blastocyst an oestrous cycle follows. The entry into quiescence of the CL and embryonic diapause, and the events of reactivation after removal of the sucking stimulus have been described in detail elsewhere (Tyndale-Biscoe 1986; Tyndale-Biscoe and Hinds 1989).

1   Strategies of breeding in Macropodoidea

Three breeding strategies can be recognized among the Macropodoidea and these reflect the adaptations of the various species to their habitats (Sharman et al. 1966; Tyndale-Biscoe et al. 1974; Tyndale-Biscoe 1989). There are continuous breeders, facultative seasonal breeders and obligate seasonal breeders (Table 7.1). Under favourable environmental conditions most species are capable of continuous breeding, with births occurring in most months of the year (Fig. 7.4). Thus, within a population, young of all ages can be found at any one time: each female may have a young at heel, another in the pouch, and a diapausing blastocyst in the uterus. This pattern is seen in all Potoroidae (rat-kangaroos) (*Bettongia* spp. *Potorous tridactylus*, *Aepyprymnus rufescens*), as well as in the desert-living kangaroos (*Macropus rufus*, *M. robustus*, *M. agilis*), the red-necked wallaby (*Macropus rufogriseus banksianus*) and mainland populations of the quokka, *S. brachyurus*. Under extreme drought conditions, species such as the red kangaroo, *M. rufus*, cease to ovulate *post-partum* and enter

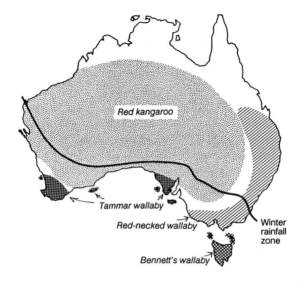

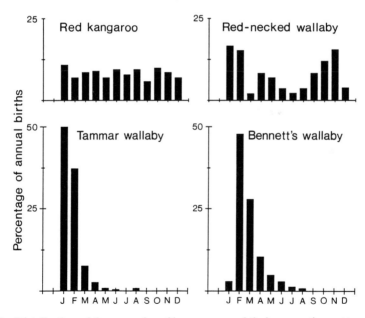

Fig. 7.4   Distribution of three species of kangaroos and their respective patterns of births throughout the year. Distributions redrawn from Strahan (1983) and breeding data for the red kangaroo, *Macropus rufus*, from Frith and Sharman (1964), for the tammar, *Macropus eugenii*, from Andrewartha and Barker (1969), and for both subspecies of *Macropus rufogriseus* (Bennett's wallaby and red-necked wallaby) from Merchant (personal communication).

lactational anoestrus. However, the pituitary–follicular axis is capable of rapid activation because, within 2 weeks of rainfall, most females have re-ovulated (Newsome 1964*a,b*; Frith and Sharman 1964; Sharman and Clark 1967).

The eastern grey kangaroo, *Macropus giganteus*, shows another variation from the opportunistic pattern most typically seen in the red kangaroo, in that it never shows post-partum oestrus, but enters lactational anoestrus after parturition. Subsequently, in late lactation, if conditions are favourable, the female ovulates and then enters lactational quiescence (Poole 1973). By contrast, the western grey kangaroo, *M. fuliginosus*, does not enter quiescence at any time, and is not only lactationally anoestrus but also seasonally anoestrus (Poole 1975; Table 7.1).

In some of the smaller species, such as quokkas on Rottnest Island off the coast at Fremantle, Western Australia, a second strategy has evolved whereby a period of true anoestrus occurs in spring. At this time both the CL and blastocyst, which have been in quiescence for several months, simultaneously disappear without reactivation. If, however, the nutritional conditions improve this seasonal anoestrus can be reversed resulting in continued breeding for that year (Shield and Woolley 1963; Wallace 1981). Two other species, *Lagostrophus fasciatus* and *Macropus irma*, also do not carry a quiescent CL and blastocyst through the non-breeding season (Tyndale-Biscoe 1965; Prince 1983). These species, then, are facultative seasonal breeders (Table 7.1).

Table 7.1   *Patterns of breeding in the Macropodoidea*

	Quiescence	Anoestrus
Lactational	Potoroidae *Macropus rufus* *M. robustus* *M. agilis* *M. rufogriseus* *banksianus* *Setonix brachyurus* (mainland population)	*(drought)* *M. giganteus* *M. antilopinus* *M. parryi* *M. parma*
Facultative seasonal	None	*S. brachyurus* (island population) *M. irma* *Lagostrophus fasciatus* *Thylogale billardierii*
Obligate seasonal	*M. eugenii* *M. rufogriseus rufogriseus*	*M. fuliginosus*

The third strategy, obligate seasonal breeding involving quiescence, is shown by two of the 47 species of Macropodoidea. For these species, the tammar wallaby, *M. eugenii*, and the Bennett's wallaby, *M. r. rufogriseus*, there is a period of several months in each annual cycle during which the quiescent CL and diapausing blastocyst remain in a viable state, but the removal or loss of the pouch young does not result in reactivation and the birth of another offspring within a month. These species occur in the most southerly parts of Australia, on Kangaroo Island and Tasmania, respectively, where winter rainfall is a predictable feature of the climate and selection for seasonal breeding would be strong. It is interesting that the mainland subspecies of *M. rufogriseus*, *M. r. banksianus*, does not show a seasonal pattern of breeding in Queensland (Fig. 7.4), although there has been no study of its breeding pattern in the more southern areas of Australia. This suggests that, at least for the Bennett's wallaby, obligate seasonal quiescence may have evolved as recently as 11 000 years ago after Tasmania became separated from mainland Australia (Tyndale-Biscoe and Hinds 1990). The rest of this review will concentrate on the research undertaken to understand the regulation of this obligate pattern of seasonal breeding in the tammar wallaby and the Bennett's wallaby.

III   GENERAL FEATURES OF THE ANNUAL CYCLE OF REPRODUCTION IN TAMMAR AND BENNETT'S WALLABIES

1   Distribution

Both species occur naturally in the winter rainfall zone in southern Australia (Fig. 7.4), with Bennett's wallaby being restricted to Tasmania. In the last century the various races of tammars were widely distributed on the mainland of South and Western Australia as well as on various islands off these coastlines (Recherche group, Kangaroo Island, South Australia; Garden Island and Houtman's Abrolhos, Western Australia). Today the largest populations of the tammar occur on Kangaroo Island off South Australia, with only small populations still occurring on the mainland and islands off Western Australia (Calaby 1971; Poole *et al.* 1991).

2   Lactational quiescence of the CL and blastocyst

The annual breeding cycle for the two species is extremely similar with slight differences in the onset and offset of the breeding season. In both species embryonic diapause lasts approximately 11 months (Berger 1966; Merchant and Calaby 1981). Breeding is more synchronized in the tammar, with births occurring within 6 weeks of the summer solstice (late January to early February in the Southern Hemisphere; Andrewartha and Barker 1969). In the captive colony maintained in Canberra (same latitude as Kangaroo Island–36°S) the mean date of birth in successive years has ranged from 26 January to 3 February (Tyndale-Biscoe *et al.* 1986). In Bennett's wallabies, births occur later, 2–6 months after the summer solstice (Catt 1977; Fleming *et al.* 1983; Curlewis 1989; Merchant, personal communication) (Fig 7.4).

After parturition tammar females enter post-partum oestrus, and ovulation occurs within 30–40 h after mating (Sutherland *et al.* 1980). In the ensuing 3–4 days the CL is formed (Sharman and Berger 1969), and if conception occurs the embryo develops to a unilaminar blastocyst of 80–160 cells within 8–9 days *post coitum* (Tyndale-Biscoe 1986; Tyndale-Biscoe and Hinds 1989). Further development of both the CL and the blastocyst is arrested by the sucking stimulus of the young (lactational quiescence).

During lactational quiescence, as in most other macropodids, loss or removal of the tammar pouch young induces reactivation of the CL and the quiescent blastocyst, and birth and post-partum oestrus occur 26–27 days later. If there is no diapausing blastocyst the female undergoes a non-pregnant or oestrous cycle, and the interval to the next oestrus is significantly longer (30–31 days) than the interval from removal of pouch young to birth (Merchant 1979; Hinds and Tyndale-Biscoe 1982*a*; Tyndale-Biscoe *et al.* 1983). Similarly, after removal of the pouch young in the Bennett's wallaby the interval to birth and post-partum oestrus is 27–28 days and to oestrus 30–31 days (Merchant and Calaby 1981; Curlewis *et al.* 1987).

Reactivation of the CL after the removal of the pouch young can be determined from concentrations of progesterone in plasma; in both species the progesterone profile is bimodal with an early peak which occurs on days 5–6 in the tammar (Hinds and Tyndale-Biscoe 1982*a*) and days 4–5 in Bennett's wallabies (Walker and Gemmell 1983; Curlewis *et al.* 1987), and a second more prolonged period of high concentrations occurring in the second half of the cycle (Fig. 7.3b). The decrease in plasma progesterone at parturition is rapid and is followed by the events of post-partum oestrus (Tyndale-Biscoe *et al.* 1983, 1988; Harder *et al.* 1985; Renfree *et al.* 1989; Hinds *et al.* 1990; see Hinds 1990 for review). The early peak invariably occurs in both pregnant and non-pregnant tammars and Bennett's wallabies. In the tammar, the interval from the early peak to birth is consistent at 21–22 days, while in the absence of birth, oestrus occurs 24 days after the peak (Hinds and Tyndale-Biscoe 1982*a*; Hinds 1989*b*). For the Bennett's wallaby, the interval from the peak to birth is 24 days (Curlewis *et al.* 1987; Curlewis and Loudon 1988*a*) but the interval from the peak to oestrus in the absence of parturition is not directly available from the literature. Because the early progesterone peak provides the first evidence of reactivation of the CL in both species, plasma progesterone concentrations serve as a precise indicator of CL activity or function and can be used to detect reactivation of the CL and then to predict the time of birth under various experimental conditions (Table 7.2).

In the tammar, during early pregnancy the presence of the CL is critical for successful reactivation and subsequent development of the blastocyst. If the CL is removed on or before day 3 after removal of the pouch young, the progesterone peak does not occur and the embryo remains in diapause, while if it is removed on day 4 the embryo reactivates but development is not sustained (Tyndale-Biscoe 1986). The CL therefore provides a critical signal to the uterus and/or embryo between days 3 and 4. Moore (1978) found that injections of progesterone induce blastocyst reactivation within 2 days, and recent studies have detected platelet-activating factor (PAF) in endometrial exudates on days 3–4 coincident with the first significant increase in peripheral progesterone concentrations (Kojima *et al.* 1993). These results suggest that the first increase in progesterone secretion from the CL may stimulate the endometrium to secrete PAF, which provides the signal for reactivation of the diapausing blastocyst. Furthermore, because injections of progesterone induce reactivation of the blastocyst within 2 days (Moore 1978), it can be concluded that the first 3 days after removal of pouch young are required for the CL to recover from the inhibition provided by the sucking stimulus. Gordon *et al.* (1988) confirmed this when they showed that returning the tammar young to the pouch within 72 h held both the CL in quiescence and the embryo in diapause. However, if the interval was longer than this, reactivation had occurred and was irreversible.

3   Seasonal quiescence of the CL and blastocyst

For the tammar the period defined as lactational quiescence occurs from January to June in the Southern Hemisphere and is followed by a period of obligate quiescence and embryonic diapause, termed seasonal quiescence, which lasts from July to December. During seasonal quiescence, seasonal factors predominate and the loss, removal or weaning of the young does not induce reactivation of the CL. The period is termed seasonal quiescence rather than seasonal anoestrus because the CL remains viable, as does the diapausing blastocyst, and the uterus appears histologically similar to uteri in lactational quiescence (Sharman and Berger 1969). Spontaneous reactivation of the CL and blastocyst occurs after the summer solstice (Berger and Sharman 1969). Thus a CL formed at the post-partum oestrus in one breeding season may remain in quiescence (lactational followed by seasonal) for up to 11 months (Berger 1966; Fig. 7.5, tammar annual cycle). As with other seasonally breeding species (sheep and red deer) transferred across the equator (Lincoln and Short 1980), tammars and Bennett's wallabies transferred to the Northern Hemisphere retain their seasonality and adjust their annual cycle so that they give birth in July or

Table 7.2 *Interval (days) to the early progesterone peak and birth after different experimental treatments in the tammar*

Treatment	Interval to			Reference
	Progesterone pulse	Birth	Pulse to birth	
Oestrus	7.0 ± 0.5	29.3 ± 0.3	22.3	Hinds and Tyndale-Biscoe (1982*a*)
Removal of pouch young	5.6 ± 0.3	27.2 ± 0.4	21.8	Hinds and Tyndale-Biscoe (1982*a*)
Bromocriptine, 5 mg/kg	5.2 ± 0.2	26.3 ± 0.3	21.1	Tyndale-Biscoe *et al.* (1986)
Hypophysectomy	7.3 ± 0.2			Hinds (1983)
Melatonin, 12 h total	8.3 ± 0.3	29.9 ± 0.3	21.6	Tyndale-Biscoe and Hinds (1992)
Photoperiod change 15L:9D to 12L:12D	10.0 ± 1.3	33.5 ± 1.1	23.5	Hinds and den Ottolander (1983)
After summer solstice, 22 December	20.6 ± 0.7	42.6 ± 0.8	22.0	Hinds and Tyndale-Biscoe (unpublished results)

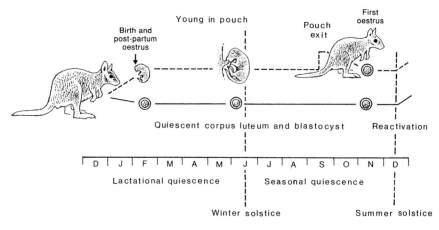

Fig. 7.5   Diagram of the annual cycle of reproduction in the tammar wallaby, *Macropus eugenii* in the Southern Hemisphere.

August after the boreal summer solstice (Berger 1970; Fleming *et al.* 1983; Dressen *et al.* 1990). It is therefore reasonable to postulate that photoperiod cues provide the proximate stimulus for the establishment and maintenance of quiescence after the winter solstice and for reactivation after the summer solstice.

IV   HORMONAL CONTROL OF REPRODUCTION IN THE TAMMAR

Three endocrine axes are important in the regulation of reproduction in the tammar: hypothalamic-hypophyseal-follicular, CL–follicular, and pineal–hypothalamic-hypophyseal–CL (Fig. 7.6). The basic interactions of the hypothalamic–hypophyseal-follicular axis appear similar to those described for other mammals, as do those of the CL–follicular axis during pregnancy and the oestrous cycle. However, an unusual influence of this latter axis is that the inhibitory effects of a newly formed CL on follicular growth extend into lactational quiescence in species showing diapause and into seasonal quiescence in the tammar and Bennett's wallabies. The interactions of the third axis (pituitary-CL) are seen only in the Macropodidae exhibiting lactational quiescence and the additional influence of the pineal gland is seen only in the two species showing seasonal quiescence. What is the role of each of these axes in the regulation of seasonality?

1 Hypothalamic–hypophyseal–follicular axis

The evidence available for the tammar indicates that the hypothalamic-hypophyseal-follicular axis functions as in other mammals during the breeding season but, in contrast to species like the sheep, it remains potentially fully functional in the non-breeding season. Throughout pregnancy, plasma concentrations of LH and follicle–stimulating hormone (FSH) remain low to undetectable, except at post-partum oestrus when there is a preovulatory surge of LH approximately 16 h after parturition (Sutherland *et al.* 1980; Tyndale-Biscoe *et al.* 1983; Harder *et al.* 1985). After day 18 of pregnancy, follicular growth resumes (see below, p.264) and before parturition a dominant Graafian follicle, secreting increasing amounts of oestradiol into the circulation, is recognizable (Harder *et al.* 1984, 1985). The increase in plasma oestradiol reaches a peak 8–12 h *post-partum.*

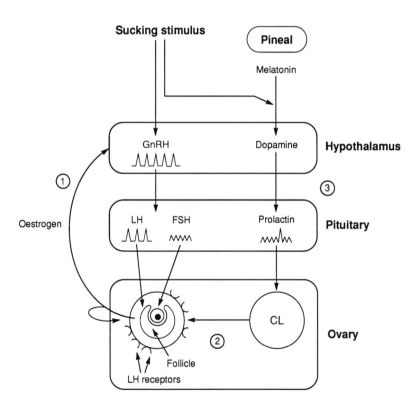

Fig. 7.6 Diagram of the three endocrine axes regulating reproduction in the tammar. (1) the hypothalamic–hypophyseal–follicular axis; (2) the CL–follicular axis; and (3) the pineal–hypothalamic–hypophyseal–CL axis.

It induces both oestrous behaviour and the preovulatory LH surge, and removal of the Graafian follicle 3 days before parturition abolishes these events (Shaw and Renfree 1984; Harder *et al.* 1985). Further, as in other mammals, removal of the ovaries results in a post-castration rise in both gonadotrophins (Evans *et al.* 1980; Tyndale-Biscoe and Hearn 1981; Hinds *et al.* 1992). Closer examination of the short-term changes in plasma LH and FSH after bilateral ovariectomy showed that there is an increase in amplitude and frequency of LH pulses, but this change in secretion is not accompanied by a marked increase in basal LH concentrations. Conversely, the pattern of release of FSH remains non-pulsatile, although basal concentrations increase (Hinds *et al.* 1992; Fig. 7.7).

Various studies in the tammar indicate that the presence of non-luteal ovarian tissue is essential for negative and positive feedback effects on

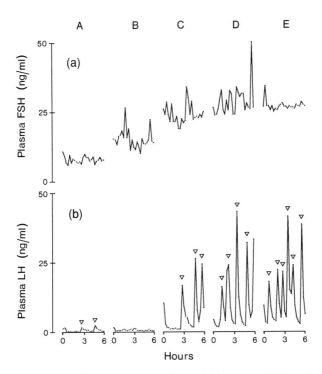

Fig. 7.7   Short-term profiles of concentrations of (a) plasma FSH and (b) plasma LH in an individual female during lactational quiescence when (A) intact; (B) 10 days after removal of the CL–bearing ovary; (C) 21 days after full ovariectomy; (D) 21 days after removal of the sucking stimulus (42 days after full ovariectomy); and in seasonal quiescence (E), 166 days after ovariectomy. ∇ indicates LH pulse.
Redrawn from Hinds, *et al.* (1992).

gonadotrophin release. LH and FSH secretion remained low in unilaterally ovariectomized females in which the ovary bearing the CL was removed, but increased when both ovaries were removed (Tyndale-Biscoe and Hearn 1981; Hinds *et al.* 1992). However, in ovariectomized females in which implants of ovarian cortex developed luteinized tissue, the concentrations of the gonadotrophins decreased to levels typical of intact females (Evans *et al.* 1980; Tyndale-Biscoe and Hearn 1981). The converse influence of the pituitary on ovarian function has also been investigated; removal of the pituitary or passive immunization against GnRH prevents follicular growth beyond the stage of small antral follicles (Hearn 1974; Panyaniti *et al.* 1985; Short *et al.* 1985).

The above responses indicate that the hypothalamic–hypophyseal–follicular axis functions in the tammar as it does in other mammals to regulate oestrus and ovulation. In contrast to the ewe, however, there is no effect of season on this axis. Injections of oestradiol-17β elicit the same negative and positive feedback responses in FSH and LH secretion in ovariectomized tammars in both periods of quiescence (Horn *et al.* 1985), as does a standard dose of GnRH (Horn and Tyndale-Biscoe, personal communication). Similarly, the short-term pattern of release of both gonadotrophins is not different between the periods of lactational and seasonal quiescence in ovariectomized females (Fig. 7.7; Hinds *et al.* 1992). Thus, unlike in the ewe, where changes in the LH pulse generator determine the annual reproductive cycle, in the tammar the hypothalamic–hypophyseal–follicular axis is essentially in breeding condition throughout the year. However, for 11 months of the year, it is prevented from being functionally active by the presence of the quiescent CL on the ovary (see below, pp.266–7).

2   CL–follicular axis

While the pituitary suppresses the activity of the CL (see below, p. 267), the CL itself inhibits follicular growth in both ovaries during the first two-thirds of pregnancy or the oestrous cycle and during lactational and seasonal quiescence; that is, the CL has both a local and systemic effect on the growth of all follicles (Tyndale-Biscoe and Hawkins 1977; Rodger *et al.* 1993). Removal of the CL during pregnancy or in either period of quiescence results in follicular growth and a return to oestrus 10–14 days later (Tyndale-Biscoe and Hawkins 1977; Evans *et al.* 1980; Renfree *et al.* 1982). The return to oestrus and ovulation are unaffected by lactation, indicating that the CL is the main agent of inhibition of follicular growth (Tyndale-Biscoe and Hawkins 1977; Hinds *et al.* 1992). After removal of the CL, injections of oestradiol alone delay the return to oestrus and ovulation suggesting that oestradiol of luteal origin may be the inhibitory agent (Evans *et al.* 1980; Renfree *et al.* 1982). However, the CL contains only negligible amounts of oestradiol, the origin of which is unknown, and the luteal tissue has no capacity to aromatize steroid precursors for the synthesis of oestrogens (Renfree *et al.* 1984). The mechanism by which the quiescent, undeveloped CL inhibits follicular growth throughout the year remains unresolved. One suggestion is that the synthesis of LH receptors on the follicles is inhibited by the CL, but this remains to be investigated. Nevertheless, when the influence of the CL is removed the hypothalamic–pituitary–follicular axis is immediately responsive and is unaffected by the season of the year (Horn *et al.* 1985: Hinds *et al.* 1992).

3 Pituitary–CL axis

In other species experiments involving removal or ablation of endocrine glands have provided many insights into their involvement in reproduction. Hypophysectomy of the tammar has been particularly revealing in establishing the relationship between the CL and pituitary. Hearn (1972*a*, 1973, 1974) found that removal of the pituitary during lactational quiescence not only caused a failure of lactation but unexpectedly allowed reactivation of the quiescent CL and the dormant blastocyst. Because the results did not discriminate between the effects of loss of the sucking stimulus and removal of the pituitary, the experiment was repeated using non-lactating females in seasonal quiescence (Hearn 1974). The same result, reactivation of the CL, was achieved. Subsequently, measurement of plasma progesterone concentrations revealed an identical profile to that described for intact females undergoing active pregnancy (Hinds 1983). This extraordinary response to hypophysectomy indicated that the activity of the CL is tonically inhibited by the pituitary, but once released, there is no requirement for luteotrophic support from the pituitary for maintenance of progesterone production during pregnancy. There is other evidence indicating that the CL functions autonomously once released from pituitary inhibition. First, the CL is devoid of LH receptors, although there are abundant prolactin receptors (Stewart and Tyndale-Biscoe 1982). Second, females passively immunized against GnRH in early pregnancy show no follicular development, and oestrus and ovulation does not occur; however, the CL grows, pregnancy is maintained and parturition occurs (Short *et al.* 1985). Thus the pituitary–ovarian axis must be intact for normal follicular maturation and ovulation, but it is not necessary for the maintenance of pregnancy or for the secretory activity of the CL. The inflence of the pineal gland on this axis is discussed in a later section.

i   The role of prolactin in lactational and seasonal quiescence
Given the involvement of the sucking stimulus in lactational quiescence, oxytocin and prolactin would be conceivable candidate(s) sustaining the inhibition of the CL. In the red kangaroo, *M. rufus*, injections of oxytocin

after removal of pouch young delayed reactivation (Sharman 1965), although some of the control animals were also delayed, presumably due to stress. Tyndale-Biscoe and Hawkins (1977) repeated this study in the tammar using injections of oxytocin or prolactin three times daily for 7 days after removal of pouch young. Both treatments resulted in a delay to birth compared with saline injected controls. Similarly, in Bennett's wallabies reactivation is also delayed after removal of the pouch young if injections of prolactin or of domperidone (a dopamine antagonist which releases prolactin in this species; Curlewis *et al.* 1986), are given (Curlewis and Loudon 1988*a*). However, these results provided no discrimination between a direct effect on the CL or an indirect effect mediated by the pituitary itself. To resolve this issue, non-lactating tammars were treated with saline, oxytocin or prolactin after hypophysectomy or anterior hypophysectomy in seasonal quiescence. Reactivation was delayed in those given prolactin but not in those given oxytocin or saline alone (Tyndale-Biscoe and Hawkins 1977). These results indicated that prolactin was the agent of inhibition of the CL, and revealed yet another function for this multipurpose hormone in mammals. More recent experiments have provided more precise information regarding the frequency of the inhibitory prolactin signal; daily intravenous injections of prolactin (producing a pulse of <2 h duration) to intact females delay reactivation after removal of pouch young. Further, if the interval between treatments was 72 or less reactivation was also delayed (Hinds and Tyndale-Biscoe, unpublished observations). These results support other findings (Gordon *et al.* 1988) that the CL is not committed to reactivation until 72 h after removal of the sucking stimulus, and also indicate that an inhibitory signal from the pituitary must occur at least once each 72 h to maintain quiescence.

Responses to the dopamine agonist bromocriptine, which inhibits prolactin release in eutherian species, provided further support for an inhibitory role for prolactin. During lactational quiescence a single intramuscular injection of bromocriptine (5 mg/kg) induced reactivation of the CL (Tyndale-Biscoe and Hinds 1984), with the progesterone peak and birth occurring at the same intervals as after removal of the pouch young (Table 7.2; Tyndale-Biscoe and Hinds 1984; Tyndale-Biscoe *et al.* 1986). Between the winter solstice and the vernal equinox, however, there was a dramatic decline in the numbers of animals which responded to treatment, and between the vernal equinox and the summer solstice, only one of 20 females reactivated (Tyndale-Biscoe *et al.* 1986). Tammars treated at the summer solstice did not respond to bromocriptine but underwent natural reactivation in the following 15 days and gave birth in late January. When these same animals were treated again in February (lactational quiescence) all reactivated (Tyndale-Biscoe and Hinds 1984). These results suggest that the pituitary becomes insensitive to dopamine between the vernal equinox and summer solstice.

The response to bromocriptine in Bennett's wallabies, although examined only at one stage in each period of quiescence, was essentially identical to the tammar. Reactivation occurred in lactational quiescence but not after the vernal equinox in seasonal quiescence (Curlewis *et al.* 1986).

Although the effects of bromocriptine in lactational quiescence were consistent with a suppression of prolactin secretion from the pituitary (e.g. lactation failed in some females), an effect on plasma prolactin concentrations measured once daily could not be detected in lactating tammars or Bennett's wallabies (Tyndale-Biscoe and Hinds 1984; Curlewis *et al.* 1986). However, during early lactation, basal prolactin concentrations are not different from those in non-lactating females (Hinds and Tyndale-Biscoe 1982*b*, 1985; Curlewis *et al.* 1986; Gordon *et al.* 1988), and so it is perhaps not surprising that no effect of bromocriptine could be demonstrated. In seasonal quiescence, prolactin concentrations also remained unchanged after bromocriptine treatment in non-lactating tammars (Tyndale-Biscoe and Hinds 1984), although in lactating Bennett's wallabies there appeared to be a significant decrease in plasma prolactin concentrations over 8 h (Curlewis *et al.* 1986). However, it could be argued that the decrease in prolactin attributed to the effect of bromocriptine is indistinguishable from the effects of the sucking stimulus which induces intermittent increases in plasma prolactin above basal concentrations during the later stages of lactation (Hinds and Tyndale-Biscoe 1985; Curlewis *et al.* 1986).

During the period of lactational quiescence, as already mentioned, the sucking stimulus is critical for the inhibition of the CL and its removal allows reactivation. The importance of the neural input in this inhibition was observed when Renfree (1979) denervated the suckled mammary gland; lactation and growth of the young were unaffected, but reactivation of the CL and blastocyst occurred. Although these results appeared inconsistent with a tonic inhibitory role for prolactin, the rate of recovery of the CL after hypophysectomy, bromocriptine treatment, and after denervation of the mammary gland was the same as after removal of the pouch young. That is, the same pathway of inhibition is likely to have been affected by all treatments. From the studies of Gordon *et al.* (1988) on the effects of temporary removal of pouch young and our studies using short duration pulses of prolactin (Hinds and Tyndale-Biscoe, unpublished observations), it appears that the inhibitory signal does not have to operate continuously. This would suggest that the most important aspect of the release of prolactin in lactational quiescence may not be basal secretion, but a short duration prolactin pulse, influenced by neural input via the sucking stimulus to the hypothalamus and pituitary at least once every 72 h, and this appears to provide the key to the inhibition of the CL.

Until the studies on the role of photoperiod were undertaken (see below, p.270) the role of prolactin in seasonal quiescence was also unresolved. On the one hand, hypophysectomy followed by injections of prolactin delayed reactivation in both periods of quiescence, but bromocriptine did not induce reactivation between the vernal equinox and the summer solstice and, despite repeated attempts, was ineffective in reducing plasma prolactin concentrations (at least in tammars). Thus either prolactin was not the common agent inhibiting the CL in lactational and seasonal quiescence, or there was an additional dopamine insensitive-component operating during seasonal quiescence. The lack of a reproductive response to bromocriptine between the vernal equinox and the summer solstice also suggested that there could be either a marked change in the sensitivity of the pituitary to dopamine release from the hypothalamus, or another photoperiod-controlled inhibitory system becomes effective during this period, or the system becomes refractory. The detection of a morning prolactin pulse during seasonal quiescence (Hinds 1989b) has furthered our understanding of the mechanisms of CL inhibition in seasonal quiescence (see below, p.280).

THE ROLE OF PHOTOPERIOD

1   Reproductive response to manipulation of photoperiod

After the winter solstice, the removal of the pouch young in both the tammar and Bennett's wallaby does not induce reactivation of the CL and blastocyst. This raises several questions about the mechanism(s) of inhibition which operates during seasonal quiescence — how does the tammar respond to photoperiod changes? What role does prolactin play? Are any other factors involved? As in other species the effects of photoperiod change on the pineal-hypothalamic-hypophyseal-gonadal axis have been examined to determine the physiological mechanisms involved. In the tammar we have assessed the response in the pineal gland by analysis of plasma melatonin profiles, the pituitary response in terms of prolactin release, and the response of the CL as it undergoes reactivation by measurement of the early progesterone pulse. What has emerged from these studies is that there is a remarkably rapid response by the tammar to photoperiod change; the first endocrine changes can be measured within 3 days.

A photoperiodic component was first implicated when tammars transferred to the Northern Hemisphere continued to give birth each year after the summer solstice (i.e. June 22; Berger 1970). Subsequently, Sadleir and Tyndale-Biscoe (1977) determined the responses of animals to experimental manipulations of the photoperiod using summer solsticial (15 h light:9 h dark; 15L:9D) and equinoctial (12L:12D) daylengths. Tammars in a state of seasonal quiescence, and maintained on 15L:9D for 40 days remained in quiescence, although not all females kept on 12L:12D did so, with six of 13 giving birth at varying intervals after the onset of this regime (Sadleir and Tyndale-Biscoe 1977; Hinds and den Ottolander 1983). However, all quiescent females, when subjected to a decrease in daylength of 3 h from 15L:9D to 12L:12D or from 12L:12D to 9L:15D, gave birth 29–36 days later. It was concluded that the reduction in daylength by 3 h, and not a particular static daylength, had induced reactivation of the CL, although the time to birth after the photoperiod change was 3–10 days longer than after removal of the pouch young in lactational quiescence. Subsequently, the standard photoregime we have used for determining the tammar's physiological response to photoperiod has been a change from 15L:9D to 12L:12D (Tyndale-Biscoe *et al.* 1986).

Similar responses to photoperiod are seen in Bennett's wallabies; quiescence is maintained when animals experience summer solsticial daylength in early seasonal quiescence, but reactivation occurs if daylength is reduced by 3 h to 12L:12D (Loudon and Curlewis 1987). Again, later in seasonal quiescence a 5 h reduction in daylength from ambient increasing daylengths to 12L:12D induces reactivation of the quiescent CL and birth occurs 28–34 days after the daylength change, an interval 2–6 days longer than after removal of pouch young (Loudon and Curlewis 1987).

Thus both species respond to a specific photoperiod change which involves a reduction in daylength, but birth is delayed, compared with the interval after removal of pouch young in lactational quiescence. In the tammar, the longer interval is due to a delay in the appearance of the early plasma progesterone peak which occurs on days 8–14 after the reduction in daylength (Fig. 7.8). The interval from the early peak to birth or oestrus remains at 21–24 days (Hinds and den Ottolander 1983). A similar delay to the progesterone peak also occurs in the Bennett's wallaby (Brinklow and Loudon 1989).

What characteristics of a photoregime determine if an animal will remain in quiescence or be reactivated? For example, as mentioned above, some tammars held on 12L:12D reactivated at various times while experiencing this daylength; such a response could suggest 12L:12D represents a critical threshold, which may be interpreted either as stimulatory or inhibitory. Analysis of the results for the various tammar studies has allowed us to define which photoregimes are stimulatory (that is induce reactivation of the CL), inhibitory (that is invoke or maintain quiescence) or permissive (Fig. 7.9); to maintain quiescence, the dark phase must be 9 h or less, to induce reactivation the dark phase must be extended to more than 9 h, while females held on 12L:12D or 9L:15D may or may not remain quiescent, i.e. show a permissive response. Experiments outlined in the next section show that in the tammar a minimum increase of 1–1$\frac{1}{2}$ h in the duration of the dark phase to a total of at least 10–10$\frac{1}{2}$ h is sufficient to induce reactivation (Tyndale-Biscoe and Hinds 1992).

Since fewer experimental manipulations of photoperiod have been done

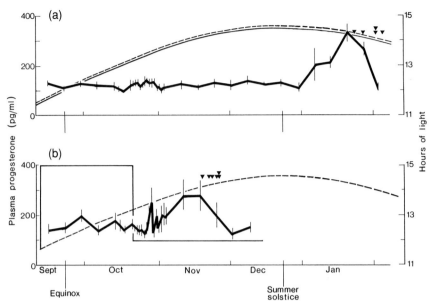

Fig. 7.8   Concentrations (mean ± SEM) of plasma progesterone in two groups of tammars experiencing artificial photoregime: (a) simulated natural increasing daylength and (b) 15L:9D to 12L:12D. Solid lines represent the artificial photoregime, dashed lines the natural photoperiod. Solid arrows indicate time of birth for individual tammars. Redrawn from Hinds and den Ottolander (1983).

with Bennett's wallabies, the minimum change in duration which induces reactivation is not as well defined. However, in early seasonal quiescence a change in duration from 10L:14D to 15L:9D did not induce reactivation, but a change in the opposite direction from 15L:9D to 12L:12D was effective (Loudon and Curlewis 1987; Fig. 7.10). An increase in the dark phase from 17L:7D to 12L:12D and from 18.25L:5.75D to 9L:15D later in seasonal quiescence also induced reactivation (Loudon and Curlewis 1987; Brinklow and Loudon 1989). At the end of lactational quiescence some Bennett's wallabies maintained on 9L:15D underwent cycles of variable lengths (Loudon and Curlewis 1987, 1989), a response which suggests that this may be a permissive photoregime (Fig. 7.10). Together these results appear to indicate that a dark phase of 9 h or less is inhibitory and that an increase in duration of the dark phase to 12 or 15 h effectively induces reactivation in Bennett's wallabies.

Do tammars and Bennett's wallabies develop refractoriness to specific photoperiods–that is, can animals change from lactational or seasonal quiescence or vice versa at any stage of the annual cycle? Although experiments have not been specifically designed to test this, several results indicate that the pineal-hypothalamic-hypophyseal-CL axis remains responsive to photoperiod manipulations throughout the year. For example, female tammars which were re-exposed to natural increasing daylength (that is, a shortening dark phase) after responding to a reduction from 15L:9D to 12L:12D, became inhibited, and did not reactivate when their young were removed (McConnell and Tyndale-Biscoe 1985). Further, tammars in a state of lactational quiescence in March (experiencing decreasing daylengths) became seasonally quiescent when re-exposed to long daylengths for a sufficient period. Six of 18 females held on 15L:9D for only 7 days before removal of pouch young reactivated, but if young were

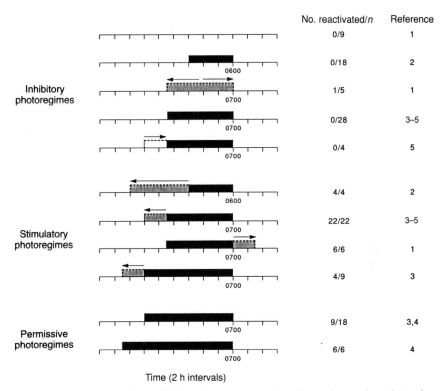

Fig. 7.9 Summary of artificial photoregimes that have been found to be stimulatory, inhibitory or permissive in tammars. Results drawn from (1) McConnell *et al.* (1986); (2) Tyndale-Biscoe and Hinds (1992); (3) Sadleir and Tyndale-Biscoe (1977); (4) Hinds and den Ottolander (1983); (5) McConnell and Tyndale-Biscoe (1985). Solid bars indicate initial photoperiod; stippled areas indicate photoperiod resulting from experimental change, with arrows indicating direction of that change.

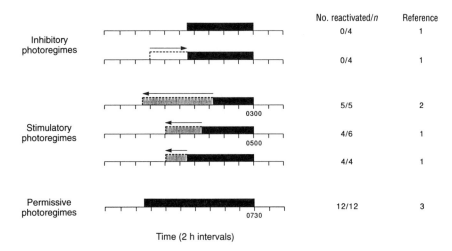

Fig. 7.10 Summary of artificial photoregimes that have been found to be stimulatory, inhibitory or permissive in Bennett's wallabies. Results drawn from: (1) Loudon and Curlewis (1987); (2) Brinklow and Loudon (1989); (3) Loudon and Curlewis (1989). Solid bars indicate initial photoperiod; stippled areas indicate photoperiod resulting from experimental change, with arrows indicating direction of that change.

removed after 14 days, there was no response (Hinds 1989*b*; Tyndale-Biscoe and Hinds 1992). Experience of a more extreme photoperiod of 18L:6D for 1 week was also inhibitory as none of the females reactivated after removal of pouch young (Tyndale-Biscoe and Hinds 1992). Thus, returning tammars to increasing or long daylengths in mid-lactational or mid-seasonal quiescence reimposes a state of inhibition.

Nevertheless, if tammars which had reactivated after experiencing a reduction in daylength from 12L:12D to 9L:15D, were held on 9L:15D they remained in a state equivalent to lactational quiescence; removal of their new pouch young induced reactivation and the early progesterone peak and birth occurred on days 6 and 28, respectively (Hinds and den Ottolander 1983), intervals equivalent to those seen after removal of pouch young in lactational quiescence.

In Bennett's wallabies, manipulations of photoperiod in early and late lactational quiescence have produced similar results, although the responses have been interpreted as a form of photorefractoriness. Lactating Bennett's wallabies transferred from decreasing natural daylengths (14L:10D), in early lactational quiescence, and held on long days (18L:6D) for 7 days, reactivated after removal of their pouch young (Curlewis and Loudon 1989). Late in lactational quiescence, 2 weeks before the winter solstice, the same experiment was repeated but the change in daylength was more extreme (approximately 9L:15D to 18L:6D). After 1 week on this photoperiod animals did not reactivate after removal of their pouch young (Curlewis and Loudon 1989). These authors argue that the results of the first experiment indicate that the animals were insensitive or refractory to inhibitory daylengths in early lactational quiescence. However, these responses could be interpreted, as for the tammar, in that 1 week was an insufficient time for the neuroendocrine axis to revert to a state equivalent to seasonal quiescence whereas the more extreme change later in the year was adequate.

For both species, it is probable that the hypothalamus and/or pituitary show changes throughout the year in their sensitivity to changing daylength. Early in lactational quiescence (or in seasonal quiescence) the system is not as sensitive to the previously inhibitory (or stimulatory) daylength — and the time to readjust the system to abrupt changes varies accordingly. Although it may be argued that these responses represent a form of refractoriness, it is clearly not total refractoriness (Nicholls *et al.* 1988) because the reproductive state can be reversed relatively rapidly.

However, not all animals need to experience a specific change in photoperiod in order to revert from seasonal quiescence to lactational quiescence. Sadleir and Tyndale-Biscoe (1977) found that at the end of seasonal quiescence, the response to different daylengths was less precise; half of the animals exposed to summer solsticial daylengths from the vernal equinox until after the summer solstice reactivated without experiencing a reduction in photoperiod, although those held on equinoctial daylength did not (Saldeir and Tyndale-Biscoe 1977). This response in some animals may be interpreted as a form of photorefractoriness which is an adaptation to ensure the onset of the next breeding season (see below, pp.288–9).

In both wallaby species, naturally increasing or artificially long days are inhibitory when these are experienced during mid-lactational quiescence or mid-seasonal quiescence, and indicate that an additional photoperiod-controlled factor, which is not present on naturally decreasing daylengths or artificial short days, influences the maintenance of the inhibition of the CL. What is the role of the pineal gland?

2   The role of the pineal gland: the melatonin message

In this section we will see that the essential melatonin message comprises a change in duration of elevated melatonin from 9 h to at least 10 or $10^1/2$ h and that this must occur for a minimum of 2–3 days for it to be interpreted by the hypothalamus and pituitary.

In seasonally breeding mammals, the pineal gland and its secretion of the hormone melatonin are important in the response to photoperiod (Lincoln and Short 1980; Goldman and Darrow 1983; Karsch *et al.* 1984;

Tamarkin *et al.* 1985). In the tammar, the role of the pineal gland has been examined by determining the effects of pinealectomy or superior cervical ganglionectomy on seasonality, and by admnistration of melatonin in intact animals at various times of the year, or under various photoperiods. The development of pineal function and its secretion of melatonin have not been determined for any marsupial and there have been no studies examining the role of the pineal in the development of puberty or seasonality in young females, although these areas pose interesting research questions.

In both Bennett's wallabies and tammars the secretion of melatonin into the circulation of adult females shows a daily rhythm with plasma concentrations being elevated at night, as in other mammals. Melatonin concentrations are low to undetectable throughout the light phase but increase shortly after the onset of the dark phase, remain elevated throughout the dark phase, and decrease to undetectable levels coincident with the onset of the next light phase (Figs 7.11 and 7.12; McConnell and Tyndale-Biscoe 1985; McConnell and Hinds 1985; McConnell 1986; Curlewis and Loudon 1988*b*). Under the arbitrary classification of Reiter (1987) this melatonin rhythm is a type C pattern (or square wave) as seen in the Djungarian hamster, sheep, and white-footed mouse. However, having this characteristic pattern does not provide any information about the component of the profile to which the animal responds. Throughout the year in both wallabies the period of elevated melatonin concentrations

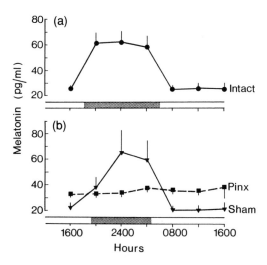

Fig. 7.11   Concentrations (Mean ± SEM) of plasma melatonin (pg/ml) measured in tammars (a) before (*n*=12) (●) and (b) after pinealectomy (*n*=5) (■) or sham pinealectomy (*n*=6) (▼). Shaded bars indicate the dark phase of each photoperiod. Redrawn from McConnell and Hinds (1985).

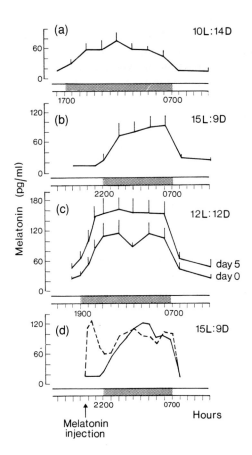

Fig. 7.12   Concentrations (Mean ± SEM) of plasma melatonin in tammars exposed to different photoperiods: (a) 10L:14D; (b) 15L:9D; (c) the response in melatonin on the first and fifth days after a change from 15L:9D to 12L:12D; and (d) the profile of melatonin (----) obtained after exogenous injection of melatonin 2.5 h before lights off on 15L:9D. Note that this profile mimics the effect of a change to 12L:12D. Shaded bars indicate the length of the dark phase. From McConnell and Tyndale-Biscoe (1985), with permission.

changes with the length of the night (McConnell and Tyndale-Biscoe 1985; McConnell 1986; Curlewis and Loudon 1988b), but only in the tammar is there any evidence for changes in amplitude (McConnell 1986; Fig. 7.12). The significance of these amplitude changes has not been investigated. Other studies indicate that duration rather than amplitude is the more important component of the melatonin message (McConnell et al. 1986; see also below, p.279).

The rhythm of melatonin secretion is abolished by pinealectomy (McConnell and Hinds 1985; Fig. 7.11) and by denervation of the pineal by bilateral cervical ganglionectomy (Renfree *et al.* 1981; McConnell 1984), as it is in other mammals (Arendt 1986). So what role does melatonin play in response to photoperiod by the tammar and Bennett's wallabies? Which component(s) are the most important for transduction of the photoperiod message?

The delay to the early progesterone peak after a change from 15L:9D to 12L:12D (see above, p.271 and Fig. 7.8; Hinds and den Ottolander 1983) could be interpreted in two ways: either the response of the pineal gland in terms of its secretion of melatonin could take some days or nights to adjust to the new night length, or the message may not be interpreted downstream in the hypothalamus and/or pituitary for several days. McConnell and Tyndale-Biscoe (1985) monitored the response of the pineal gland by measuring melatonin secretion in the first 5 days after a photoperiod change from 15L:9D to 12L:12D. The increase in melatonin occurred on the first night after the change, coincident with the onset of the new dark phase, which occurred 3 h earlier in the evening (1900 h; Fig. 7.12). Thus, the pineal responds rapidly to a photoperiod change and the delay to reactivation is more likely due to a delay in the transduction of this altered secretion of melatonin and its subsequent interpretation in the hypothalamus and pituitary.

If melatonin is the mediator of the message between the pineal and hypothalamus then treatment with exogenous melatonin should induce the same response as a stimulatory photoperiod change, and it does. Females held on an inhibitory daylength (15L:9D) and injected for 15 days with melatonin $2\frac{1}{2}$ h in advance of the beginning of the dark phase (Fig. 7.12), responded as if the photoperiod had been altered to 12L:12D; births occurred 30–32 days later (McConnell and Tyndale-Biscoe 1985). Implants or injections of melatonin in seasonal quiescence also induced reactivation in Bennett's wallabies (Loudon *et al.* 1985; Loudon and Curlewis 1987). These results for both species indicated that melatonin mediated the message between the pineal gland and the hypothalamus–pituitary gland, but the component of the melatonin profile inducing the response was not apparent.

It has been argued that the amplitude, duration and/or phase of elevated melatonin is used to measure photoperiod in other seasonally breeding species. There is scant evidence for a role of amplitude except in the ram (Lincoln *et al.* 1981). For most species either duration or phase has been implicated (Reiter 1987, 1990). In species such as the white-footed mouse (Margolis and Lynch 1981) and the Syrian hamster (Reiter *et al.* 1976; Tamarkin *et al.* 1976; Watson-Whitmyre and Stetson 1983) the response to photoperiod is thought to be due to exposure to melatonin at a particular, sensitive phase of each 24-h period. However, more recent evidence for the Syrian hamster indicates duration may also be important (Karp *et al.*

1991; Pitrosky *et al.* 1991). In the sheep (Bittman *et al.* 1983; Karsch *et al.* 1984) and the Djungarian hamster (Carter and Goldman 1983; Goldman and Elliott 1988) the absolute duration of the elevation in concentrations of melatonin was thought initially to be critical. However, most species are likely to use a combination of the absolute duration and the direction of the change in duration to monitor daylength (Reiter 1987, 1990). In the sheep the latter component has now been recognized (Robinson and Karsch 1987; Wayne *et al.* 1988; Karsch and Woodfill 1990), as too has the presence of an endogenous rhythm of breeding apparent in pinealectomized ewes which can be re-synchronized by just 70 days of an appropriate melatonin treatment (Robinson and Karsch 1988; Karsch *et al.* 1989).

For the tammar, several experiments have provided evidence that change in duration of elevated melatonin is the critical component. When tammars were maintained under conditions of 24L:0D the melatonin profile was abolished and the animals remained in quiescence (McConnell *et al.* 1986). If, on this regime of 24L:0D, a single injection of melatonin was given at 1900 h, animals also remained in quiescence. However, if the exogenous treatment was continued when the animals were returned to a photoperiod of 15L:9D, reactivation occurred (McConnell *et al.* 1986). Further, if the dark phase was extended by 3 h in the morning rather than in the evening the animals also reactivated (McConnell *et al.* 1986). That is, reactivation occurred in animals independent of extension of the 9 h dark period in the morning or the evening, indicating that the change in duration of elevated melatonin is more important than the time of day at which it is elevated. Specific experiments have not been done in Bennett's wallabies to determine if change in duration of elevated melatonin provides the critical message which leads to reactivation.

Nevertheless, in the tammar, a change in duration alone is not sufficient to induce reactivation. As mentioned, tammars which experienced a decrease in duration from 12L:12D to 15L:9D or to 18L:6D became inhibited (Hinds 1989*b*; Tyndale-Biscoe and Hinds 1992). Similarly, a simulated increase in duration from 18L:6D to 15L:9D by injection of exogenous melatonin does not induce a response (Tyndale-Biscoe and Hinds 1992). However, injection of melatonin at 6 h and again at 3 h in advance of a 6 h dark phase mimicked an increase to 12 h duration of melatonin and induced reactivation in all treated females. Reactivation also occurred in three of five animals injected only at 6 h in advance of the 6 h dark phase. For both groups, plasma melatonin was not elevated throughout the entire 12 h, as concentrations returned to basal level between each injection; that is, the animals did not need to experience elevated melatonin throughout the entire 12 h for it to be read as a stimulatory change in duration (Tyndale-Biscoe and Hinds 1992).

Clearly, change in duration is important, but what is the minimum change required to induce reactivation? By giving melatonin injections between

1/2 and 2^1/2 h in advance of 9 h dark phase, it has become clear that the minimum requirement for most animals is a change in duration of the dark phase from 9 to 10^1/2 h (Tyndale-Biscoe and Hinds 1992).

How long must the tammars see this critical change in duration before an irreversible response is ensured? Treatment of tammars with melatonin for consecutive nights revealed that the minimum number of days of treatment required to induce reactivation was 2 or 3 (Tyndale-Biscoe and Hinds 1992). This response is very rapid. It is more rapid than in the sheep, for example, in which there is a delay of 50–60 days before a change in LH levels is detected after an appropriate photoperiod change or melatonin infusion (Bittman *et al.* 1983; Bittman and Karsch 1984). The rapid response in the tammar is also seen in the Bennett's wallaby, as births occur 30–34 days after photoperiod change or melatonin implants (Loudon *et al.* 1985; Loudon and Curlewis 1987).

Although many of these experiments should be repeated using pinealectomized tammars and exogenous melatonin treatments to confirm the results discussed above, we can summarize the essential components of the melatonin message as derived from the experiments with intact tammars. Clearly, a change in duration of elevated melatonin is essential to evoke either an inhibitory or stimulatory response. To be stimulatory there must be an increase in duration of elevated melatonin to greater than 10 or 10^1/2 h, and it must occur for a minimum of 2–3 days for it to be interpreted downstream in the hypothalamus and pituitary. Conversely, to be an inhibitory response the duration must change to 9 h or less and last for at least 7–14 days (Hinds 1989*b*; Tyndale-Biscoe and Hinds 1992). The critical components of the melatonin message in the Bennett's wallaby have not been determined but would appear to be similar.

3   Pituitary response to photoperiod change

In the sheep, the melatonin message affects GnRH release from the hypothalamus and thus LH release from the pituitary (Karsch *et al.* 1984; Goodman 1988). In the tammar, melatonin affects the release of prolactin from the pituitary. From the earlier studies described above (Tyndale-Biscoe and Hawkins 1977; Tyndale-Biscoe and Hinds 1984), there was substantial, though often paradoxical, evidence for a role for prolactin in the inhibition of the CL in lactational and seasonal quiescence. Hinds and den Ottolander (1983) showed a decrease in prolactin concentrations after a stimulatory photoperiod change, Tyndale-Biscoe and Hinds (1984) described higher concentrations of plasma prolactin in seasonal quiescence than in lactational quiescence, but such changes were not supported by further studies (Hinds and Tyndale-Biscoe 1985; McConnell and Hinds 1985). These apparent inconsistencies were resolved when an early morning plasma prolactin peak was detected in animals held on inhibitory photoperiods in seasonal quiescence (McConnell *et al.* 1986). This peak of 1–2 h duration occurred coincident with the end of the dark phase in animals held on 15L:9D (Fig. 7.13) and was absent within 5 days after the photoperiod was altered to 12L:12D (McConnell *et al.* 1986). The between-study discrepancies in plasma prolactin concentrations were now explicable; the infrequency of samples as well as inappropriate sampling times in relation to the occurrence of the morning prolactin peak meant that it was not detected in the earlier studies.

The short duration prolactin peak also provided a candidate for the critical, non-basal component of prolactin secretion which could be influenced by the sucking stimulus in lactational quiescence and by changing photoperiod in seasonal quiescence. Could this prolactin peak maintain the inhibition of the CL during seasonal quiescence while prolactin was at basal concentrations for the rest of the day? When the photoperiod was altered from 15L:9D to 12L:12D the peak was present until day 3 but absent in most animals by the fourth morning (Hinds 1989*b*; Fig. 7.14),

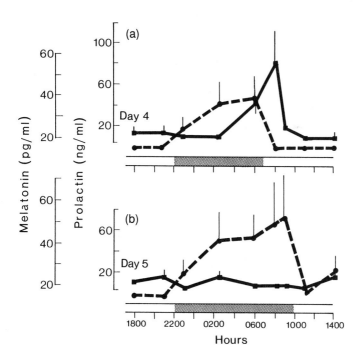

Fig. 7.13   Concentrations of plasma melatonin (pg/ml ●--●) and prolactin (ng/ml ■——■) in 6 tammars sampled while on (a) an inhibitory photoperiod of 15L:9D and (b) 5 days after a change to 12L:12D. Shaded bars indicate length of the dark phase. From McConnell *et al.* (1986).

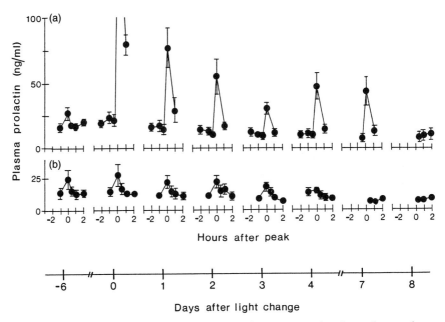

Fig. 7.14   Concentrations of plasma prolactin on successive days after a photo-period change from 15L:9D to 12L:12D in tammars injected each morning with (a) TRH or (b) saline. From Hinds (1989*b*).

about 5 days in advance of the plasma progesterone peak. If, after a photoperiod change from 15L:9D to 12L:12D, animals are treated with thyrotrophin-releasing hormone (TRH) to induce an endogenous peak of prolactin, or if exogenous prolactin was given intravenously at the time of lights on for 8 or 10 days, quiescence was maintained for the period of treatment (Figs 7.14 and 7.15). The early progesterone peak in TRH- and prolactin-treated animals occurred 5 days after the last injection (Fig. 7.15; Hinds 1989*b*). In the converse of the above experiment, tammars were maintained on an inhibitory daylength (15L:9D) and, for 5 successive mornings, treated with bromocriptine (5 mg/kg) approximately an hour before the endogenous pulse of prolactin would be expected to occur. Half of the treated animals lost their prolactin peak, reactivated and gave birth 28–29 days later, indicating that, at least in some tammars, the pituitary is sensitive to a dopamine agonist between the vernal equinox and the summer solstice (Hinds 1989*b*).

Similarly, in Bennett's wallabies subjected to a stimulatory photo-period change, both domperidone, which induces prolactin release, and exogenous prolactin, delay reactivation (Brinklow and Loudon 1989).

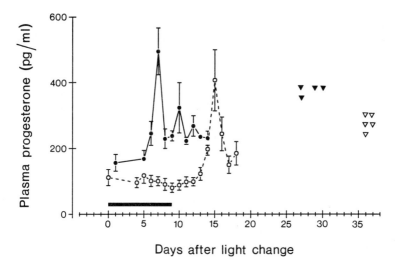

Fig. 7.15   Concentrations of plasma progesterone for four tammars treated with saline (●) and five tammars treated with prolactin (○) for 10 consecutive mornings (solid bar) after a photoperiod change from 15L:9D to 12L:12D. Births in control tammars, ▼; births in prolactin-treated tammars, ▽. From Hinds (1989*b*).

Because prolactin concentrations remained elevated between treatments, the exact nature of the prolactin signal in Bennett's wallabies has not been determined. To date a morning pulse of prolactin has not been detected (Pearce-Kelly *et al.* 1992), although in none of the studies have samples been collected frequently enough to establish unequivocally that a prolactin pulse is absent in this species. Nevertheless, prolactin is important (Brinklow and Loudon 1989) and there are pronounced changes with time in the release of prolactin in response to a standard challenge with domperidone; throughout seasonal quiescence prolactin release is higher than in lactational quiescence, and coincident with the onset of the breeding season the response declines significantly (Loudon and Brinklow 1990; Pearce-Kelly *et al.* 1992). Bennett's wallabies receiving melatonin implants which prevented the normal perception of day-night secretion of melatonin, showed low responses to domperidone and continued to undergo oestrous cycles throughout the normal period of seasonal quiescence (Loudon and Brinklow 1990). Another group of females which had entered seasonal quiescence before receiving the melatonin implants initially showed an elevated release of prolactin after domperidone treatment. After insertion of melatonin implants the prolactin response to domperidone declined significantly, the animals reactivated and also continued to undergo oes-trous cycles as indicated by elevations in plasma progesterone. These results indicate that melatonin affects prolactin secretion, possibly via the dopaminergic system, at the beginning and end of the breeding season (Loudon and Brinklow 1990). Such findings are consistent with those in the tammar in which the presence of the morning prolactin peak is also influenced by changes in the duration of melatonin secretion.

For both species it is clear that prolactin plays a major role in the maintenance of quiescence, and the small differences in mechanisms are probably due to convergent adaptations and the more recent evolution of the pattern of seasonal breeding in Bennett's wallabies compared with the tammar in which seasonality evolved before the Australian mainland and Kangaroo Island became separated. Nevertheless, the prolactin peak is not just a feature of reproduction in the female tammar. Although male tammars are not seasonal breeders (Catling and Sutherland 1980; Tyndale-Biscoe 1980; Inns 1982), a short duration prolactin peak is seen at the end of the dark phase in males experiencing 15L:9D and it is absent shortly after a photoperiod change to 12L:12D (Hinds and Tyndale-Biscoe, unpublished observations). Thus the morning prolactin peak has not evolved independently in the sexes, but a specific use of the hormone has evolved in the females: the luteal cells of the CL have become sensitive to its influence, which, unlike eutherians, is to inhibit its activity.

4   Mechanism of response to photoperiod change: a summary

The events which may occur in the tammar in response to a stimulatory photoperiod change can now be summarized (Fig. 7.16). A change in photoperiod from 15L:9D to 12L:12D causes an immediate change in the pattern of secretion of melatonin and if this is experienced for 3 consecutive nights the stored information leads to abolition of the prolactin pulse on the fourth morning. If the CL does not experience a prolactin peak for the next 72 h it undergoes irreversible reactivation. Once the CL has reactivated the following events occur: within 1 day of reactivation the CL stimulates the endometrium to produce a signal, possibly PAF, which releases the blastocyst from diapause; 1–2 days later there is an early peak of progesterone which is essential for the stimulation of the endometrium for the maintenance of the rest of pregnancy; birth of a new young occurs 21–22 days after the early progesterone peak (Fig. 7.16).

The rapidity of the tammar's response to photoperiod provides us with an ideal model to investigate the way in which the hypothalamus processes the melatonin message. This centre(s) should be active between the third and fourth morning, prior to the abolition of the prolactin peak, and could be located more readily in this species than in others such as the ewe or hamster where the time to detect a response in the pituitary is 3–6 weeks.

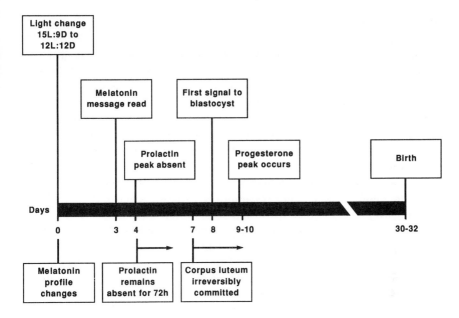

Fig. 7.16   Summary of sequence of events that may occur in the tammar after a stimulatory photoperiod change from 15L:9D to 12L:12D.

VI   EFFECTS OF PINEALECTOMY ON SEASONALITY

At first it was believed that the role of the pineal gland in mammals was primarily to inhibit reproductive activity. This view arose because pinealectomized male Syrian hamsters remained in reproductive condition even when exposed to daylengths which were normally inhibitory (Hoffman and Reiter 1965). However, this antogonadal role was not supported for other species in which further studies indicated a role for melatonin in biological time-keeping. For example, in Turkish hamsters, which like Syrian hamsters are long day breeders, pinealectomy induced testicular regression (Carter *et al.* 1982). In the sheep, the response varies with the time of year that surgery is performed, but generally there is a disorganization of the timing of the annual reproductive cycle. However, the animals do not become locked into one reproductive state because alternating periods of reproductive activity and anoestrus still persist although not in synchrony with pineal-intact sheep (Karsch and Woodfill 1990). What are the effects of pinealectomy in the tammar?

As in eutherian species the effect of pinealectomy on seasonality in the tammar is complex, with the time of year that surgery is performed and the time since surgery affecting the response. Furthermore, experimental manipulations such as treatment with bromocriptine or photoperiod challenge resulted in equally varied responses among pinealectomized animals.

Removal of the pineal gland or its denervation by superior cervical ganglionectomy abolishes the secretion of melatonin in the tammar (Renfree *et al.* 1981; McConnell and Hinds 1985; Fig. 7.11). Renfree *et al.* (1981) examined the effects of removal of the superior cervical ganglia from lactating tammars in early lactational quiescence. The tammars remained in quiescence until their pouch young were removed at the vernal equinox; reactivation was induced immediately, with births occurring 27–28 days later. A similar though not so uniform response occurred in tammars pinealectomized between April and June; when their pouch young were removed at the vernal equinox in September, half of the animals gave birth several weeks in advance of the control animals in which births occurred at the normal time after the summer solstice. The remaining pinealectomized animals also gave birth at the normal time (McConnell 1984; McConnell and Hinds 1985). These results suggest that some tammars need an intact pineal to interpret the increasing daylengths after the winter solstice in order to enter seasonal quiescence whereas others do not.

Pinealectomy in seasonal quiescence does not immediately affect the neuroendocrine reproductive axis in all tammars. Only one-third of the females pinealectomized or ganglionectomized in seasonal quiescence (July to December), gave birth in advance of sham-operated animals. Because most of the operated animals gave birth at the normal time, it would appear that an intact pineal is not necessary for initiation of the breeding season. In the following year, sham-operated, pinealectomized, and ganglionectomized animals were treated with bromocriptine after the vernal equinox, when this drug was known to be ineffective in inducing reactivation in intact females (Tyndale-Biscoe and Hinds 1984; Tyndale-Biscoe *et al.* 1986). Again one-third of the pinealectomized females and half of the ganglionectomized females, but no sham-operated females reactivated. A similar array of responses was seen when this group of animals was challenged with an artificial photoperiod change (15L:9D to 12L:12D); some operated females responded, as did all sham-operated controls, while other treated females remained quiescent. That is, under each of these manipulations some pinealectomized animals were aseasonal while others retained seasonality (McConnell 1984; Tyndale-Biscoe and Hinds, unpublished observations). The absence of a melatonin message has clearly disrupted downstream responses in the hypothalamus and pituitary in some but not all pinealectomized animals. By contrast, while melatonin implants also deprive intact animals of a changing melatonin message the reproductive response is more uniform; both tammars and Bennett's wallabies reactivate (Renfree *et al.* 1981; Loudon and Brinklow 1990).

The long-term effects of the loss of the pineal gland in the tammar have also been determined. In the 3 years after surgery, half of the pinealectomized animals retained seasonality (i.e. gave birth after the summer solstice each year), while the remaining animals became aseasonal or anoestrus. The smaller number of ganglionectomized animals all lost seasonality after the second year and about half became anoestrus (Tyndale-Biscoe, Hinds, and McConnell, unpublished observations). Again this mixed response is in contrast to the response in pinealectomized sheep which all continue to undergo periods of breeding and non-breeding activity for over 4 years (Karsch and Woodfill 1990). The responses in the tammar suggest that for some females there is an endogenous rhythm which may be synchronized by the prevailing photoperiod, probably via the secretion of melatonin. The changing pattern of melatonin acts on the neuroendocrine reproductive axis so that the breeding season begins and ends at the appropriate time. However, for other females the loss of an intact pineal gland leads to a cessation of breeding activity.

VII   CONTROL OF INITIATION AND END OF THE BREEDING SEASON

Although we now have substantial evidence regarding the mechanisms involved in the response to photoperiod change, particularly in the tammar (Fig. 7.16), the control of the beginning and end of each breeding season has not been fully resolved for either species. In photoperiodic eutherian species and birds, the beginning of the breeding season occurs when a certain 'threshold' of daylength has been attained. If the species was responding just to a particular daylength, for example 12L:12D at the vernal equinox, then one would predict that when this daylength is experienced again at the autumnal equinox that breeding would cease. However, in most species breeding ceases before the supposed 'threshold' is experienced again. This phenomenon, wherein a species no longer responds to a daylength which was once stimulatory (or inhibitory), has been termed photorefractoriness (see Nicholls et al. 1988 for review). The mechanism(s) underlying refractoriness remain largely unresolved. It could involve either a disruption in the secretion of melatonin from the pineal gland or a disruption in the post-pineal processing of the photoperiodic message (Nicholls et al. 1988). The latter mechanism seems to apply in the ewe, for example, in which the breeding season is initiated and ends because the animals become refractory to previously inhibitory (long) (Robinson et al. 1985b) or stimulatory (short) days, in the absence of a specific change in the pattern of melatonin secretion (Robinson and Karsch 1984).

In the Bennett's wallaby, the evidence indicates that entry into seasonal quiescence does not involve development of refractoriness although initiation of the next breeding season does appear to be due to refractoriness. Merchant and Calaby (1981) observed that most female Bennett's wallabies were in seasonal quiescence shortly after the winter solstice. In later studies, in which progesterone profiles were determined, Curlewis *et al.* (1987) found that lactational quiescence extended into August for a minority of animals, but that the majority of the females were in seasonal quiescence within 4–6 weeks after the winter solstice. What triggers the entry into seasonal quiescence? Loudon and Curlewis (1989) have experimentally determined whether development of refractoriness is a factor. Using the experimental protocol of Robinson and Karsch (1984), Bennett's wallabies were maintained on winter solsticial daylength, natural increasing daylengths or simulated natural increasing daylengths. The breeding season was extended by several months in females retained on winter solsticial daylengths, but animals experiencing natural or simulated natural photoperiod ceased to show periods of elevated progesterone indicative of oestrous cycles shortly after the winter solstice (Loudon and Curlewis 1989). Animals implanted with melatonin before the winter solstice also did not become seasonally quiescent but continued to undergo successive reproductive cycles (Loudon and Brinklow 1990); that is, in the absence of a change in the pattern of melatonin secretion they remained reproductively active. These results indicate that Bennett's wallabies do not become refractory to short days but must experience increasing daylengths after the winter solstice in order to enter seasonal quiescence.

The end of lactational quiescence and entry into seasonal quiescence has not been assessed experimentally for the tammar. However, examination of the animal breeding records for the captive colony in Canberra (1974–91) show that the majority of tammars do not reactivate after removal of pouch young in May, several weeks before the winter solstice. In March and April 80% of animals give birth after removal of pouch young. Because only 80% of females conceive at a post-partum oestrus this figure represents 100% reactivation. However, in May only 27% of animals give birth after removal of pouch young, in June 22.5%, in July 6%, and no births were recorded for the remainder of the year (September to December). These results could be interpreted as development of refractoriness to decreasing daylengths in May and June, but there has been no experimental examination of this possibility. However, if they do become refractory to decreasing daylengths then it is difficult to explain why female tammars continue to respond to injections of bromocriptine between the winter solstice and the vernal equinox.

Tammars and Bennett's wallabies show differences in the timing of entry into seasonal quiescence, and this may reflect different mechanisms. By contrast, they may use the same mechanism to initiate reactivation after the summer solstice because both species may become refractory to long daylengths (Sadleir and Tyndale-Biscoe 1977; Curlewis and Loudon 1989). If Bennett's wallabies were held on summer solsticial daylength beginning 1 week before the summer solstice, they commenced breeding at the same time as control animals held on natural photoperiods. Thus the animals do not need to experience decreasing daylengths to initiate the onset of the season, i.e. they become refractory to long days. However, when animals were placed on summer solsticial daylength in early seasonal quiescence, they gave birth approximately 1 month in advance of control females experiencing natural photoperiod. Curlewis and Loudon (1989) interpreted this in two ways. If, as in the starling, the development of refractoriness to long days is proportional to the absolute daylength (Dawson and Goldsmith 1983), then early exposure to long days would advance the onset of the breeding season. Alternatively, if winter solstice daylengths are seen as short days then the early experience of long days may have accelerated the onset of refractoriness. However, these interpretations must be viewed with some caution for two reasons. First, when tammars experienced an accelerated increase in photoperiod such that the summer solstice was advanced by about 5 weeks, births occurred at the normal time (Hearn 1972*b*). Second, the results of Sadleir and Tyndale-Biscoe (1977) suggest that the difference in the responses could be because the control Bennett's wallabies were not maintained on a simulated natural photoperiod. When tammars experienced summer solsticial daylength from the vernal equinox until after the summer solstice, eight of 12 females gave birth or re-ovulated at the same time as control females held on simulated natural photoperiod, although this occurred approximately 27 days in advance of females under natural photoperiod (Sadleir and Tyndale-Biscoe 1977). Thus while in both species a decrease in daylength may not be necessary for the initiation of the breeding season, none of the present studies satisfactorily addresses the possibility that the rate of development of photorefractoriness occurs in proportion to absolute daylength (Nicholls *et al.* 1988). Another interpretation of the above results is that the animals show an endogenous rhythm of reproduction, which is normally synchronized by the prevailing photoperiod.

VIII   CONCLUSIONS

For most eutherian species regulation of seasonal breeding involves responses to photoperiod change via the pineal gland, hypothalamus, and pituitary, which results in an intimate interplay between melatonin, the GnRH pulse generator, and the gonadotrophins. The catecholamines and opioids also interact in this system. Subsequently the gonads become involved. For the two macropodid species, the tammar and Bennett's wallaby, showing seasonal quiescence of the CL, the same axis is involved in responses to photoperiod, but the interaction of melatonin via the hypothalamus affects pituitary release of prolactin, rather than gonadotrophin secretion. The hypothalamic–hypophyseal–follicular axis remains potentially active throughout the year, its function being blocked by the presence of the quiescent CL, which is itself inhibited by the pituitary and its release of prolactin.

The basic pattern of opportunistic breeding seen in the red kangaroo has been adapted by the tammar and Bennett's wallabies for seasonal breeding in the more southern areas of Australia. The use of prolactin during seasonal quiescence represents a logical extension of its use during lactational quiescence. Since both island and mainland tammars are obligate seasonal breeders, seasonality probably evolved at an earlier time than in Bennett's wallabies, which have only been separated from mainland Australia for about 11 000 years. Thus some differences in emphasis or sensitivity to prolactin might be expected, reflecting the evolution of slightly different regulatory mechanisms. Many members of the various Macropodid and Potoroid sub-Families have developed morphologically identical or closely similar adaptations, perhaps in response to similar environmental challenges or opportunities (Flannery 1989); the mechanisms of seasonality described here probably represent another example of convergent evolution in the kangaroos.

Can we now propose a model which describes the regulation of the annual reproductive cycles of the tammar and Bennett's wallabies? Together the results for lactational and seasonal quiescence in the tammar suggest that transient pulses of prolactin seen at least once in 72 h maintain quiescence. The response to photoperiod manipulations and to exogenous melatonin treatment varies depending on the stage of the annual cycle at which these are imposed. Thus the transition from one state to the next could involve a change in the frequency of these pulses, this frequency change could reflect changes in the sensitivity of the pituitary lactotrophs to the release of dopamine from the hypothalamus. During lactational quiescence, when daylengths are decreasing, the sucking stimulus predominates and no other factor appears to induce prolactin release. However, on increasing daylengths, daily pulses of prolactin are secreted due to changes in the pattern of secretion of melatonin. These pulses override any decreases in pulse frequency that would be expected as lactation ceases after the vernal equinox. Between the vernal equinox and the summer solstice the prolactin pulse is driven solely by a photoperiodic factor which supplants the dopamine-sensitive component of pituitary prolactin release in most females. At or near the summer solstice, the frequency or amplitude of the prolactin pulses decreases until the critical threshold of 72 h is passed at which time reactivation occurs. At the onset of the breeding season, this change in pulse frequency could be induced by neural mechanisms in the hypothalamus (i.e. development of refractoriness) or to a reduced sensitivity of the luteal cells to these pulses. If prolactin pulse frequency is affected by hypothalamic input under the influence of changing photoperiod and the pattern of melatonin secretion, the male tammar, which is non-seasonal, could be used to investigate such an hypothesis. Under inhibitory photoperiods (15L:9D) males show a morning prolactin pulse which is lost after a change to 12L:12D. It might be predicted that males would not show a peak of prolactin between the summer solstice and the winter solstice, as daylengths were decreasing, but would do so between the winter and summer solstices when daylengths were increasing.

In female Bennett's wallabies the development of similar changes in neural mechanisms within the hypothalamus in terms of dopamine release or sensitivity of the pituitary to this release may occur at the summer solstice and provide the trigger for reactivation. At the end of the breeding season, coupled with the changes in melatonin as the daylength increases, the sensitivity of the dopaminergic system is also altered to maintain inhibition of the CL via prolactin.

It only remains to be said that the study of the endocrine regulation of breeding in marsupials provides alternative insights into the various and complex mechanisms used by different species to achieve the same goal — successful production and survival of another generation. For the marsupials, the adoption of a reproductive strategy which emphasizes lactation rather than pregnancy, has meant that different uses for the hormone prolactin have evolved. For the tammar and Bennett's wallabies this includes the exploitation of prolactin in the regulation of the annual breeding cycle. Furthermore, because the response of both species to photoperiod manipulation is so rapid they offer special opportunities for investigation of the hypothalamic centre that is computing the melatonin signal. This neural centre must store the information and compare it with prior information and then act on it. In the tammar the centre must be active within 3–4 days after a stimulatory photoperiod change or melatonin treatment, so finding and defining the site should be more feasible than for any eutherian species in which interpretation of the signal occurs over a longer period of several weeks.

ACKNOWLEDGEMENTS

I thank Drs Hugh Tyndale-Biscoe, Peter Janssens, and John Clarke for their comments and constructive criticisms of the manuscript.

REFERENCES

Andrewartha, H.G. and Barker, S. (1969). Introduction to a study of the ecology of the Kangaroo Island wallaby, *Protemnodon eugenii* (Desmarest) within

Flinders Chase, Kangaroo Island, South Australia. *Transactions of the Royal Society of South Australia*, **93**, 127–32.

Arendt, J. (1986). Role of the pineal gland and melatonin in seasonal reproductive function in mammals. In *Oxford Reviews of Reproductive Biology*, Vol. 8 (ed. J.R. Clarke), pp. 266–320. Clarendon Press, Oxford.

Bartke, A. and Steger, R. W. (1992). Seasonal changes in the function of the hypothalamic–pituitary–testicular axis in the Syrian hamster. *Proceedings of the Society for Experimental Biology and Medicine*, **199**, 139–48.

Berger, P.J. (1966). Eleven-month 'Embryonic Diapause' in a marsupial. *Nature*, **211**, 435–6.

Berger, P.J. (1970). Reproductive biology of the tammar wallaby, *Macropus eugenii*. Ph.D. thesis, Tulane University USA., *Dissertation Abstracts International*, **31B**, 3760–1.

Berger, P.J. and Sharman, G.B. (1969). Progesterone-induced development of dormant blastocysts in the tammar wallaby, *Macropus eugenii* Desmarest: Marsupialia. *Journal of Reproduction and Fertility*, **20**, 201–10.

Bittman, E.L. and Karsch, F.J. (1984). Nightly duration of pineal melatonin secretion determines the reproductive response to inhibitory day length in the ewe. *Biology of Reproduction*, **30**, 585–93.

Bittman, E.L., Dempsey, R.J., and Karsch, F.J. (1983). Pineal melatonin secretion drives the reproductive response to daylength in the ewe. *Endocrinology*, **113**, 2276–83.

Brinklow, B.R. and Loudon, A.S.I. (1989). Effect of exogenous prolactin and bromocriptine on seasonal reproductive quiescence in the Bennett's wallaby (*Macropus rufogriseus rufogriseus*). *Journal of Endocrinology*, **120**, 189–93.

Calaby, J.H. (1971). The current status of Australian Macropodidae. *Australian Journal of Zoology*, **16**, 17–29.

Carter, D.S. and Goldman, B.G. (1983). Antigonadal effects of timed melatonin infusion in pinealectomized male Djungarian hamsters (*Phodopus sungorus sungorus*): duration is the critical parameter. *Endocrinology*, **113**, 1261–7.

Carter, D.S., Hall, V.D., Tamarkin, L., and Goldman, B.D. (1982). Pineal is required for testicular maintenance in the Turkish hamster (*Mesocricetus brandti*). *Endocrinology*, **111**, 863–71.

Catling, P.C. and Sutherland, R.L. (1980). Effect of gonadectomy, season, and the presence of females on plasma testosterone, luteinizing hormone, and follicle stimulating hormone levels in male tammar wallabies (*Macropus eugenii*). *Journal of Endocrinology*, **86**, 25–33.

Catt, D.C. (1977). The breeding biology of Bennett's wallaby (*Macropus rufogriseus fruticus*) in South Canterbury, New Zealand. *New Zealand Journal of Zoology*, **4**, 401–11.

Curlewis, J.D. (1989). The breeding season of Bennett's wallaby (*Macropus rufogriseus rufogriseus*) in Tasmania. *Journal of Zoology*, **218**, 337–9.

Curlewis, J.D. and Loudon, A.S.I. (1988a). Experimental manipulations of prolactin following removal of pouch young or bromocriptine treatment during lactational quiescence in the Bennett's wallaby. *Journal of Endocrinology*, **119**, 405–11.

Curlewis, J.D. and Loudon, A.S.I. (1988b). Effects of photoperiod on the 24-hour melatonin profiles of the Bennett's wallaby (*Macropus rufogriseus rufogriseus*). *Journal of Pineal Research*, **5**, 373–83.

Curlewis, J.D. and Loudon, A.S.I. (1989). The role of refractoriness to long daylength in the annual reproductive cycle of the female Bennett's wallaby (*Macropus rufogriseus rufogriseus*). *Journal of Experimental Zoology*, **252**, 200–6.

Curlewis, J.D., White, A.S., Loudon, A.S.I., and McNeilly, A.S. (1986). Effects of lactation and season on plasma prolactin concentrations and response to bromocriptine during lactation in the Bennett's wallaby (*Macropus rufogriseus rufogriseus*). *Journal of Endocrinology*, **110**, 59–66.

Curlewis, J.D., White, A.S., and Loudon, A.S.I. (1987). The onset of seasonal quiescence in the female Bennett's wallaby (*Macropus rufogriseus rufogriseus*). *Journal of Reproduction and Fertility*, **80**, 119–24.

Dawson, A. and Goldsmith, A.R. (1983). Plasma prolactin and gonadotrophins during gonadal development and the onset of photorefractoriness in male and female starlings (*Sturnus vulgaris*) on artificial photoperiods. *Journal of Endocrinology*, **97**, 253–60.

Dressen, W., Grun, H., and Hendrichs, H. (1990). Radio telemetry of heart rate in male tammar wallabies (Marsupialia:Macropodidae): temporal variations and behavioural correlates. *Australian Journal of Zoology*, **38**, 89–103.

Evans, S.M., Tyndale-Biscoe, C.H., and Sutherland, R.L. (1980). Control of gonadotrophin secretion in the female tammar wallaby (*Macropus eugenii*). *Journal of Endocrinology*, **86**, 13–23.

Fadem, B.H., Trupin, G.L., Maliniak, E., VandeBerg, J.L., and Hayssen, V. (1982). Care and breeding of the gray, short-tailed opossum (*Monodelphis domestica*). *Laboratory Animal Science*, **32**, 405–9.

Flannery, T.G. (1989). Phylogeny of the Macropodoidea; a study in convergence. In *Kangaroos, wallabies and rat-kangaroos* (ed. G. Grigg, P. Jarman, and I. Hume), pp. 1–46. Surrey Beatty and Sons Pty Ltd., Sydney, New South Wales.

Fleming, D., Cinderey, R.N., and Hearn, J.P. (1983). The reproductive biology of Bennett's wallaby (*Macropus rufogriseus rufogriseus*) ranging free at Whipsnade Park. *Journal of Zoology*, **201**, 283–91.

Frith, H.J. and Sharman, G.B. (1964). Breeding in wild populations of the red kangaroo, *Megaleia rufa*. *CSIRO Wildlife Research*, **9**, 86–114.

Gemmell, R.T. (1987). Effect of melatonin and removal of pouch young on the seasonality of births in the marsupial possum, *Trichosurus vulpecula*. *Journal of Reproduction and Fertility*, **80**, 301–7.

Gemmell, R.T. (1990). Influence of daylength on the initiation of the breeding season of the marsupial possum, *Trichosurus vulpecula*. *Journal of Reproduction and Fertility*, **88**, 605–9.

Godfrey, G.K. (1969). The influence of increased photoperiod on reproduction in the dasyurid marsupial, *Sminthopsis crassicaudata*. *Journal of Mammalogy*, **50**, 132–3.

Goldman, B.D. and Darrow, J.M. (1983). The pineal gland and mammalain photoperiodism. *Neuroendocrinology*, **37**, 386–96.

Goldman, B.D. and Elliot, J.A. (1988). Photoperiodism and seasonality in hamsters: role of the pineal gland. In *Processing of environmental information in vertebrates*. (ed. M.H. Stetson), pp. 203–18. Springer-Verlag, Berlin.

Goodman, R.L. (1988). Neuroendocrine mechanisms mediating the photoperiodic control of reproductive function in sheep. In *Processing of environmental information in vertebrates*, (ed. M.H. Stetson), pp. 179–202. Springer-Verlag, Berlin.

Gordon, K., Fletcher, T.P., and Renfree, M.B. (1988). Reactivation of the quiescent corpus luteum and diapausing embryo after temporary removal of the sucking stimulus in the tammar wallaby (*Macropus eugenii*). *Journal of Reproduction and Fertility*, **83**, 401–6.

Harder, J.D., Hinds, L.A., Horn, C.A., and Tyndale-Biscoe, C.H. (1984). Oestradiol in follicular fluid in utero-ovarian venous and peripheral plasma during parturition and post-partum oestrus in the tammar, *Macropus eugenii*. *Journal of Reproduction and Fertility*, **72**, 551–8.

Harder, J.D., Hinds, L.A., Horn, C.A., and Tyndale-Biscoe, C.H. (1985). Effects of removal in late pregnancy of the corpus luteum, Graafian follicle or ovaries on plasma progesterone, oestradiol, LH, parturition and post-partum oestrus in the tammar wallaby, *Macropus eugenii. Journal of Reproduction and Fertility*, **75**, 449–59.

Hartman, C.G. (1923). The oestrous cycle in the opossum. *American Journal of Anatomy*, **32**, 353–421.

Hearn, J.P. (1972a). The pituitary gland and reproduction in the marsupial, *Macropus eugenii* (Desmarest). Ph.D. thesis. Australian National University, Canberra, ACT, Australia.

Hearn, J.P. (1972b). Effect of advanced photoperiod on termination of embryonic diapause in the marsupial *Macropus eugenii*, (Macropodidae). *Australian Mammalogy*, **1**, 40–2.

Hearn, J.P. (1973). Pituitary inhibition of pregnancy. *Nature*, **241**, 207–8.

Hearn, J.P. (1974). The pituitary gland and implantation in the tammar wallaby, *Macropus eugenii. Journal of Reproduction and Fertility*, **39**, 235–41.

Hinds, L.A. (1983). Progesterone and prolactin in marsupial reproduction. Ph.D. thesis. Australian National University, Canberra, ACT, Australia.

Hinds, L.A. (1989a). Plasma progesterone through pregnancy and the oestrous cycle in the eastern quoll, *Dasyurus viverrinus. General and Comparative Endocrinology*, **75**, 110–17.

Hinds, L.A. (1989b). Morning prolactin pulse maintains seasonal quiescence in the tammar wallaby, *Macropus eugenii. Journal of Reproduction and Fertility*, **87**, 735–44.

Hinds, L.A. (1990). The control of pregnancy, parturition and luteolysis in marsupials. *Reproduction, Fertility and Development*, **2**, 535–52.

Hinds, L.A. and den Ottolander, R.C. (1983). Effects of changing photoperiod on peripheral plasma prolactin and progesterone concentrations in the tammar wallaby, *Macropus eugenii. Journal of Reproduction and Fertility*, **69**, 631–9.

Hinds, L.A. and Smith, M.J. (1992). Evidence from plasma progesterone concentrations for male-induced ovulation in an Australian marsupial, the brushtailed bettong, *Bettongia penicillata. Journal of Reproduction and Fertility*, **95**, 291–302.

Hinds, L.A. and Tyndale-Biscoe, C.H. (1982a). Plasma progesterone levels in the pregnant and non-pregnant tammar, *Macropus eugenii. Journal of Endocrinology*, **93**, 99–107.

Hinds, L.A. and Tyndale-Biscoe, C.H. (1982*b*). Prolactin in the marsupial, *Macropus eugenii*, during the estrous cycle, pregnancy and lactation. *Biology of Reproduction*, **26**, 391–8.

Hinds, L.A. and Tyndale-Biscoe, C.H. (1985). Seasonal and circadian patterns of circulating prolactin during lactation and seasonal quiescence in the tammar, *Macropus eugenii*. *Journal of Reproduction and Fertility*, **74**, 173–83.

Hinds, L.A., Tyndale-Biscoe, C.H., Shaw, G., Fletcher, T.P., and Renfree, M.B. (1990). Effects of prostaglandin and prolactin on luteolysis and parturient behaviour in the non-pregnant tammar, *Macropus eugenii*. *Journal of Reproduction and Fertility*, **88**, 323–33.

Hinds, L.A., Diggle, P.J., and Tyndale-Biscoe, C.H. (1992). Effects of the ovary, sucking stimulus and season on the pattern of LH and FSH release in the female tammar, *Macropus eugenii*. *Reproduction, Fertility and Development*, **4**, 303–12.

Hoffman, R.A., and Reiter, R.J. (1965). Pineal gland: influence on gonads of male hamsters. *Science*, **148**, 1609–10.

Horn, C.A., Fletcher, T.P., and Carpenter, S. (1985). Effects of oestradiol-17β on peripheral plasma concentrations of LH and FSH in ovariectomized tammars (*Macropus eugenii*). *Journal of Reproduction and Fertility*, **73**, 585–92.

Inns, R.W. (1982). Seasonal changes in the accessory reproductive system and plasma testosterone levels of the male tammar wallaby, *Macropus eugenii*, in the wild. *Journal of Reproduction and Fertility*, **66**, 675–80.

Karp, J.D., Hastings, M.H., and Powers, J.B. (1991). Melatonin and the coding of daylength in male Syrian hamsters. *Journal of Pineal Research*, **10**, 210–17.

Karsch, F.J. and Woodfill, C.J.I. (1990). Neuroendocrinology of seasonal breeding: mode of action of melatonin. In *Neuroendocrine regulation of reproduction* (ed. S.S.C. Yen and W.W. Vale), pp. 9–17. Serono Symposia, Norwell, MA.

Karsch, F.J., Bittman, E.L., Foster, D.L. Goodman, R.L., Legan, S.J., and Robinson, J.E. (1984). Neuroendocrine basis of seasonal reproduction. *Recent Progress In Hormone Research*, **40**, 185–232.

Karsch, F.J., Robinson, J.E., Woodfill, C.J.I., and Brown, M.B. (1989). Circannual cycles of luteinizing hormone and prolactin secretion in ewes during prolonged exposure to a fixed photoperiod: evidence for an endogenous reproductive rhythm. *Biology of Reproduction*, **41**, 1034–46.

Kojima, T., Hinds, L.A., Muller, W.J, O'Neill, C., and Tyndale-Biscoe, C. H. (1993). *In vitro* production and secretion of progesterone and presence of platelet activating factor (PAF) in early pregnancy of the marsupial, *Macropus eugenii*. *Reproduction, Fertility and Development*, **5**, 15–25.

Lee, A.K. and Cockburn, A. (1985). *The evolutionary ecology of marsupials*. Cambridge University Press, Cambridge.

Legan, S.J., Karsch, F.J., and Foster, D.L. (1977). The endocrine control of seasonal reproductive function in the ewe: a marked change in response to the negative feedback action of estradiol on luteinizing hormone secretion. *Endocrinology*, **101**, 818–24.

Lincoln, G.A. and Short, R.V. (1980). Seasonal breeding: Nature's contraceptive. *Recent Progress in Hormone Research*, **36**, 1–52.

Lincoln, G.A., Almeida, O.F.X., and Arendt, J. (1981). Role of melatonin and circadian rhythms in seasonal reproduction in rams. *Journal of Reproduction and Fertility*, **30**, (Suppl.) 23–31.

Loudon, A.S.I. and Brinklow, B.R. (1990). Melatonin implants prevent the onset of seasonal quiescence and suppress the release of prolactin in response to a dopamine antagonist in the Bennett's wallaby (*Macropus rufogriseus rufogriseus*). *Journal of Reproduction and Fertility*, **90**, 611–18.

Loudon, A.S.I. and Curlewis, J.D. (1987). Refractoriness to melatonin and short daylengths in early seasonal quiescence in the Bennett's wallaby (*Macropus rufogriseus rufogriseus*). *Journal of Reproduction and Fertility*, **81**, 543–52.

Loudon, A.S.I. and Curlewis, J.D. (1989). Evidence that the seasonally breeding Bennett's wallaby (*Macropus rufogriseus rufogriseus*) does not exhibit short-day photorefractoriness. *Journal of Reproduction and Fertility*, **87**, 641–8.

Loudon, A.S.I., Curlewis, J.D., and English, J. (1985). The effect of melatonin on the seasonal embryonic diapause of the Bennett's wallaby (*Macropus rufogriseus rufogriseus*). *Journal of Zoology*, **206**, 35–9.

McConnell, S.J. (1984). The pineal gland and reproduction in the tammar wallaby, *Macropus eugenii*. Ph.D. thesis. Australian National University, Canberra, ACT, Australia.

McConnell, S.J. (1986). Seasonal changes in the circadian plasma profile of the tammar, *Macropus eugenii*. *Journal of Pineal Research*, **3**, 119–25.

McConnell, S.J. and Hinds, L.A. (1985). Effect of pinealectomy on plasma melatonin, prolactin and progesterone concentrations during seasonal reproductive quiescence in the tammar, *Macropus eugenii*. *Journal of Reproduction and Fertility*, **7**, 433–40.

McConnell, S.J. and Tyndale-Biscoe, C.H. (1985). Response in peripheral plasma melatonin to photoperiod change and the effects of exogenous melatonin on seasonal quiescence in the tammar, *Macropus eugenii*. *Journal of Reproduction and Fertility*, **73**, 529–38.

McConnell, S.J., Tyndale-Biscoe, C.H., and Hinds, L.A. (1986). Change in duration of elevated concentrations of melatonin is the major factor in photoperiod response of the tammar, *Macropus eugenii*. *Journal of Reproduction and Fertility*, **77**, 623–32.

McNeilly, A.S. and Land, R.B. (1979). Effect of suppression of plasma prolactin on ovulation, plasma gonadotrophins and corpus luteum function in LH-RH-treated anoestrous ewes. *Journal of Reproduction and Fertility*, **56**, 601–9.

Margolis, D.J. and Lynch, G.R. (1981). Effects of daily melatonin injections on reproduction in the white-footed mouse, *Peromyscus leucopus*. *General and Comparative Endocrinology*, **44**, 530–7.

Merchant, J.C. (1979). The effect of pregnancy on the interval between one oestrus and the next in the tammar wallaby, *Macropus eugenii*. *Journal of Reproduction and Fertility*, **56**, 459–63.

Merchant, J.C. and Calaby, J.H. (1981). Reproductive biology of the red-necked wallaby (*Macropus rufogriseus banksianus*) and Bennett's wallaby (*M.r. rufogriseus*) in captivity. *Journal of Zoology*, **194**, 203–17.

Moore, G.P.M. (1978). Embryonic diapause in the marsupial *Macropus eugenii*. Stimulation of nuclear RNA polymerase activity in the blastocyst during resumption of development. *Journal of Cellular Physiology*, **94**, 31–6.

Newsome, A.E. (1964a). Anoestrus in the red kangaroo *Megaleia rufa* (Desmarest). *Australian Journal of Zoology*, **12**, 9–17.

Newsome, A.E. (1964b). Oestrus in the lactating red kangaroo, *Megaleia rufa* (Desmarest), in Central Australia. *Australian Journal of Zoology*, **13**, 735–59.

Nicholls, T.J., Goldsmith, A.R., and Dawson, A. (1988). Photorefractoriness in birds and comparison with mammals. *Physiological Reviews*, **68**, 133–76.

Panyaniti, W., Carpenter, S.M., and Tyndale-Biscoe, C.H. (1985). Effects of hypophysectomy on folliculogenesis in the tammar wallaby *Macropus eugenii* (Marsupialia: Macropodidae). *Australian Journal of Zoology*, **33**, 303–11.

Pearce-Kelly, A.S., Loudon, A.S.I., and Curlewis, J.D. (1992). Seasonal and lactational effects on the prolactin response to a dopamine antagonist and TRH in the Bennett's wallaby (*Macropus rufogriseus rufogriseus*). *General and Comparative Endocrinology*, **86**, 323–31.

Pilton, P.E. and Sharman, G.B. (1962). Reproduction in the marsupial *Trichosurus vulpecula*. *Journal of Endocrinology*, **25**, 119–36.

Pitrosky, B., Massonpevet, M., Kirsch, R., Vivienroels, B., Canguilhem, B., and Pevet, P. (1991). Effects of different doses and durations of melatonin infusions on plasma melatonin concentrations in pinealectomized Syrian hamsters — consequences at the level of sexual activity. *Journal of Pineal Research*, **11**, 149–55.

Poole, W.E. (1973). A study of breeding in grey kangaroos, *Macropus giganteus* Shaw, *M. fuliginosus* (Desmarest), in central New South Wales. *Australian Journal of Zoology*, **23**, 333–53.

Poole, W.E. (1975). Reproduction in the two species of grey kangaroos, *Macropus giganteus* Shaw, and *M. fuliginosus* (Desmarest). II. Gestation, parturition and pouch life. *Australian Journal of Zoology*, **23**, 333–53.

Poole, W.E. (1976). Breeding biology and current status of the grey kangaroos, *Macropus fuliginosus fuliginosus*, of Kangaroo Island, South Australia. *Australian Journal of Zoology*, **23**, 169–87.

Poole, W.E. and Catling, P.C. (1974). Reproduction in the two species of grey kangaroo, *Macropus giganteus* Shaw and *M. fuliginosus* (Desmarest). I. Sexual maturity and oestrus. *Australian Journal of Zoology*, **22**, 277–302.

Poole, W.E., Wood, J.T., and Simms, N.G. (1991). Distribution of the tammar *Macropus eugenii*, and the relationships of populations as determined by cranial morphometrics. *Wildlife Research*, **18**, 625–39.

Prince, R.I.T. (1983). Banded hare wallaby, *Lagostrophus fasciatus*. In *The Australian Museum complete book of Australian mammals* (ed. R. Strahan), pp. 201–2. Angus and Robertson, Sydney.

Reiter, R.J. (1987). The melatonin message: duration versus coincidence hypotheses. *Life Sciences*, **46**, 2119–31.

Reiter, R.J. (1990). Mechanisms of reproductive regulation by the pineal gland. In *Neuroendocrine regulation of reproduction* (ed. S.S.C. Yen and W.W. Vale), pp. 105–11. Serono Symposia, Norwell, MA.

Reiter, R.J., Blask, D.E., Johnson, L.Y., Rudeen, P.K., Vaughan, M.K., and Waring, P.J. (1976). Melatonin inhibition of reproduction in the male hamster: its dependency on time of day of administration and on intact and sympathetically innervated pineal gland. *Neuroendocrinology*, **22**, 107–16.

298   Lyn A. Hinds

Renfree, M.B. (1979). Initiation of development of diapausing embryo by mammary denervation during lactation in a marsupial. *Nature*, **278**, 549–51.

Renfree, M.B. (1983). Marsupial reproduction: the choice between placentation and lactation. *Oxford Reviews of Reproductive Biology*, Vol 5. (ed. J.R. Clarke), pp. 1–29. Clarendon Press, Oxford.

Renfree, M.B., Lincoln, D.W., Almeida, O.F.X., and Short, R.V. (1981). Abolition of seasonal embryonic diapause in a wallaby by pineal denervation. *Nature*, **293**, 138–9.

Renfree, M.B., Wallace, G.I., and Young, I.R. (1982). Effects of progesterone, oestradiol-17β and androstenedione on follicular growth after removal of the corpus luteum during lactational and seasonal quiescence in the tammar wallaby. *Journal of Endocrinology*, **92**, 397–403.

Renfree, M.B., Flint, A.P.F., Green, S.W., and Heap, R.B. (1984). Ovarian steroid metabolism and luteal oestrogens in the corpus luteum of the tammar wallaby. *Journal of Endocrinology*, **101**, 231–40.

Renfree, M.B., Fletcher, T.P., Blanden, D.R., Lewis, P.R., Shaw, G., Gordon, K. *et al.* (1989). Physiological and behavioural events around the time of birth in macropodid marsupials. In *Kangaroos, wallabies and rat-kangaroos* (ed. G. Grigg, P. Jarman, and I. Hume), pp. 323–37. Surrey Beatty and Sons Pty Ltd, Sydney, New South Wales.

Robinson, J.E. and Karsch, F.J. (1984). Refractoriness to inductive day lengths terminates the breeding season of the Suffolk ewe. *Biology of Reproduction*, **31**, 656–63.

Robinson, J.E. and Karsch, F.J. (1987). Photoperiodic history and a changing melatonin pattern can determine the neuroendocrine response of the ewe to daylength. *Journal of Reproduction and Fertility*, **80**, 159–65.

Robinson, J.E. and Karsch, F.J. (1988). Timing the breeding season of the ewe: what is the role of daylength? *Reproduction Nutrition and Development*, **28**, 365–74.

Robinson, J.E., Radford, H.M., and Karsch, F.J. (1985a). Seasonal changes in pulsatile luteinizing hormone (LH) secretion in the ewe: relationship of frequency of LH pulses to day length and response to estradiol negative feedback. *Biology of Reproduction*, **33**, 324–34.

Robinson, J.E., Wayne, N.L., and Karsch, F.J. (1985b). Refractoriness to inhibitory day lengths initiates the breeding season of the Suffolk ewe. *Biology of Reproduction*, **32**, 1024–30.

Rodger, J.C., Cousins, S.J., Mate, K.E., and Hinds, L.A. (1993). Ovarian function and its manipulation in the tammar wallaby, *Macropus eugenii*. *Reproduction, Fertility and Development*, **5**, 27–38.

Russell, E.M. (1982). Patterns of parental care and parental investment in marsupials. *Biological Reviews*, **57**, 423–86.

Sadleir, R.M.F.S. and Tyndale-Biscoe, C.H. (1977). Photoperiod and the termination of embryonic diapause in the marsupial *Macropus eugenii*. *Biology of Reproduction*, **16**, 605–8.

Sharman, G.B. (1954). Reproduction in marsupials. *Nature*, **173**, 302–3.

Sharman, G.B. (1955a). Studies on marsupial reproduction. II. The oestrous cycle of *Setonix brachyurus*. *Australian Journal of Zoology*, **3**, 44–55.

Sharman, G.B. (1955b). Studies on marsupial reproduction. III. Normal and delayed pregnancy in *Setonix brachyurus*. *Australian Journal of Zoology*, **3**, 56–70.

Sharman, G.B. (1955c). Studies on marsupial reproduction. IV. Delayed birth in *Protemnodon eugenii*. *Australian Journal of Zoology*, **3**, 156–61.

Sharman, G.B. (1965). The effects of the suckling stimulus and oxytocin injection on the corpus luteum of delayed implantation in the red kangaroo. *Excerpta Medica*, **83**, 669–74.

Sharman, G.B. and Berger, P.J. (1969). Embryonic diapause in marsupials. *Advances in Reproductive Physiology*, **4**, 211–40.

Sharman, G.B. and Clark, M.J. (1967). Inhibition of ovulation by the corpus luteum in the red kangaroo, *Megaleia rufa*. *Journal of Reproduction and Fertility*, **14**, 129–37.

Sharman, G.B., Calaby, J.H., and Poole, W.E. (1966). Patterns of reproduction in female diprotodont marsupials. *Symposium of the Zoological Society of London*, **15**, 205–32.

Shaw, G. and Renfree, M.B. (1984). Concentrations of oestradiol-17β in plasma and corpora lutea throughout pregnancy in the tammar, *Macropus eugenii*. *Journal of Reproduction and Fertility*, **72**, 29–37.

Shield, J.W. and Woolley, P. (1963). Population aspects of delayed birth in the quokka (*Setonix brachyurus*). *Proceedings of the Zoological Society of London*, **141**, 783–90.

Short, R.V., Flint, A.P.F., and Renfree, M.B. (1985). Influence of passive immunization against GnRH on pregnancy and parturition in the tammar wallaby, *Macropus eugenii*. *Journal of Reproduction and Fertility*, **75**, 567–75.

Smith, M.J. (1992). Evidence from the oestrous cycle for male-induced ovulation in the brushtailed bettong, *Bettongia penicillata* (Marsupialia: Potoroidae). *Journal of Reproduction and Fertility*, **95**, 283–9.

Smith, P.J., Bennett, H.J., and Chesson, C.M. (1978). Photoperiod and some factors affecting reproduction of female *Sminthopsis' crassicaudata* (Gould) (Marsupialia: Dasyuridae). *Australian Journal of Zoology*, **26**, 449–63.

Stewart, F. and Tyndale-Biscoe, C.H. (1982). Prolactin and luteinizing hormone receptors in marsupial corpora lutea: relationship to control of luteal function. *Journal of Endocrinology*, **92**, 63–72.

Strahan, R. (ed.) (1983). *The Australian Museum complete book of Australian mammals*. Angus and Robertson, Sydney.

Sutherland, R.L., Evans, S.M., and Tyndale-Biscoe, C.H. (1980). Macropodid marsupial luteinizing hormone: validation of assay procedures and changes in plasma levels during the oestrous cycle in the female tammar wallaby (*Macropus eugenii*). *Journal of Endocrinology*, **86**, 1–12.

Tamarkin, L., Westrom, W.K. Hamill, A.I., and Goldman, B.D. (1976). Effect of melatonin on the reproductive systems of male and female Syrian hamsters: a diurnal rhythm in sensitivity to melatonin. *Endocrinology*, **99**, 1534–41.

Tamarkin, L., Baird, C.J., and Almeida, O.F.X. (1985). Melatonin: a coordinating signal for mammalian reproduction? *Science*, **227**, 714–20.

Tyndale-Biscoe, C.H. (1965). The female urogenital system and reproduction of the marsupial *Bettongia lesueur* (Quoy and Gaimard). *Australian Journal of Zoology*, **13,**, 255–67.

Tyndale-Biscoe, C.H. (1980). Photoperiod and the control of seasonal reproduction in marsupials. In *Proceedings of the VI International Congress of Endocrinology* (ed. I.A. Cumming, J.W. Funder, and F.A.O. Mendelsohn), pp. 277–82. Australian Academy of Science, Canberra, ACT.

Tyndale-Biscoe, C.H. (1986). Embryonic diapause in a marsupial: roles of the corpus luteum and pituitary in its control. In *Comparative endocrinology: developments and directions* (ed. C.L. Ralph), pp. 137–55. Alan R. Liss New York.

Tyndale-Biscoe, C.H. (1989). The adaptiveness of reproductive processes. In *Kangaroos, wallabies and rat-kangaroos* (ed. G. Grigg, P. Jarman, and I. Hume), pp. 277–85. Surrey Beatty and Sons Pty Ltd, Sydney, New South Wales.

Tyndale-Biscoe, C.H. and Hawkins, J. (1977). The corpora lutea of marsupials, aspects of function and control. In *Reproduction and evolution* (ed. J.H. Calaby and C.H. Tyndale-Biscoe), pp. 245–52. Australian Academy of Science, Canberra, ACT.

Tyndale-Biscoe, C.H. and Hearn, J.P. (1981). Pituitary and ovarian factors associated with seasonal quiescence of the tammar wallaby, *Macropus eugenii. Journal of Reproduction and Fertility*, **63**, 225–30.

Tyndale-Biscoe, C.H. and Hinds, L.A. (1984). Seasonal patterns of circulating progesterone and prolactin and response to bromocriptine in the female tammar, *Macropus eugenii. General and Comparative Endocrinology*, **53**, 58–68.

Tyndale-Biscoe, C.H. and Hinds, L.A. (1989). The hormonal milieu during development in marsupials. In *Development of preimplantation embryos and their environment* (ed. K. Yoshinaga and T. Mori), pp. 237–46. Alan R. Liss, New York.

Tyndale-Biscoe, C.H. and Hinds, L.A. (1990). Control of seasonal reproduction in the tammar and Bennett's wallabies. In *Progress in comparative endocrinology* (ed. A. Epple, C.G. Scanes, and M.H. Stetson), pp. 659–67. Wiley-Liss, Inc., New York.

Tyndale-Biscoe, C.H. and Hinds, L.A. (1992). Components of the melatonin message in the response to photoperiod of the tammar wallaby, *Macropus eugenii. Journal of Pineal Research*, **12**, 155–166.

Tyndale-Biscoe, C.H. and Janssens, P.A. (1988). *The developing marsupial. Models for biomedical research.* Springer-Verlag, Berlin.

Tyndale-Biscoe, C.H. and Renfree, M.B. (1987). *Reproductive physiology of marsupials.* Cambridge University Press, Cambridge.

Tyndale-Biscoe, C.H., Hearn, J.P., and Renfree, M.B. (1974). Control of reproduction in macropodid marsupials. *Journal of Endocrinology*, **63**, 589–614.

Tyndale-Biscoe, C.H., Hinds, L.A., Horn, C.A., and Jenkin, G. (1983). Hormonal changes at oestrus, parturition and post-partum oestrus in the tammar wallaby (*Macropus eugenii*). *Journal of Endocrinology*, **96**, 155–61.

Tyndale-Biscoe, C.H., Hinds, L.A., and McConnell, S.J. (1986). Seasonal breeding in a marsupial: opportunities of a new species for an old problem. *Recent Progress in Hormone Research*, **42**, 471–512.

Tyndale-Biscoe, C.H., Hinds, L.A., and Horn, C.A. (1988). Fetal role in the control of parturition in the tammar, *Macropus eugenii. Journal of Reproduction and Fertility*, **82**, 419–28.

Walker, M.T. and Gemmell, R.T. (1983). Plasma concentrations of progesterone, oestradiol-17β and 13, 14-dihydro-15-oxo-prostaglandin F2α in the pregnant wallaby (*Macropus rufogriseus rufogriseus*). *Journal of Endocrinology*, **97**, 369–77.

Wallace, G.I. (1981). Uterine factors and delayed implantation in macropodoid marsupials. *Journal of Reproduction and Fertility*, **29**, (Suppl.) 173–81.

Walton, J.S., McNeilly, J.R., McNeilly, A.S., and Cunningham, F.J. (1977). Changes in concentrations of follicle-stimulating hormone, luteinizing hormone, prolactin and progesterone in the plasma of ewes during the transition from anoestrus to breeding activity. *Journal of Endocrinology*, **75**, 127–36.

Watson-Whitmyre, M. and Stetson, M.H. (1983). Simulation of peak pineal melatonin release restores sensitivity to evening melatonin injections in pinealectomized hamsters. *Endocrinology*, **112**, 763–5.

Wayne, N.L., Malpaux, B., and Karsch, F.J. (1988). How does melatonin code for daylength in the ewe: duration of nocturnal melatonin release or coincidence of melatonin with a light-entrained sensitive period? *Biology of Reproduction*, **39**, 66–75.

8 Isolation of X- and Y-bearing sperm for sex preselection

LAWRENCE A. JOHNSON

I INTRODUCTION

This review deals primarily with the most recent methods that have been reported in the scientific literature to isolate X- and Y-chromosome-bearing sperm of mammals, with emphasis on mammals of agricultural importance.

Although an effort will be made to touch on many of the methods that have appeared in the literature, the primary emphasis of this chapter will be on those methods that have been based on the only known difference between X and Y sperm, chromosomal constitution.

The greatest body of literature regarding the separation of X and Y sperm deals with physical and immunological methods that have been devised in order to separate X- from Y-bearing sperm. Since virtually all the physical and immunological separation methods lack reproducibility, their importance for this review is greatly diminished. The lack of repeatability in these methods is due in large measure to the fact that the methods are designed to take advantage of a real or a perceived difference of negligible size that may or may not be specific to X and Y sperm. For those who desire to study the many methods that have been put forth, see detailed reviews by Kiddy and Hafs (1971), Amann and Seidel (1982), Levin (1987), and Gledhill (1988).

1   Historical perspective

Since the beginning of history, human beings have been interested in controlling the sex of their own offspring and of the animals that they have dominion over. This interest has generally focused on intervention prior to or after coitus or insemination as well as *in vitro* manipulation of sperm. Manipulation of sperm has attracted the greatest interest since sex determination in mammals is based on the chromosome content of the fertilizing sperm. There have been many theories advanced for controlling sex ratio. Many of these theories originated with the Greek philosophers. For example, there were those who believed that females developed on the left side of the uterus, males on the right. Even as late as the eighteenth century some believed that sperm from the right testes produced more male young while sperm from the left testes produced more females.

One of the first serious scientific studies on the control of prenatal sex was reported by J.L. Lush in 1925. The basis for separation of X and Y sperm was the possible differential density of X- and Y-bearing sperm in the rabbit. The progeny from the inseminations made with sperm separated by centrifugation failed to show an altered sex ratio. This method and a host of others put forth over the past 60–70 years can be grouped under the broad heading of 'physical separation' methods. They are based on actual or perceived differences in the weight, density, size, motility, or surface charge of sperm. In many of the reports, positive results were documented, but scientific validation was lacking.

Sex chromosomes were described in mammals by Guyer (1910). Avery *et al.* (1944) established DNA as the carrier of genetic information. This ultimately led to an interest in using new and rapidly developing single cell technology for studying the age-old question of how to control sex

Table 8.1   *Flow cytometric analysis of X- and Y-bearing sperm for DNA difference*

Species	Percentage difference
Turkey	0
Human	2.9
Rabbit	3.0
Swine	3.6
Cattle	3.8
Dog	3.9
Horse	4.1
Sheep	4.2
Chinchilla	7.5
Microtus oregoni	9.2[a]
Microtus oregoni	12.5[a]

[a] From Oregon and Washington, respectively; no X chromosome is carried on sperm in the vole (Johnson and Clarke 1990).

ratio. Gledhill *et al.* (1976) were among the first to see the advantages of using flow cytometric measurement of DNA in the individual sperm as a potential means of monitoring mutagenic events caused by various forms of environmental radiation. These initial studies laid the groundwork for the eventual flow cytometric separation of X- and Y-chromosome-bearing sperm.

2   Basis for X and Y sperm separation

Exploiting some aspect of X- and Y-bearing sperm differences is the ideal method on which sex preselection should be based. Only then is gender development in the fertilized egg known. It also allows for the greatest economy in developing sexed embryos, since all manipulation takes place prior to fertilization. In mammals, the male is the heterogametic sex and, thereby, determines the sex of the progeny. Since sperm carry an X or a Y chromosome, union with the egg results in an XX (female) or XY (male). In birds, the situation is markedly different in that all the sperm contain similar chromosome complements and similar DNA content and the female is heterogametic and determines the sex of progeny. The DNA content of the nucleus of X-bearing and Y-bearing mammalian sperm nearly always differs. To our knowledge, the X chromosome is always larger than the Y and always carries more DNA than the Y sperm (Table 8.1), thus it can be used as a differential marker. The X chromosome of the human contains approximately 2.8–3.0% more DNA than the Y chromosome (Sumner and Robinson 1976; Johnson *et al*. 1993). In other mammals the difference between X and Y sperm DNA is generally greater (cattle, 3.8%; swine, 3.6%; horse, 4.1%; chinchilla, 7.5%; creeping vole, 9.1–12.5%; Johnson 1992). On the basis of visual estimation of karyotypes from more than 500 species Moruzzi (1979) found 24 species with greater than 6% difference in axial chromatid length between X and Y chromosomes and suggested they might be the best candidates for sperm sexing research. This difference was an estimate of chromatin content difference between X and Y sperm. We have found good agreement between Moruzzi's values and our own values based on flow cytometry (Table 8.1).

Since production of spermatozoa occurs in the seminiferous tubules, it follows that one would expect equality of X and Y sperm production. They are created by a division of the normal chromosome complement. No experimental data exist to demonstrate that sperm development may be different for X sperm than for Y sperm. It is generally assumed that survival and maturation rates of all sperm are the same. The fact that sperm differ according to their chromosome constitution provides the basis for separation by DNA content. However, intracellular sperm DNA does not lend itself to surface measurement. If one were to choose an ideal means of differentiating X from Y sperm, it would be a specific surface marker (antigenic protein). Such a protein would lend itself to the production of an antibody and subsequent adaptation to large scale preparation of sexed semen. However, no such marker exists at the present time. Thus, DNA is the only sex-specific marker known, measured, and validated.

II   SEPARATION METHODS

1   Physical

In addition to the earlier work (Lush *et al*. 1925), more recent studies have been published claiming the successful separation of sperm based on density using Percoll gradients (Kaneko *et al*. 1983). However, validation of these results has not been forthcoming. Bull sperm fractions collected from Percoll gradients were subjected to DNA flow cytometric analysis to determine proportions of X and Y sperm in the separated fractions. There was no change in the sperm sex ratio based on the flow cytometric DNA evaluation (Upreti *et al*. 1988).

According to Roberts (1972), there is a theoretical basis for a difference in size of X- and Y-bearing sperm, since X sperm carry the larger chromosome. However, no experimental evidence has been presented to substantiate the hypothesis. Morphologically, there has never been any data to back up the claim by some (Shettles 1961), that the X sperm head is morphologically any different than the Y sperm head. Theoretically, it is possible that one can separate sperm on the basis of dry weight since DNA mass differs between the X and Y sperm (for review see Johnson 1992). However, the difference in mass between X and Y sperm would be no more than 1%. Thus it would require exceptional measuring technology to accurately measure a 1% difference in the weight of X and Y sperm. The technique of free-flow electrophoresis has given some indication of being useful in differentiating between two populations of sperm based on a difference in sperm surface charge. However, there has been no confirmation of these results by others. Surface charge was the basis for other attempts to separate sperm. Bhattacharya *et al.* (1977) used the principle of thermal convection and counter-streaming sedimentation in combination with galvanization to attempt to move sperm at different velocities. Using their application the lighter Y sperm would lag and accumulate at some distance below the X sperm. The claimed success for this process has not been confirmed by other workers. Sephadex gel filtration as described by Steeno *et al.* (1975) has received some attention, particularly with human sperm. The ability of the sephadex method to separate X and Y sperm has not been confirmed (Lobel *et al.*, 1993).

Albumin column separation has been used on many species (Ericsson *et al.* 1973), but especially in humans. Although it is used in numerous clinics world-wide, there is still a good deal of controversy concerning its efficacy. The method seeks to enrich the Y sperm population, which is based on the view that the Y sperm swim faster and thus can be collected off the column prior to the arrival of the X sperm at the bottom of the column. Clinical data put forth by Beernink *et al.* (1993) claim 72% males in 1034 births (data collected over many years) after the use of the method. A study designed to test the method through the use of isolated sperm for hamster sperm penetration experiments and subsequent karyotyping of the fertilized embryo resulted in no confirmation of the Ericsson data (Brandriff *et al.* 1986). In fact, the results showed a tendency to a greater number of females from the separated semen.

2   Immunological

Attempts to inactivate the Y sperm by immunological methods have focused on the H–Y antigen (Goldberg *et al.*, 1971). Other reports have suggested the usefulness of H–Y antibody as a means of selecting only the X-bearing rabbit sperm (Zavos 1985). Recently, several investigators have shown that a diverse number of genes are expressed by the haploid genome. However, it is still not evident that haploid gene expression occurs on sex chromosomes (see review by Bradley 1989). Hoppe and Koo (1984) suggested that X- and Y-bearing sperm probably share the same surface antigen due to their origin in the same testicular milieu, also suggesting that sex specific expression of H–Y antigen on sperm is questionable. The data of Hendriksen *et al.* (1993) appear to confirm that there is no preferential expression. Experiments were conducted to determine if H–Y antigen is preferentially expressed on Y-chromosome-bearing sperm that have been flow cytometrically sorted into X or Y populations based on DNA. No difference in binding between the sorted X and Y sperm populations of the boar and bull was found. Flow cytometric evaluation of bull semen samples resulting from treatment with antibodies to H–Y antigen and subsequent removal of dead sperm showed no deviation from the expected 50:50 sex ratio (Johnson 1988).

3   Sperm separation markers: long arm of Y chromosome; DNA

Identification of Y sperm based on the long arm of the Y chromosome (human) was reported by Barlow and Vosa (1970). This resulted in the quinacrine hydrochloride fluorescent technique being used to monitor sperm separation. The validity and repeatability of the quinacrine technique has been questioned because of the variability of sperm preparation procedures and the difficulty of consistently reading the fluorescent signal.

Alternately, the application of flow cytometric analysis to measure individual sperm DNA content in order to determine the proportions of X- and Y-bearing sperm in a sample of semen has served as a reliable check on procedures purported to show the sex ratio (Pinkel *et al.* 1985; Johnson 1988; Upreti *et al.* 1988). In no case has there been evidence to support the claims made for various 'physical separation' procedures or other non-DNA-based separation procedures mentioned in the previous section. The remainder of this chapter will deal with the successful differentiation and separation of X and Y sperm using DNA as a marker.

III   FLOW CYTOMETRY (FCM) FOR X AND Y SPERM DNA ANALYSIS

Modern FCM/cell sorting technology was first developed by Fulwyler (1965) and Kamentsky and Melamed (1967). Flow systems were commercialized in the 1970s and have developed rapidly in conjunction with the computer revolution in the 1980s and continues in the 1990s. The primary application has been in medical research and diagnosis with respect to blood and tumour cells. FCM is an effective tool for use with many types of cell suspensions. Since sperm are produced in suspension, they are readily adapted to FCM analysis and sorting resulting in a classic illustration of the usefulness of

FCM for measuring the relative DNA content of individual sperm at a very rapid rate.

1 Hydrodynamic orientation

The flat shape of the mammalian sperm head and the fact that there was a mismatch in refractive index between sperm and their liquid suspension was investigated by Gledhill *et al.* (1976). This mismatch resulted in a non-Gaussian fluorescence distribution of sperm DNA when sperm were analysed in FCM with orthogonal geometry. This group showed that when they stained sperm from several species with acriflavine–Feulgen the observed fluorescence was orientation-dependent. This dependency has come to be known as an orientation artefact (Gledhill *et al.* 1976), due in large measure to the fact that sperm nuclei contain highly packed chromatin which results in brighter fluorescence emitted from the edge of the sperm head compared with the more transparent flat side. Without controlling orientation, the variability associated with orientation-dependent fluorescence overshadows any difference in DNA content fluorescence between the X and Y sperm populations.

Dean *et al.* 1978 looked at methods of controlling the orientation of sperm heads in flow. They found that the bevelled needle as originally designed by Fulwyler (1977) and Stovel *et al.* (1978) was effective in subjecting the cells to planer hydrodynamic forces as the cells flowed out the end of the needle. The sample stream is pushed into a more ribbon-shaped stream which tends to similarly orientate the flat-headed sperm cell. The broad dimension of the sperm head coincides with the broad dimensions of the ribbon stream. Pinkel *et al.*, 1982*b* used this principle to adapt an experimental flow cytometer (orienting flow cytometer) of orthogonal geometry to orient sperm. They were successful in differentiating between the X and Y sperm of the mouse. They also used the system to compare the DNA of normal mouse sperm with sperm from mice that had the Cattanach translocation (a piece of chromosome number 7 relocated to the X; Cattanach 1961). The latter was verification of the apparent bimodal differentiation in the putative X and Y sperm populations of normal mouse sperm. The DNA difference in normal mouse sperm is small (3.2%). The difficulty in separation is made more significant due to the irregular mouse sperm head shape. Cattanach mouse sperm, on the other hand, have a 4.9% difference (Pinkel *et al.* (1982*b*) which makes differentiation comparably easier.

A schematic diagram of the bevelled needle and its effect on orientation (Johnson and Pinkel 1986) of the sperm is shown in Fig. 8.1. The design shown has been used in all the sperm DNA analysis and sorting research that this author has published over the past 10 years.

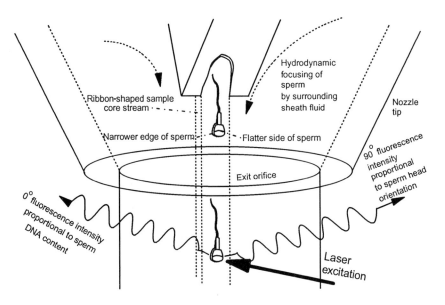

Fig. 8.1   Illustration of sperm orientation in an orthogonal flow system to enable resolution of X- and Y-chromosome-bearing sperm populations. As the sample core stream exits the bevelled injection needle two opposing sides of the stream are compressed by the surrounding sheath fluid while laminar flow is maintained. This produces a sample core stream flattened on two sides, giving it a shape more like a ribbon than standard cylindrical sample core streams. Consequently, the flattened, paddle-shaped sperm head tends to align its broad dimension in the broad dimension of the sample core stream. In this 'properly oriented' position the laser excitation strikes the flatter side of the sperm head. Fluorescence emission at 90° to the incident laser passes out the edge of the sperm. Since the edge has more fluorescent intensity than the flatter face these properly oriented sperm can be selected by gating on the brightest 90° population. Simultaneous analysis of fluorescence emitted at 0° to the excitation laser enables resolution of sperm bearing the X chromosome (2.8–7.5% brighter, depending on species) from those sperm bearing the Y chromosome.

2   Epi-illumination flow cytometer

The optical design of this epi-illumination type of flow cytometer is relatively insensitive to the orientation of cells during measurement. This is best typified by the ICP22 (Ortho Instruments, Westwood, MA, USA; no longer commercially built). The original unit was called the Phywe and was designed by Dittrich and Gohde (1969). The system is essentially an epifluorescence microscope focussed on to a flow cell and using an arc lamp rather than a laser as an excitation source (Fig.

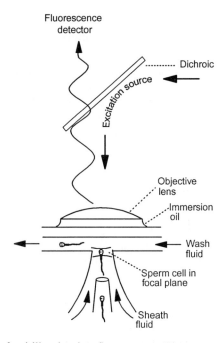

Fig. 8.2   Schematic of epi-illumination flow system. These systems also give good resolution of X and Y sperm since the optical artefact created by orientation, angle of detection, and refractive index is minimized. Excitation and emission are in the same plane and efficient optical coupling can be achieved by oils or gels of the same refractive index as the excitation and collection optics. However, these original systems were not designed for cell sorting so while they can resolve X- and Y-bearing sperm they cannot physically separate them (ICP22, Westwood, MA, USA).

8.2). Hydrodynamic forces orient the cells so their longitudinal axes are parallel to the direction of flow which flows directly into the sensing path of a microscope objective. This system was also effective at differentiating between X- and Y-chromosome-bearing sperm of mice (Pinkel *et al.* 1982*b*).

Determining the proportion of X and Y sperm in semen of domestic animals by epi-illumination FCM was reported by Garner, *et al.* (1983). In this study bovine, porcine, ovine, and the rabbit sperm were subjected to analysis for DNA using the ICP22 instrument. This study demonstrated the utility of DNA differentiation for assessing the proportions of X and Y sperm. It was subsequently used for monitoring the proportions of X and Y sperm in samples of semen prepared by various physical separation methods (Pinkel *et al.* 1985).

3   Modified orthogonal flow cytometer/cell sorter (MFCM)

High resolution flow cytometric DNA analysis of sperm nuclei is difficult when compared to other cells because of the inordinate compactness of chromatin in the morphologically flat, paddle-shaped sperm head that is characteristic of many mammals. The dense packing of the chromatin causes a high index of refraction as noted earlier. The difference in refractive index between the sperm nuclei and the surrounding medium, coupled with the flat shape of the sperm head, results in preferential emission of light in the plane of the cell (from the edge).

Although the epi-illumination flow systems offered advantages for the flat paddle-shaped sperm, in that one need not orient the sperm in flow, they had the drawback of being unable to sort cells. In 1986 we reported the modification of a commercial flow sorting system (MFCM) to control orientation and to analyse and sort sperm of mammals based on DNA content (Fig. 8.3; Johnson and Pinkel 1986). The basic orthogonal system (Epics V and Epics 753 series) was built by Coulter Corporation of Hialeah, FL, USA. Our modification consisted of several parts, the first being a bevelled sample injection needle substituted for the standard cylindrical needle. The needle was bevelled on the outlet end at an angle of 25–30° (Fig. 8.1). We replaced the light scatter detector at the 0° position with a fluorescence detector. The additional 0° fluorescence detector allowed the 90° detector to distinguish oriented sperm from misoriented sperm. By electronic gating, the forward detector could then be used to see only oriented sperm and have the ability to differentiate between the DNA of X and Y sperm (Johnson and Pinkel 1986). The signal received by the 0° detector was transmitted by fibre-optic cable to an existing photomultiplier tube and was thereby incorporated directly into the standard instrument data acquisition system. During the past year, similar modifications have also been made on a FacStar Plus, another flow cytometer/cell sorter of orthogonal geometry (Becton Dickinson, Mountain View, CA, USA) with equally good results (Cran et al, 1993).

The fluorescent signal emitted from the sperm head edge (90° angle from laser emission) is brighter than from the flat side (forward detector). The edge emission is used to characterize the orientation of the sperm heads as they pass the laser beam. The light emitted from the edge is collected by the 90° detector (Fig. 8.3). Misorientated sperm give off less light and therefore can be electronically gated out of the analysis. In general, about 65–95% of the sperm heads are properly oriented. Intact sperm are more difficult to orient (30–50%, Fig. 8.4) because of the presence of the tail. This percentage is species and individual sample dependent. Of the paddle-shaped sperm that we have analysed, the rabbit has proven the easiest to orientate, for reasons still unknown.

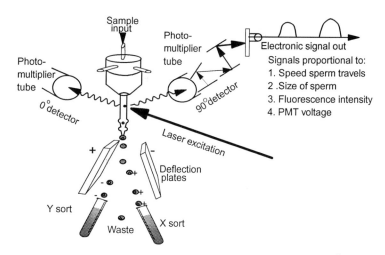

Fig. 8.3   System schematic of an orthogonal sperm sorting system. Unlike standard cell sorters this system also utilizes an optical detector and a photomultiplier tube (PMT) for the O° detection of fluorescence which is proportional to the DNA content of properly oriented sperm. The PMTs collect the fluorescence emitted by the sperm and convert this optical signal into a proportional electronic signal. This small signal is also amplified by the PMT for further signal processing and graphic display. With the bevelled injection needle and 0° optical detector, modification of the X- and Y- chromosome-bearing sperm populations can be resolved. By placing electronic sort windows around the resolved populations, X and Y sperm can be physically separated. For example, if at the point of analysis (intercept of laser and stream) the sperm fluoresces with an intensity indicative of it containing a Y chromosome, a timing circuit is started for the predetermined time it takes the sperm to travel from the analysis point to where the stream is just about to break into uniform droplets. When the Y sperm reaches this point a negative charge is put on the entire stream. The droplet containing the Y sperm then carries a negative charge as it completely breaks off from the stream. The negatively charged droplet then passes through an electrostatic field and is deflected toward the positively charged plate. Thus the Y sperm are deposited in an appropriate left sort vessel. Conversely, X-bearing sperm are deflected to the right.

IV   VERIFICATION OF X AND Y SPERM SEPARATION

Enrichment for X or Y sperm in a sample of semen can be monitored directly using FCM analysis for DNA content of the individual sperm (Johnson 1988; Upreti *et al*. 1988). This type of verification is critical to the development of authentic methods of separating X from Y sperm.

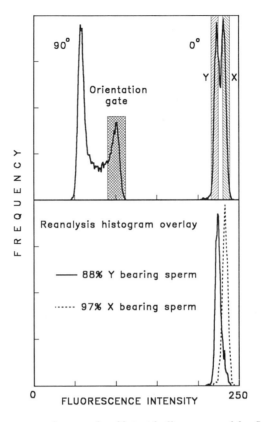

Fig. 8.4   A representative sample of intact bull sperm used for flow sorting (upper histograms) and the resulting purified populations of X and Y sperm as determined by flow cytometric reanalysis (lower histograms). The upper left-hand histogram (90°) shows the orientation distribution (boar and rabbit are very similar). The brightest peak are those sperm whose edge is toward the 90° detector and thus have their flatter face toward the 0° detector. These properly oriented sperm are selected by an electronic gate (orientation gate) for simultaneous DNA content analysis. The upper-right bimodal distribution is produced by the 0° fluorescence collected from the gated sperm. The slightly dimmer left population is comprised of sperm containing the lesser total DNA content (Y chromosome bearing) and conversely the slightly brighter right population are sperm bearing the X chromosome. Electronic sort windows select these X and Y populations and the flow cytometer sorts them from one another. The X and Y populations are recovered at rates of 0.5–1 million per hour. About 100 000 sperm aliquots of the sorted samples are prepared for reanalysis by removing the sperm tails by sonication. Improved resolution is gained by restaining to original stain concentrations. The sorted samples are analyzed and the resulting histograms are shown in the lower figure. A Gaussian distribution computer model is fitted to these curves for an estimation of purity. Purities have also been confirmed by live births, embryo sexing, and the polymerase chain reaction.

Historically, the only verification available was to determine the sex ratio of progeny. Not only is that costly, it usually takes the duration of the species gestation to determine the sex of the offspring. With DNA FCM evaluation of the purportedly sexed sperm, the planning of a fertility trial to determine sex ratio among progeny can be delayed until one is virtually certain of some success. This reduces costs and increases precision.

1 Sperm analysis for DNA content

Sperm which are to be analysed to determine the DNA content can be stained with either DAPI or Hoechst 33342 (Johnson *et al.* 1987*a*). Both dyes preferentially bind to the adenine–thymine regions (Muller and Gautier 1975) of the DNA helix. We have found no other vital dyes to be as effective at uniformly staining the sperm DNA for X and Y sperm separation. In our hands, even though DAPI is effective, Hoechst 33342 is the superior fluorochrome for sperm DNA analysis and sorting (Johnson *et al.* 1987*a*, 1989) due to its greater permeability and vital characteristics. The stained sperm heads are introduced under pressure into the MFCM in liquid suspension. The sperm enter the sample insertion tube and tend to orient as they exit the bevelled end of the sample tube (Fig. 8.1) into sheath fluid. The integrity of the ribbon-shaped sample stream (laminar flow) is maintained as the two fluids (sample and sheath) pass out through the 76 µm orifice of the flow cell nozzle. The stream containing the sperm intersects a laser beam generated by an Innova 90–5 Argon-ion laser. The lasing occurs in the ultraviolet (351, 364 nm) at up to 200 MW of power. The stained nuclei are individually excited by the laser beam, giving off fluorescent signals proportional to the amount of DNA that has been bound by the dye. These signals are detected by the photomultiplier tubes, amplified, converted from analogue to digital format and displayed as a frequency distribution.

The sperm DNA content of mammalian sperm as determined by flow cytometric analysis results in a bimodal distribution (Fig. 8.5) representing X- and Y-chromosome-bearing sperm populations. Data collected from a single sample can be transferred to another computer and fitted to a pair of Gaussian distributions whose means, relative areas, and coefficients of variation are adjusted to give the best least squares fit to the data (Johnson *et al.* 1987*a*). The per cent separation of the two populations is calculated by the difference $= 100 [x-y/0.5)] (x+y)$, where x and y are the respective channel means for the two peaks.

Most agriculturally important animals have a DNA difference of 3.0–4.2% between X- and Y-bearing sperm. It is difficult to consistently measure DNA difference of that magnitude without the specific modifications described. The MFCM has the resolving power to the repeatably distinguish X- from Y-bearing sperm (Johnson *et al.* 1987*a*; Johnson and

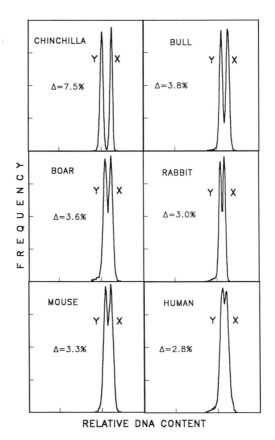

Fig. 8.5   Histograms of DNA content from six different species are shown. All DNA distributions are placed in the same position on their respective histogram, therefore, total DNA content between species is not comparable. However, this does allow for comparative separation of the X and Y peaks between species. The difference (Δ) in DNA content varies considerably between X-and Y-chromosome-bearing sperm of the profiled species. Chinchilla X and Y sperm can be almost completely resolved ($\Delta = 7.5\%$) while human X and Y sperm have considerable overlap with an approximate DNA difference of only 2.8%.

Clarke 1988). However, this can only be done with efficiency and consistency if one is using the bevelled tip and forward detector as modifications to the standard orthogonal flow cytometer (Figs 8.1 and 8.3). Because sperm can be held in suspension, preparations from virtually any species, including birds, can be analysed by flow cytometry (Fig. 8.5). Sperm may be washed prior to staining to remove seminal plasma, although it is not essential. The sperm are then subjected to sonication for about 10 s to break off the tails

(Johnson *et al.* 1987*a*). The sperm are stained with 9 μm Hoechst 33342. The mechanism of the Hoechst 33342 stain attachment is external binding rather than intercalation. Hoechst 33342 is non-toxic to sperm. The effect on viability is not important relative to simple flow cytometric DNA analysis of sperm nuclei (Johnson *et al.* 1987*a*). However, when one is interested in maintaining viability for insemination of intact sperm we have found Hoechst 33342 to be essential (Johnson *et al.* 1989; Johnson 1991; Cran *et al.* 1993).

Flow cytometric analysis to determine the sperm sex ratio of semen is currently routine procedure for most mammals excluding humans (Johnson 1992). Human sperm are characterized by a more angular or bullet-shaped head making orientation more problematic and less adaptable to reanalysis of DNA of sorted sperm. Also, the DNA difference between X-and Y-bearing human sperm is less than 3.0%; (Johnson and Welch 1991; Johnson *et al.* 1993) which necessitates increased precision in the analysis. However, we have recently found fluorescence *in situ* hybridization procedures to be effective in documenting the purity of sorted X and Y human sperm (Johnson *et al.* 1993), thus overcoming the problem of a difficult reanalysis. Recently, we have adapted single sperm sorting to 96 well plates for analysis using the polymerase chain reaction to assess the purity of sorting single sperm (Welch *et al.* 1993).

2   Measurement for proportions of X and Y sperm

Many samples of semen that have been prepared by one of numerous physical separation or immunological procedures have been received in my laboratory from various commercial and academic sources for verification of the proportions of X and Y sperm. The samples have been prepared as sperm nuclei and flow cytometrically analysed for sperm sex ratio. More than 300 samples have been analysed and none have shown anything other than a 50:50 ratio of Y to X sperm (Johnson 1988; unpublished data). This ratio of Y to X sperm technique allows precise prediction of the ultimate outcome if one were to use the 'separated' sperm. The DNA analysis has also been shown to be a good predictor of the ultimate sex of offspring conceived by sorted viable sperm (Johnson *et al.* 1989; Johnson 1991; Cran *et al.* 1993).

V   Flow cytometric sorting of sperm

The first flow sorting of sperm for purposes of isolating X from Y sperm was reported by Pinkel *et al.* (1982*a*). The work was done on a specially built orienting flow cytometer using sperm from the creeping vole (*Microtus oregoni*). Validation of the separation was conducted on a Phywe flow analyser by reanalysis of sorted sperm for DNA content. The creeping vole was selected for these experiments because of the wide difference (9.1%) in DNA content between 'X' and Y sperm. This results from the fact that half the sperm do not carry a sex chromosome (X) due to selective non-disjunction (Ohno *et al.* 1963). We also chose the *Microtus oregoni* because of its ease of sorting the sperm with a wide DNA difference and because the germ cells have 17 YO and 18 XX chromosomes. This gave us the ability to achieve high purity in flow-sorted sperm populations and to easily differentiate between a developing male or female in metaphase spreads (17 *vs* 18). It was in the process of conducting these studies (Libbus *et al.* 1987; Libbus and Johnson 1988) that we discovered a subspecies of the original *Microtus oregoni* which had a 12.5% DNA difference in the sperm (Johnson and Clarke 1990). However, the need for the vole as an animal model became less important once we established procedures for sorting viable sperm (Johnson *et al.* 1989).

Chinchilla sperm has become a standard in our FCM studies. Chinchilla has a 7.5% difference in X and Y DNA (Fig. 8.5) making it more useful than the vole because it is easily maintained in colonies. We attained purities of greater than 90% for X and Y sperm populations (Johnson *et al.* 1987*b*) which has led to its use by us as a reference standard for flow sorting and flow analysis of sperm based on DNA content using the MFCM.

1   Flow sorting process on the MFCM

Flow sorting involves vibrating the flow chamber, causing the descending stream carrying the sperm away from the laser to break into small, uniform droplets. At the flow rates commonly used for sperm about 7.0% of the droplets actually contain a sperm. Once the DNA mass in a particular sperm head is identified by its respective pulse height as clearly belonging in the left (Y) or right (X) peak of the frequency distribution, timing and charging circuits are activated, and droplets carrying sperm are electrically charged positive (X) or negative (Y) according to the DNA content of the sperm within the droplet (Fig. 8.3). The charged droplet (as well as uncharged droplets containing no sperm) then pass through an electrostatic field of about 2000 volts. The droplets carrying X-bearing sperm (given positive charge) are deflected towards the negative pole of the high voltage, plate, while the droplets carrying the Y-bearing sperm (given negative charge) are deflected towards the positive pole of the high voltage plate. The deflected droplets carrying X and Y sperm fall into their respective collection tubes (Fig. 8. 3). The droplets containing no sperm or sperm with an undesired pulse height (outside or between the sort windows) that have been set on each peak are left uncharged and thus fall into the discard tube. The purities of the collected X and Y sperm populations are then determined by flow cytometric reanalysis of respective samples for DNA content. Resultant histograms are computer fitted to double Gaussian peaks as described earlier and as illustrated in Fig. 8.5.

2 Flow sorting sperm nuclei

Since sperm nuclei (without tails) are the easiest to orient with respect to the laser beam, they sort with highest purity. We have sorted sperm nuclei from bull, boar, ram, chinchilla, creeping voles (Johnson et al. 1987b; Johnson and Clarke 1988), rabbit (Johnson et al. 1989), human (Johnson et al. 1993), stallion, and dog (unpublished data). Purities ranged from 80% to greater than 95% with domestic animal species. Purity determinations were made through cytometric reanalysis of sorted samples for DNA. The initial viability testing of the sorted sperm nuclei was determined through microinjection of sperm heads into hamster eggs (Johnson and Clarke 1988) and culturing the eggs through the pronuclear stage. Sorted sperm have also been microinjected into sheep eggs and subsequently cultured, some to the 16 cell stage (Clarke et al. 1988).

3 Flow sorting intact viable sperm

A certain percentage of intact sperm randomly orient to the laser in the proper manner if a standard cylindrical sample needle is used. Bevelling the insertion tube enhances sorting efficiency for intact sperm as well as for nuclei (Fig. 8.1 and 8.4). Ultimately it was found that hydrodynamic orientation varied according to species and flow rate (Johnson and Welch 1991). Viable sperm preparation and staining is handled differently than for sperm nuclei. The key differences are the absence of sonication required for nuclei and the added step of incubation at 35°C, to effect rapid stain penetration in the intact viable sperm. Maintaining sperm viability after sorting over an extended period (Johnson et al. 1989; Johnson 1991) was a key factor in achieving successful sorting and production of offspring. There has been one other report of sorting viable sperm. In that study, an unmodified commercial flow cytometer/cell sorter was used to sort Hoechst 33342 treated bull and rabbit sperm (Morrell et al. 1988). However, without an MFCM, reanalysis for proportions of X-or Y-bearing sperm could not be performed. The fertility results obtained in the study showed some alteration of the sex ratio of offspring but the sex ratios were not significantly altered from 50:50 (Morrell et al. 1988).

4 Flow sorting sperm for production of offspring by insemination or *in vitro* fertilization (IVF)

Flow rates for sorting sperm nuclei for most species generally ranges from 700 to 1500 sperm/s. This results in a deflecting of about 100 sperm/s in each direction. Effectively this results in about 400 000 sperm/h in each collection tube. Intact sperm on the other hand flow at a higher rate of 2500 sperm/s, also resulting in approximately 100 sperm/s deflected. However, since orientation characteristics affect efficiency, there is a difference in X and Y sort efficiency. Therefore, an hour sorting of intact sperm also results in about 400 000 intact sperm. Sorted sperm generally average 75% motility for most species which results in about 300 000 viable sperm for each hour sort. Several variables affect sort rates. Uniformity of staining and individual ejaculate characteristics and the particular species are the primary variables. With the advent of more advanced computers and flow cytometer/cell sorters, it may be possible to increase sorting rate to 1 000 000 sperm/h (unpublished data).

Sorting of sperm is a relatively slow production technique when compared with the number of sperm normally required for artificial insemination in most species. Surgical insemination to minimize the number of sperm per fertilization has been useful in overcoming that problem in the pig and rabbit. Even fewer numbers of sperm need to be sorted for IVF since about 2000 to 4000 sperm per egg are needed to fertilize cattle and swine ova. Sorted sperm are washed and concentrated after sorting and coincubated with ova.

VI   FERTILITY RESULTS USING FLOW-SORTED SPERM

Boar and rabbit X and Y sperm have been flow sorted and surgically inseminated into sows and does, respectively (Johnson *et al*. 1989; Johnson 1991). Bull sperm have been sorted and used for IVF experiments and subsequent embryo transfer (Cran *et al*. 1993). Boar sperm have also been sorted and used for IVF (Rath *et al*. 1993). Ram sperm have been sorted and used for surgical insemination into the oviduct of ewes (Johnson and Rexroad, unpublished).

1   Rabbit

Sperm from mature bucks were sorted into X- and Y-bearing populations and surgically inseminated into the tip of the uterine horn of New Zealand White does. Does had been primed with human chorionic gonadotrophin and were scheduled to ovulate approximately 4 h post-surgery. A summary of the actual and predicted sex of offspring are shown in Fig. 8.6. There was a significant deviation in the sex ratio of progeny produced from the inseminations with sorted sperm. Offspring born from X sorted sperm were 94% female, and from Y sorted sperm were 86% male (Johnson *et al*. 1989).

2   Swine

Sperm from mature boars was sorted into X- and Y-bearing sperm populations, and surgically inseminated into the isthmus of the oviduct of gilts
whose oestrous cycles had been synchronized to control ovulation time.
Surgery was performed using mid-ventral laparotomy 2–4 h after expected
ovulation (Johnson 1991). A summary of the results in terms of deviation
from the theoretical 50:50 sex ratio (ranging from 68Y to 74X%) is shown in
Fig. 8.6. These results represent 19 litters. There was a significant deviation
in the sex ratio of progeny produced from either X or Y sperm populations
when compared with the theoretical 50:50 ratio. The sorting process has also
shown promise for producing embryos via IVF in pigs (Rath *et al.* 1993).

3   Cattle

Recent results using sorted bull sperm for IVF of cow eggs have shown the
feasibility of sorting bull sperm into X- and Y-bearing sperm populations for

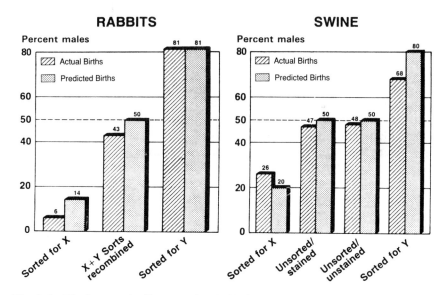

Fig. 8.6   The bar graphs illustrate the similarity of predicted percentages of male
progeny based on reanalysis of aliquots from the sorted samples of X- and Y-bearing
sperm for DNA when compared with the actual sex of the offspring born for rabbits
and swine (Johnson *et al.* 1989; Johnson 1991). The values are based on 13 litters of
rabbits and 19 litters of swine. The values in the bars represents the percentage for
each group (Johnson 1992). Reprinted by permission, *Journal of Animal Science*,
Champaign, IL.

the application of IVF and subsequent embryo transfer. Embryos have been produced and transferred to recipients resulting in the birth of calves whose sex was predetermined by reanalysis for DNA by embryo sexing (Cran *et al*. 1993). Optimum conditions for sperm number, composition, and dilution to suit the specific needs of IVF and sorted sperm are still being refined and improved.

VII   CONCLUSIONS

Sperm separated with flow cytometric cell sorting have several limitations if one considers them useful for artificial insemination. These limitations include the low number of sperm sorted in any given time period resulting in the problem of maintaining sperm viability for 3–4 days of sorting bull sperm just for a single insemination. Swine insemination doses are even larger (3×10^9), increasing the required sorting time substantially. However, by using IVF with sorted sperm the requirement of high sperm number is significantly reduced. In fact, only 2000–4000 sperm per egg are necessary for IVF. The methods discussed have potential for providing sexed sperm for specific production situations in domestic animals. Other factors that will impact the use of X and Y sperm are the feasibility of cryopreservation of sorted sperm, and the possible impact of reduced birth rates from sorted sperm inseminations.

The more cost-effective means of producing sexed sperm would be to utilize a sex specific sperm surface membrane marker that would lend itself to antibody development. Such a marker, if present, has not yet been identified. Efforts to identify such a marker using sorted sperm have begun, but to date there has been no success (Fenner *et al*. 1992). The promise of such a procedure is hard to minimize in that antibody tagged sperm (applied immediately after semen collection) could be readily separated utilizing micromagnetic beads or fluids or by using affinity chromatographic procedures. This would make the technology efficient and quite easy to adapt to semen production schemes, especially in cattle. If a process could be accomplished in this way, it would likely be quite cost effective.

ACKNOWLEDGEMENTS

The author gratefully acknowledges the special assistance of Glenn R. Welch in preparing this manuscript and thanks Joan Tishue for the typescript.

REFERENCES

Amann R.P., and Seidel, G.E. (ed.) (1982). *Prospects for sexing mammalian sperm.* Colorado University Press, Boulder, Co.

Avery, O.T., MacLeod, C.M., and McCarty, M. (1944). Studies on the chemical nature of the substance inducing transformation of pneumococcal types. Induction of transformation by a deoxyribonucleic acid fraction isolated from pneumococus Type III. *Journal of Experimental Medicine*, **79**, 137.

Barlow, P. and Vosa, C.G. (1970). The Y-chromosome in human spermatozoa. *Nature*, **226**, 961–2.

Beernink, F.J., Dmowski, W.P., and Ericsson, R.J. (1993). Sex preselection through albumin separation of sperm. *Fertility and Sterility.* **59**, 382–6.

Bhattacharya, B.C., Shome, P., Gunther, A.H., and Evans, B.M. (1977). Successful separation of X and Y and spermatozoa in human and bull semen. *International Journal of Fertility*, **22**, 30–5.

Bradley, M.P. (1989). Immunological sexing of mammalian semen: Current status and future options. *Journal of Dairy Science*, **72**, 3372–80.

Brandriff, B.F., Gordon, L.A., Haendal, S., Singer, S., Moore, D.H., and Gledhill, B.L. (1986). Sex chromosome ratios determined by karyotypic analysis in albumin-isolated human sperm. *Fertility and Sterility*, **46**, 678–85.

Cattanach, B.M. (1961). A chemically induced variegated-type position effect in the mouse. *Zeitschrift fur Vererbungslehre*, **92**, 165–82.

Clarke, R.N., C.E. Rexroad, C.E., A.M. Powell, A.M., and Johnson. L.A. (1988). Microinjection of ram spermatozoa into homologous and heterologous oocytes. *Biology of Reproduction*, **38**, (Suppl. 1), 75.

Cran, D.G., Johnson, L.A., Miller, N.G.A., Cochrane, D., and Polge, C. (1993). Production of bovine calves following separation of X-and Y-chromosome bearing sperm and *in vitro* fertilisation. *Veterinary Record*, **132**, 40–1.

Cran, D.G., Johnson, L.A., Miller, N.G.A., Cochrane, D., and Polge. C. (1993*b*). Gender preselection: Altered sex ratio in bovine embryos after IVF using X- and Y-chromosome bearing flow cytometrically sorted sperm. *Journal of Reproduction and Fertility.*

Dean, P.N., Pinkel, D., and Mendelsohn, M.L. (1978). Hydrodynamic orientation of sperm heads for flow cytometry. Biophysical Journal **23**, 7–13.

Dittrich, W. and Gohde, W. (1969). Impulse fluorometry in single cells in suspension. *Zeitschrift fur Naturforschung*, **24b**, 360–1.

Ericsson, R.J., Langevin, C.N., and Nishino, M. (1973). Isolation of fractions rich in human Y sperm. *Nature*, **246**, 421–4.

Fenner, G.P., Johnson, L.A. Hruschka, W.R., and Bolt, D.J. (1992). Two-dimensional electrophoresis and densistometric analysis of solubilized bovine sperm plasma membrane proteins detected by silver staining and radio-iodination. *Archives of Andrology.* **29**, 21–32.

Fulwyler, M.J. (1965). Electronic separation of biological cells by volume. *Science*, **150**, 910.

Fulwyler, M.J. (1977). Hydrodynamic orientation of cells. *Journal of Histochemistry and Cytochemistry* **25**, 781–3.

Garner, D.L., Gledhill, B.L., Pinkel, D., Lake, S., Stephenson, D., Van

Dilla, M.A., and Johnson, L.A. (1983). Quantification of the X and Y chromosome-bearing spermatozoa of domestic animals by flow cytometry. *Biology of Reproduction*, **28**, 312–21.

Goldberg, E.H., Boyse, E.A., Bennett, D., Scheid, M., and Carswell, E.A. (1971). Serological demonstrations of H–Y (male) antigen on mouse sperm. *Nature*, **232**, 478–80.

Gledhill, B.L. (1988). Gender preselection: historical, technical, and ethical perspective. *Seminars in Reproductive Endocrinology*, **6**, 385–395.

Gledhill, B.L., Lake, S., Steinmetz, L.L., Gray, J.W., Crawford, J.R., Dean, P.N., and Van Dilla, M.A. (1976). Flow Microfluorometric analysis of sperm DNA content: effect of cell shape on the fluorescence distribution. *Journal of Cellular Physiology*. **87**, 367–76.

Guyer, M.F. (1910). Accessory chromosomes in man. *Biological Bulletin (Woods Hole Massachusetts)*, **19**, 219.

Hendriksen, P.J.M., Tieman, M., van der Lende, T., and Johnson L.A. (1993). Binding of anti–H–Y monoclonal antibodies to separated X and Y chromosome bearing porcine spermatozoa. *Molecular Reproduction Development*, **35**, 189–96.

Hoppe, P.C. and Koo, G.C. (1984). Reacting mouse sperm with monoclonal H–Y antibodies does not influence sex ratio of eggs fertilized *in vitro*. *Journal of Reproduction and Immunology*, **6**, 1–9.

Johnson, L.A. (1988). Flow cytometric determination of sperm sex ratio in semen purportedly enriched for X or Y bearing sperm. *Theriogenology*, **29**, 265.

Johnson, L.A. (1991). Sex preselection in swine: altered sex ratios in offspring following surgical insemination of flow sorted X- and Y-bearing sperm. *Reproduction in Domestic Animals*, **26**, 309–14.

Johnson, L.A. (1992). Gender preselection in domestic animals using flow cytometrically sorted sperm. *Journal of Animal Science*, **70**, (Suppl. 1), 8–18.

Johnson, L.A. and Clarke, R.N. (1988). Flow sorting of X and Y chromosome-bearing mammalian sperm: Activation and pronuclear development of sorted bull, boar and ram sperm microinjected into hamster oocytes. *Gamete Research*, **21**, 335–43.

Johnson, L.A. and Clarke, R.N. (1990). Sperm DNA and sex chromosome differences between two geographical populations of the creeping vole, *Microtus oregoni*. *Molecular Reproduction Development* **27**, 159–62.

Johnson, L.A. and Pinkel, D. (1986). Modification of a laser-based flow cytometer for high resolution DNA analysis of mammalian spermatozoa. *Cytometry*, **7**, 268–73.

Johnson, L.A., and Welch, G.R. (1991). Flow sorting of X and Y sperm based on DNA: influence of viability and head shape on hydrodynamic orientation. *Cytometry*, Suppl. **5**, 86.

Johnson, L.A., Flook, J.P., and Look, M.V. (1987*a*). Flow analysis of X and Y chromosome-bearing sperm for DNA using an improved preparation method and staining with Hoechst 33342. *Gamete Research*, **17**, 203–12.

Johnson, L.A., Flook, J.P., Look, M.V., and Pinkel, D. (1987*b*). Flow sorting of X and Y chromosome-bearing spermatozoa into two populations. *Gamete Research*, **16**, 1–9.

Johnson, L.A., Flook, J.P., and Hawk, H.W. (1989). Sex preselection in rabbits: Live births from X and Y sperm separated by DNA and cell sorting. *Biology of Reproduction*, **41**, 199–203.

Johnson, L.A., Welch, G.R., Keyvanfar, K., Dorfmann, A., Fugger, E.F., and Schulman, J.D. (1993). Gender preselection in humans? Flow cytometric separation of X and Y sperm for the prevention of X-linked diseases. *Human Reproduction*, **8**, 1733–9.

Kamentsky, L.A. and Melamed, R.R.. (1967). Spectrophotometric cell sorter. *Science*, **156**, 1364–5.

Kaneko, S., Yamaguchi, J., Kobayashi, T., and Iizuka, R. (1983). Separation of human X- and Y-bearing sperm using Percoll density gradient centrifugation. *Fertility and Sterility*, **40**, 661–5.

Kiddy, C.A. and Hafs, H.D. (ed.) (1971). *Sex ratio at birth–prospects for control. Journal of Animal Science*, **31**, Symposium Suppl. 104 pp).

Levin, R.J. (1987). Human sex pre-selection. In *Oxford reviews of reproductive biology* Vol. 9 (ed. J.R. Clarke), pp.161–91. Oxford University Press, Oxford.

Libbus, B.L. and Johnson, L.A. (1988). The creeping vole, *Microtus oregoni*: karyotype and sex-chromosome differences between two geographical populations. *Cytogenetics and Cell Genetics*, **47**, 181–4.

Libbus, B.L., Perreault, S.D., Johnson, L.A., and Pindel, D. (1987). Incidence of chromosome aberrations in mammalian sperm stained with Hoechst 33342 and UV-laser irradiated during flow sorting. *Mutation Research*, **182**, 265–74.

Lobel, S.M., Pomponio, R.J., and Mutter, G.L. (1993). The sex ratio of normal and manipulated human sperm quantitated by the polymerase chain reaction. *Fertility and Sterility*, **59**, 387–92.

Lush, J.L. (1925). The possibility of sex control by artificial insemination with centrifuged spermatozoa. *Journal of Agricultural Research*, **30**, 893–913.

Morrell, J.M., Keeler, K.D., Noakes, D.E., Mackenzie, N.M., and Dresser, D.W. (1988). Sexing of sperm by flow cytometry. *Veterinary Record*, **122**, 322–4.

Moruzzi, J.F. (1979). Selecting a mammalian species for the separation of X- and Y-chromosome-bearing spermatozoa. *Journal of Reproduction and Sterility*, **57**, 319–23.

Muller, W. and Gautier, F. (1975). Interactions of heteroaromatic compounds with nucleic acids. *European Journal of Biochemistry*, **54**, 358.

Ohno, S., Jainchill, J., and Steinius, C. (1963). The creeping vole (*Microtus oregoni*) as a gonosomic mosaic. I. The OY/XY constitution of the male. *Cytogenetics*, **2**, 232–9.

Pinkel D., Gledhill, B.L., Lake, S., Stephenson, D., and Van Dilla, M.A. (1982a). Sex preselection in mammals? Separation of sperm Bearing Y and 'O' chromosomes in the vole *Microtus oregoni*. *Science*, **218**, 904–6.

Pinkel, D., Gledhill, B.L., Van Dilla, M.A., Stephenson, D., and Watchmaker, G. (1982b). High resolution DNA measurements of mammalian sperm. *Cytometry*, **3**, 1–9.

Pinkel, D., Garner, D.L., Gledhill, B.L., Lake, S., Stephenson, D., and Johnson, L.A. (1985). Flow cytometric determination of the proportions of X and Y chromosome-bearing sperm in samples of purportedly separated bull sperm. *Journal of Animal Science*, **60**, 1303–7.

Rath, D., Johnson, L.A., and Welch, G.R. (1993). *In vitro*, culture of porcine embryos: Development to blastocysts after *in vitro*, fertilization (IVF) with flow cytometrically sorted and unsorted semen. *Theriogenology*, **39**, 293.

Roberts, A.M. (1972). Gravitational separation of X-and Y-spermatozoa. *Nature*, **262**, 223–35.

Shettles, L.B. (1961). Human spermatozoan shapes in relation to sex ratios. *Fertility and Sterility*, **12**, 502–8.

Steeno, O., Adimoelja, A., and Steeno, J. (1975). Separation of X-and Y-bearing human spermatozoa with the sephadex gel-filtration method. *Andrologia*, **7**, 95–7.

Stovel, R.T., Sweet, R.G., and Herzenberg, L.A. (1978). A means for orienting flat cells in flow systems. *Biophysical Journal*, **23**, 1–5.

Sumner, A.T. and Robinson, J.A. (1976). A difference in dry mass between the heads of X- and Y-bearing spermatozoa. *Journal of Reproduction and Fertility*, **48**, 9–15.

Upreti, G.C., Riches, P.C., and Johnson, L.A. (1988). Attempted sexing of bovine spermatozoa by fractionation on a percoll density gradient. Gamete Research. **20**, 83–92.

Welch, G.R., Jackson, J., Waldbeiser, G., Wall, R.J., and Johnson, L.A. (1993). Single cell sorting and PCR to confirm, separations of X- and Y-chromosome bearing sperm. *Cytometry*, **6**, 26.

Zavos, P. (1985). Sperm separation attempts via the use of albumin gradients in rabbits. *Theriogenology*, **23**, 875.

INDEX

α_2-macroglobulin 45, 56, 57
ACTH (adrenocorticotrophic hormone) 190
adrenal glands 176–7
ageing, reproductive 215–39
  changes in female sexual behaviour 233–4
  female rodents 221–9
  hypothalamus 218–20
  pituitary 220–1
  testicular function and male sexual behaviour 234–7
  women 229–32
albumin column sperm separation 307
allogeneic pregnancy 183–5
altered self 196
androstenedione 232
angiogenesis 115, 116, 120
anoestrus 258
anti-aromatase test 148–50
antibody tagged sperm 322
antigens, of paternal and fetal origin 181, 195–6
antigen-specific T cells 195–6
anti-Müllerian hormone (AMH) 139–57
  cloning and sequencing of AMH gene 142–3
  effect on adult ovary 154
  effects upon female genital primordia 147–50
  expression and ontogeny 143–6
  expression *in vitro* 146
  hormonal regulation 146–7
  mapping bioactive site 150–3
  PMDS 154–6
  protein purification 140–2
  testicular descent 153–4
  testicular differentiation 153
apical surface of uterine epithelium 89, 91, 92–3, 97
ascorbic acid 63–4
assisted reproduction 60–1, 319–20, 322
attachment, embryo
  cell adhesion molecules, *see* cell adhesion processes
  CG 12, 16

basal lamina 34, 36
  invasion by trophoblast 93, 103–17, 118–20

ovary 38, 47, 48, 49
testis 37, 39
basement membranes 34–5, 65
  ovary 48, 54
  testis 39–40, 41, 45
Bennett's wallaby 257, 259, 286, 289–91
  general features of reproductive cycle 259–63
  photoperiod 271–84 *passim*
  photorefractoriness 287–9
blastocyst, diapausing 259–63
bromocriptine 268–70, 282
brush tail possums 252, 255

cadherins 100–1, 104, 118, 120
calcium, intercellular 10
CAM-105 101, 118, 120
catecholamines (CAs) 217, 218–19, 220, 238
cattle sperm 321–2
cell adhesion processes 87–122
  cellular aspects of implantation 90–4
  classification of cell adhesion molecules 118–19
  implantation models 94–6
  molecules influencing embryo attachment 96–103
  ovarian steroids and uterine receptivity 88–90
  trophoblast invasion of basal lamina and decidua 103–17
chondroitin sulphate proteoglycans 51–3, 60–1, 114
chorionic gonadotrophin (CG) 2, 11–24
  establishment of early pregnancy 18–24
  first secretion 13–16
  immunization against 16
  regulation of secretion 16–18
  see also human chorionic gonadotrophin
CL–follicular axis 263, 264, 266–7
cloprostenol 3, 5, 6, 8, 18–19
  see also PGF$_{2\alpha}$
collagen
  cell adhesion processes in implantation 113–14, 120
  ovary 47, 63–4
  testis 38–9, 41
collagenase 55–7, 59, 61
collagen-induced arthritis (CIA) 197–8
continuous breeding 256–8
cord development 36–8

330 Index

Royal Society of
Medicine Library
WITHDRAWN STOCK